GENETIC ENHANCEMENT OF *RABI* SORGHUM – ADAPTING THE INDIAN DURRAS

GENETIC ENHANCEMENT OF *RABI* SORGHUM – ADAPTING THE INDIAN DURRAS

P. Sanjana Reddy
Directorate of Sorghum Research, Rajendranagar, Hyderabad, India

J.V. Patil
Directorate of Sorghum Research, Rajendranagar, Hyderabad, India

ELSEVIER

AMSTERDAM • BOSTON • HEIDELBERG • LONDON
NEW YORK • OXFORD • PARIS • SAN DIEGO
SAN FRANCISCO • SINGAPORE • SYDNEY • TOKYO
Academic Press is an imprint of Elsevier

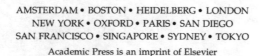

AP

Academic Press is an imprint of Elsevier
32 Jamestown Road, London NW1 7BY, UK
525 B Street, Suite 1800, San Diego, CA 92101-4495, USA
225 Wyman Street, Waltham, MA 02451, USA
The Boulevard, Langford Lane, Kidlington, Oxford OX5 1GB, UK

ISBN: 978-0-12-801926-9

British Library Cataloguing-in-Publication Data
A catalogue record for this book is available from the British Library

Library of Congress Cataloging-in-Publication Data
A catalog record for this book is available from the Library of Congress

For information on all Academic Press publications
visit our website at http://store.elsevier.com/

Typeset by MPS Limited, Chennai, India
www.adi-mps.com

Printed and bound in USA

Publisher: Nikki Levy
Senior Acquisition Editor: Nancy Maragioglio
Editorial Project Manager: Billie Jean Fernandez
Production Project Manager: Julia Haynes
Designer: Ines Cruz

Contents

Preface

Sorghum (*Sorghum bicolor* (L.) Moench) is the fifth important cereal crop in the world, primarily grown for grain production on about 35 m ha in about 106 countries and a staple food crop for over 500 million people in Africa, Asia, Oceania, and the Americas. The crop is mainly grown in tropical and subtropical areas which are marginal and stress prone. The postrainy sorghums or the winter sorghums are specialized sorghums of India. They are agronomically and physiologically distinct from the rainy sorghums and adapt well to the drought conditions as they have been grown and selected under receding moisture conditions occurring after cessation of rains. Unlike rainy sorghum where the grain quality is affected due to grain mold infection, the winter sorghum/postrainy sorghum is the main source of food as it is free from grain mold. It also serves as an important source of fodder since fodder from other crops is not available during this season. As not much progress has been made in the improvement of the postrainy sorghums, the landraces with tolerance to shoot fly, terminal drought and charcoal rot and with bold, round and lustrous grain dominate the winter sorghum tracts. Rapid productivity enhancement has not been possible for winter sorghum unlike rainy season grown sorghum due to lack of success with hybrid technology. However winter sorghum is an important crop in lieu of climate change and also as a health crop. Though several researchers worked for winter sorghum improvement in Indian Council of agricultural research (ICAR), India and International Crops Research Institute for the Semi Arid Tropics (ICRISAT), the information is scattered and not compiled. Hence, an attempt is made to pool the available literature to give a glimpse on what has gone through in decades of research on winter sorghum improvement.

This book has eight chapters. The first five chapters deal with crop biology, origin and taxonomy, genetic variability, genetics and cytogenetics. These do not tell about actual breeding, but a knowledge about these is useful in enhancing the effectiveness of a breeding programme. The sixth chapter deals with history of winter sorghum improvement in India. Winter sorghum research did not receive much emphasis until nineties and the varieties or hybrids bred and released could not match M 35-1, a landrace variety that is grown for about seven and a half decade, in yield or quality. However, lot of strategic research is required to develop new varieties and hybrids for post-rainy season adaptation that can break the yield plateau. The history of past research for winter sorghum improvement will help to plan future strategies to bring in real impact at the farmers level. The seventh chapter on conventional and molecular breeding for sorghum improvement deals with strategies for breeding for yield and adaptation, tolerance to abiotic stresses influencing winter sorghum (terminal drought and mid-season cold stress) and tolerance to biotic stresses limiting yields of winter sorghum (charcoal rot among diseases and shootfly, aphids and shoot bug among insects). The eighth chapter exposes the reader on alternate and industrial uses of sorghum.

Sorghum's place in most of its traditional environments will remain important, with great potential in non-traditional environments. We hope that the "Genetic Enhancement of *rabi* sorghum: Adapting the Indian Durras" will make a valuable reference book for students, teachers, researchers interested in winter sorghum research and development.

Editors

CHAPTER

1

Introduction

OUTLINE

Sorghum [*Sorghum bicolor* (L.) Moench] is cultivated as a major food crop in several countries in South Asia, Africa, and Central America. The crop is mainly grown in tropical and subtropical areas that are marginal and stress prone. Sorghum is known as *guinea-corn, dawa,* or *sorgho* in West Africa, *durra* in the Sudan, *mshelia* in Ethiopia and Eritrea, *mtama* in East Africa, *kaffircorn, mabele,* or *amabele* in southern Africa, and as *jowar, jonna, cholam,* or *jola* in the Indian subcontinent (Bantilan et al., 2004).

Sorghum is grown mainly for animal feed in the United States, Australia, and South America, while it is mainly grown as food in Africa and India. In India, sorghum is grown in two seasons, the *kharif* (or rainy) season and the *rabi* (or postrainy) season. During *rabi*, it is sown between September and end of October in the Deccan Plateau between 10° and 20°N latitude (Seetharama et al., 1990), and it accounts for 45% of the total sorghum area under cultivation and 32% of the total production (Sajjanar et al., 2011). In India, sorghum grown in the rainy season is mainly utilized as feed, as the grain is often caught in rains prevailing during harvesting and the grain quality is affected due to molds. However, postrainy sorghum is primarily used as a food owing to its good grain quality, and it also serves as a main source of fodder, especially during dry seasons, and is an important crop in lieu of food security. Most of the *rabi* sorghum varieties belong to *durra* or intermediates of the *durra* race, while the *kharif* cultivars that are being grown belong to *caudatum* and

Genetic Enhancement of rabi *sorghum – Adapting the Indian Durras.*
DOI: http://dx.doi.org/10.1016/B978-0-12-801926-9.00001-7

FIGURE 1.1 *Rabi* sorghum panicle.

kafir races (Reddy et al., 2003). The majority of *rabi* germplasm samples, which mostly belonged to the *durra* race, were collected by the National Bureau of Plant Genetic Resources (NBPGR) and International Crops Research Institute for the Semi-Arid Tropics (ICRISAT; Mathur et al., 1993). From the breeding point of view, *kafir*, *caudatum*, and *durra*, having genes contributing to yield, have been extensively utilized in breeding programs across the globe. Despite some major differences in both environment and cultural practices, *rabi* sorghum grown in India is similar to sorghum grown on residual moisture elsewhere in Africa. However, two major differences exist: African postrainy sorghums are grown in low soil fertility conditions, as they are cropped on receding flood plains after burning vegetation; and low plant density is employed (Seetharama et al., 1990) (Figure 1.1).

1.1 PRODUCTION STATISTICS

Sorghum [*S. bicolor* (L.) Moench] is the fifth most important crop in acreage next only to wheat, rice, maize, and barley in the world (Dillon et al., 2007). It was cultivated on 35 m ha in 106 countries in 2011 and a staple food crop for over 500 million people in Africa, Asia, Oceania, and the Americas. In 2011, India was the world's largest producer of sorghum (7.0 million metric tons), followed by Nigeria (6.9), Mexico (6.4), the United States (5.4), and Argentina (4.5). India is also the largest cultivator of sorghum (7.4 m ha), followed by Nigeria (4.9), Niger (2.9), and Ethiopia (2.2) (FAOSTAT, 2013).

Crop	Area (m ha)	Production (m tons)
Wheat	220.39	704.08
Rice	164.12	722.76
Maize	170.40	883.46
Barley	48.60	134.28
Sorghum	35.48	54.20

Source: FAOSTAT (2013).

During the last 30 years (1981–2011), the annual world production of sorghum has decreased from 73.3 to 54.2 million tons, and the area planted decreased from 45.9 to 35.5 million ha. However, the average yields in 1981 and 2011 were almost similar (1598 and 1527 kg/ha). However, wide variations exist between these global figures and figures at the national level. In India, for example, between 1981 and 2011, the area planted to sorghum fell from 16.6 to 7.4 million ha (Figure 1.2), and annual production fell from 12.1 to 7.0 million tons (Figure 1.3), but yields increased by 31% from 727 (in 1981) to 949 kg/ha (in 2011) (Figure 1.4). There exists a wide gap between average yield levels of India compared to the world yield levels.

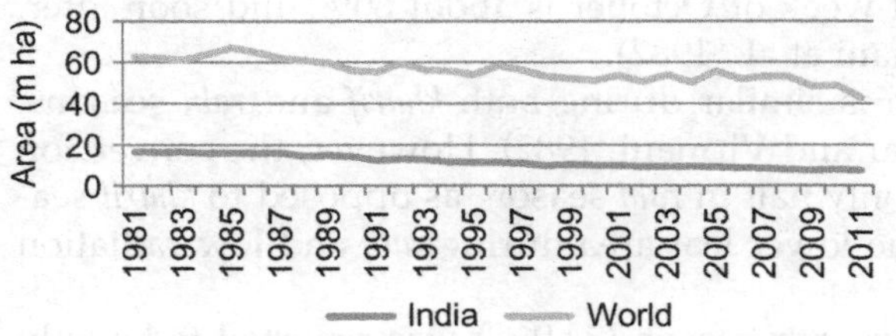

FIGURE 1.2 Trends in area under sorghum for three decades.

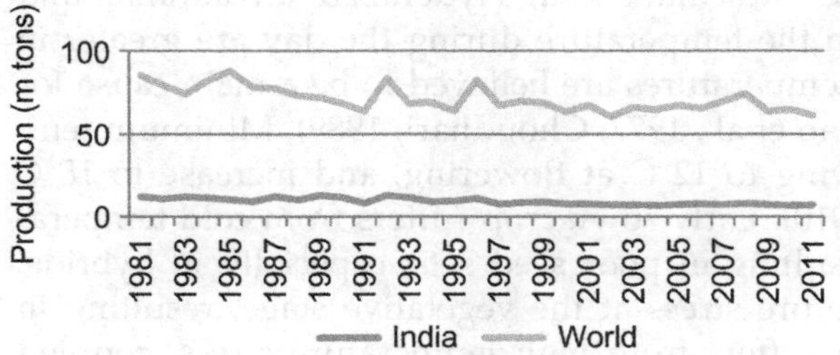

FIGURE 1.3 Trends in sorghum production for three decades.

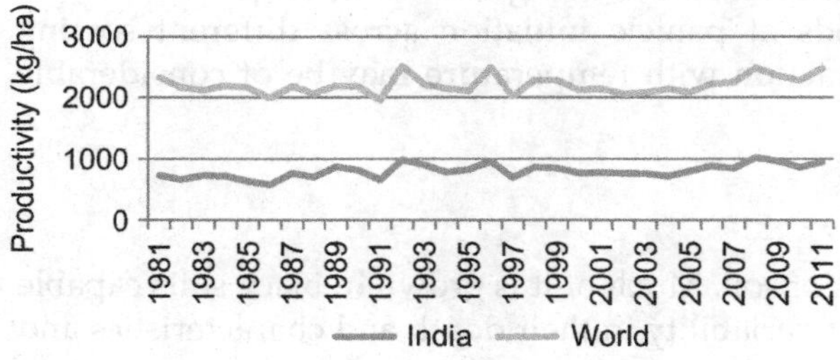

FIGURE 1.4 Trends in sorghum productivity for three decades.

Sorghum is well adapted to hot and dry (semiarid) agro-ecologies, where it is difficult to grow other food grains. Since sorghum is grown in drought-prone areas, it is cultivated with minimal inputs; in conditions of low soil fertility; under rainfed conditions with erratic and inadequate rainfall using traditional cultivars, which are poor yielders; and also exposed to a variety of disease and pest problems that affect the crop yields (Rai et al., 1999). The lower average yields are primarily due to the hot, dry conditions where sorghum is mostly grown, rather than the plant's own capability. However, sorghum has high yield potential, comparable to rice, wheat, and maize (House, 1985). Where moisture is not a limiting factor, sorghum had yielded up to 11,000 kg/ha at the field level, with average yields ranging from 7000 to 9000 kg/ha. In traditional sorghum-cultivating areas, yields of 3000–4000 kg/ha are obtained under better management conditions, dropping to 300–1000 kg/ha when moisture and soil fertility become limiting factors (House, 1985).

1.2 ENVIRONMENTAL FACTORS LIMITING *RABI* SORGHUM PRODUCTIVITY

1.2.1 Climatic Factors

Rainfall: The *rabi* crop requires about 175 mm of water (Tarhalkar, 1986). Most *rabi* sorghum is grown in areas receiving annual rainfall of about 600–800 mm. The probability of receiving 10 mm of rainfall during the first week of October is about 60%, and soon after, it decreases to about 2–5% per week (Virmani et al., 1982).

Solar radiation: The mean daily radiation is similar during both *kharif* and *rabi* seasons, being only 6% less in *rabi* season (Sivakumar and Virmani, 1982). However, the conversion of incident solar radiation to dry matter is only half in *rabi* season, as opposed to *kharif* season (Sivakumar and Huda, 1985), due to the lower leaf area during *rabi* and low radiation use efficiency (Seetharama et al., 1982).

Temperature: The mean temperature during *rabi* season (24.9°C) was reported to be only marginally less than *kharif* season (27.9°C) at Patancheru, Hyderabad (Sivakumar and Virmani, 1982). However, the variations in the temperature during the day are greater in *rabi* than *kharif* season, and the low night temperatures are believed to be a main cause for the reduction in growth and grain yield (Rao et al., 1977; Choudhari, 1989). Minimum temperatures decline from about 20°C at sowing to 12°C at flowering, and increase to 18°C during grain-filling period (Mukri et al., 2010). Early-sown crop suffers from cold temperature stress during the flowering stage, resulting in poor seed sets, especially in hybrids. Late-sown crop suffers from cold temperature stress at the vegetative stage, resulting in profuse tillering. The late-sown crop also suffers from high temperature stress, coupled with low soil moisture and high evaporative demand at the grain-filling stage.

Photoperiod: The change in photoperiods at panicle initiation across different sowing dates is less in *rabi* sorghum, but its interaction with temperature may be of considerable significance (Seetharama et al., 1990).

1.2.2 Edaphic Factors

Rabi sorghum is grown in varied depths of soil. Much of it is grown in black soils capable of holding moisture. The soils show a great variability in their depth and characteristics and are grouped as shallow, medium, and deep black soils. The depth of shallow soils varies from 0.0 to 22.5 cm, and about 20–22% of the black soils are classed as shallow soils. The soil depth of medium soils is about 22.5–90 cm. About 65% of the area is comprised of medium black soils. Deep black soils are more fertile than medium black soils, have a granular structure in the surface layer, and become cloddy at lower depths. The clay content of black soils ranges from 40% to 60% but may be as high as 70%. The differences in grain yield between shallow and deep vertisols is around 1.0 t/ha (Tandon and Kanwar, 1984).

1.3 USES

Sorghum grain is mostly consumed directly for food (55%) in the form of porridge (thick or thin) and flatbread. It is also an important source of feed grain (33%), especially in Australia and the Americas. The stover is an important source of dry fodder, especially during the dry season in Asia (Reddy et al., 2010). Sorghum has great potential as a fodder resource due to its quick and rapid growth, high green fodder yield, and good quality. Of late, sweet sorghum is emerging as an important biofuel crop, making sorghum a unique crop with multiple advantages as food, feed, fodder, fuel, and fiber. Hence, it is popularly known as a *smart crop*. In addition to these uses, sorghum crop residues and green plants provide building material, and fuel for cooking, particularly in the dry land areas (Chandel and Paroda, 2000), and in paper and cardboard, jaggery, and ethanol production. Sorghum has great potential to provide raw material for industries. Industrial application of sorghum makes its cultivation economically viable for marginal farmers. The grain is used for industrial purposes, such as potable alcohol, malt, beer, liquids, gruels, starch, adhesives, core binders for metal casting, ore refining, and grits as packaging material (Reddy et al., 2006) (Figure 1.5).

FIGURE 1.5 Uses of sorghum.

1.4 NUTRITIONAL STATUS

Sorghum grain contains 8–15% protein, 5–15% sugar, and 32–57% starch. It is relatively rich in micronutrients (mg/kg) iron (41–127), zinc (14–35), phosphorus (1498–3797), Ca (207–447), K (1150–2569), Mn (10–24), Na (12–54), and Mg (750–1506) (Shegro et al., 2012). Tannins, found in red-grained sorghums, contain antioxidants that protect against cell damage, a major cause of disease and aging. The protein and starch in sorghum grain are comparatively slowly digested than other cereals. This aspect is highly beneficial for people with diabetes; hence, sorghum is considered health food. Sorghum is a good alternative to wheat flour for individuals suffering from celiac disease, as the starch from sorghum grain is gluten free.

1.5 CHALLENGES FOR GENETIC ENHANCEMENT

Sorghum is cultivated by poor farmers and grown under subsistence conditions. Hence, they cannot take advantage of high yield potential, as the growers are unable to follow improved management practices. Higher yields can be obtained by growing varieties/hybrids with improved tolerance to drought, heat, and low soil fertility, as well as resistance to pests and diseases. Pest problems comprise one of the major constraints for achieving higher yields in sorghum grown in tropical areas. Immense losses are caused by insect pests attacking sorghum at all stages of growth, the important ones being shoot fly and aphids in winter sorghum. Pests like the shoot fly and stem borer attack in the very early stages of sorghum, resulting in the complete loss of the crop. Aphids attack from the flowering to milk stages and cause both qualitative and quantitative yield losses. The midge and ear-head bugs attack at the grain-filling stage leading to the losses up to 100%.

1.6 CHARACTERISTICS AND CULTIVARS

Winter sorghums are characterized by their response to shorter day lengths (photoperiod sensitivity), flowering, and maturity (occurring more or less at the same time irrespective of temperature fluctuations) and sowing dates (thermo-insensitivity within the postrainy season varieties). They are tolerant to terminal moisture stress and resistant to stalk rot/charcoal rot. As the fodder is as important as grain, the varieties are selected to produce high biomass (grain and stover) and have high lustrous grain with semi-corneous endosperm. Tolerance to shoot fly, lodging (mechanical), and rust are also required (Sanjana Reddy et al., 2012). All these characters exist in M 35-1, a variety selected from a local landrace nearly 75 years ago at Mohol in Maharashtra, which produces high stable yields of grain and stover across different sowing dates. As a result, M 35-1, a landrace selection developed in 1937 still dominates the postrainy season sorghum areas in India (Sanjana Reddy et al., 2009). In the past, focused breeding efforts on *rabi* sorghum led to the development of several *rabi* sorghum varieties such as CSV8R, Swati, CSV14R, CSV 18R, CSV216R, and CSV22R. Heterosis breeding led to the release of hybrids like

CSH8R, CSH12R, CSH13R, CSH15R, and CSH19R. However, these varieties have become more popular compared to hybrids, a situation quite opposite to *kharif*. Though the hybrids are heterotic for grain yield, they have poor grain quality and are vulnerable to biotic and abiotic stresses (Shinde et al., 2010).

References

Bantilan, M.C.S., Deb, U.K., Gowda, C.L.L., Reddy, B.V.S., Obilana, A.B., Evenson, R.E., 2004. Introduction. In: Bantilan, M.C.S., Deb, U.K., Gowda, C.L.L., Reddy, B.V.S., Obilana, A.B., Evenson, R.E. (Eds.), Sorghum Genetic Enhancement: Research Process, Dissemination and Impacts. International Crops Research Institute for the Semi-Arid Tropics, Patancheru, Andhra Pradesh, India, pp. 5–18.

Chandel, K.P.S., Paroda, R.S., 2000. Status of plant genetic resources conservation and utilization in Asia-Pacific region – Regional synthesis report Asia-Pacific Association of Agricultural Research Institutions. FAO Regional Office for Asia and the Pacific, Bangkok, Thailand, 158 p.

Choudhari, S.D., 1989. Production technology and a new line of work to improve "Rabi sorghum production". Paper Presented During the Working Group Meeting on Production Technology of Rabi Sorghum. 19 September 1989, CRIDA, Hyderabad.

Dillon, S.L., Shapter, F.M., Henry, R.J., Cordeiro, G., Izquierdo, L., Slade, L.L., 2007. Domestication to crop improvement: genetic resources for sorghum and saccharum (Andropogoneae). Ann. Bot. (Lond.). 100, 975–989.

FAOSTAT, 2013. Agricultural data. Available from: <http://apps.fao.org> (accessed May 2013).

House, L.R., 1985. A Guide to Sorghum Breeding, second ed. International Crops Research Institute for the Semi-Arid Tropics, Patancheru, India.

Mathur, P.N., Gopal Reddy, V., Prasada Rao, K.E., Mengesha, M.H., 1993. Collection of *rabi* sorghum germplasm I. Northern Karnataka and adjoining areas of Andhra Pradesh. Indian J. Plant Genet. Resour. 6, 1–8.

Mukri, G., Biradar, B.D., Sajjanar, G.M., 2010. Effect of temperature on seed setting behavior in *rabi* sorghum (*Sorghum bicolor* (L.) Moench). Electron. J. Plant Breed. 1 (4), 776–782.

Rai, K.N., Murty, D.S., Andrews, D.J., Bramel-Cox, P.J., 1999. Genetic enhancement of pearl millet and sorghum for the semi-arid tropics of Asia and Africa. Genome. 42, 617–628.

Rao, N.G.P., Vidyabhushanam, R.V., Rana B.S., 1977. Recent development in sorghum breeding in India; Pages 13 to 18 in Section 7, Plant Breeding Papers, Third International Congress of the Society for the Advancement of Breeding Researches in Asia and Oceania (SABRAO), February, 1977, Canberra, Australia.

Reddy, B.V.S., Sanjana, P., Ramaiah, B., 2003. Strategies for improving post-rainy season sorghum: a case study for landrace hybrid breeding approach. Paper presented in the Workshop on Heterosis in Guinea Sorghum, Sotuba, Mali, pp. 10–14.

Reddy, B.V.S., Ramesh, S., Reddy, P.S., 2006. Sorghum genetic resources, cytogenetics, and improvement. In: Singh, R.J., Jauhar, P.P. (Eds.), Genetic Resources Chromosome Engineering and Crop Improvement, Cereals, vol. 2. CRC Press, Taylor & Francis Group, Boca Raton, FL, pp. 309–363.

Reddy, B.V.S., Ashok Kumar, A., Sanjana Reddy, P., 2010. Recent advances in sorghum improvement research at ICRISAT. Kasetsart J. (Natural Science). 44, 499–506.

Sajjanar, G.M., Biradar, B.D., Biradar, S.S., 2011. Evaluation of crosses involving *rabi* landraces of sorghum for productivity traits. Karnataka J. Agric. Sci. 24 (2), 227–229.

Sanjana Reddy, P., Reddy, B.V.S., Ashok Kumar, A., 2009. M 35-1 derived sorghum varieties for cultivation during the postrainy season. E-J. SAT Agric. Res. 7 (1–4).

Sanjana Reddy, P., Patil, J.V., Nirmal, S.V., Gadakh, S.R., 2012. Improving post-rainy season sorghum productivity in medium soils: does ideotype breeding hold a clue? Curr. Sci. 102, 904–908.

Seetharama, N., Reddy, B.V.S., Peacock, J.K., Bidinger, F.R., 1982. Sorghum improvement for drought resistance. In: Drought Resistance in Crop Plants with Emphasis on Rice, IRRI, Philippines, pp. 317–356.

Seetharama, N., Singh, S., Reddy, B.V.S., 1990. Strategies for improving postrainy sorghum productivity. Proc. Indian Natl. Sci. Acad. 56 (5&6), 455–467.

Shegro, A., Shargie, N.G., van Biljon, A., Labuschagne, M.T., 2012. Diversity in starch, protein and mineral composition of sorghum landrace accessions from Ethiopia. J. Crop Sci. Biotechnol. 15 (4), 275–280.

Shinde, D.G., Biradar, B.D., Salimath, P.M., Kamatar, M.Y., Hundekar, A.R., Deshpande, S.K., 2010. Studies on genetic variability among the derived lines of B × B, B × R and R × R crosses for yield attributing traits in *rabi* sorghum (*Sorghum bicolor* (L.) Moench). Electron. J. Plant Breed. 1 (4), 695–705.

Sivakumar, M.V.K., Huda, AKS, 1985. Solar energy utilization by tropical sorghums. Agric. Forest Meteorol. 35, 47–57.

Sivakumar, M.V.K., Virmani, S.M., 1982. The physical environment; in Sorghum in the Eighties. International Crops Research Institute for the Semi-Arid Tropics, Patancheru, India, pp. 83–100.

Tandon, H.L.S., Kanwar, J.S., 1984. A Review of Fertilizer Use Research on Sorghum in India. Research Bulletin No. 8. International Crops Research Institute for the Semi-Arid Tropics, Patancheru, A.P. India.

Tarhalkar, P.P., 1986. Agronomical investigations on rabi sorghum: a brief review. Presented at the AICSIP Annual Workshop. 14–16 May 1986, Andhra Pradesh Agricultural University, Hyderabad.

Virmani, S.M., Sivakumar, M.V.K., Reddy, S.J., 1982. Rainfall probability estimates for selected locations of semi-arid India. International Crops Research Institute for the Semi-Arid Tropics (ICRISAT), Patancheru, Andhra Pradesh, India (Research Bulletin No. 1), 170 pp.

CHAPTER

2

Taxonomy and Origin

OUTLINE

2.1 TAXONOMY

Sorghum was first described by Linnaeus in 1753 under the name *Holcus*. In 1794, Moench distinguished the genus *Sorghum* from genus *Holcus* (Celarier, 1959; Clayton, 1961) and brought all the sorghums together under the name *Sorghum bicolor* (House, 1978; Clayton, 1961). Subsequently, several authors have discussed the systematics, origin, and evolution of sorghum since Linnaeus (de Wet and Huckabay, 1967; de Wet and Harlan, 1971; Doggett, 1988). Dahlberg (2000) provides an excellent overview of the present-day classification to describe the variation found within cultivated sorghums.

Sorghum is classified under the family Poaceae, tribe Andropogoneae, subtribe Sorghinae, and genus *Sorghum* Moench (Clayton and Renvoize, 1986). Garber (1950) and Celarier (1959) further divided the genera into five subgenera: sorghum, chaetosorghum, heterosorghum, parasorghum, and stiposorghum. *S. bicolor* was further broken down into three subspecies: *S. bicolor* subsp. *bicolor*, *S. bicolor* subsp. *drummondii*, and *S. bicolor* subsp. *verticilliflorum*. The cereal sorghums were found to consist of four wild races and five cultivated races (Harlan et al., 1976). The four wild races of *S. bicolor* are *arundinaceum*, *virgatum*, *aethiopicum*, and *verticilliflorum*. They are placed in *S. bicolor* subsp. *verticilliflorum*, formerly subsp. *arundinaceum*. Cultivated sorghums are classified as *S. bicolor* subsp. *bicolor* and are represented by several agronomic types, such as grain sorghum, sweet sorghum, sudangrass, and broomcorn (Berenji and Dahlberg, 2004). Additionally, there are two weedy sorghums widespread in temperate zone; that is, Johnsongrass and spontaneous sorghum (shattercane).

S. bicolor subsp. *bicolor* contains cultivated sorghum races. The cultivated races that are presently conceived are *bicolor*, *guinea*, *kafir*, *caudatum*, and *durra*. Intermediates that are caused by hybridization of these races exhibit characters of both parents (Smith and

Genetic Enhancement of rabi sorghum – Adapting the Indian Durras.
DOI: http://dx.doi.org/10.1016/B978-0-12-801926-9.00002-9

Frederiksen, 2000). Harlan and de Wet (1972) classified *S. bicolor* (L.) Moench, subsp. *bicolor* into 5 basic and 10 hybrid races as depicted next.

Basic races (5)	**1.** Race *bicolor* (B)
	2. Race *guinea* (G)
	3. Race *caudatum* (C)
	4. Race *kafir* (K)
	5. Race *durra* (D)
Intermediate races (10)	**1.** *Guinea-bicolor* (GB)
	2. *Caudatum-bicolor* (CB)
	3. *Kafir-bicolor* (KB)
	4. *Durra-bicolor* (DB)
	5. *Guinea-caudatum* (GC)
	6. *Guinea-kafir* (GK)
	7. *Guinea-durra* (GD)
	8. *Kafir-caudatum* (KC)
	9. *Durra-caudatum* (DC)
	10. *Kafir-durra* (KD)

The 15 races of cultivated sorghum are identified by mature spikelets, and this classification is based on five fundamental spikelet types (Harlan and de Wet, 1972). The International Plant Genetic Resources Institute Advisory Committee on Sorghum and Millets Germplasm has accepted and recommended this classification for describing sorghum germplasm (IBPGR and ICRISAT, 1993). These races are known to differ significantly not only for grain quality traits, but also for yield potential. The interracial hybrids were found to have greater heterosis than those of intraracial hybrids. Knowledge of the racial characteristics helps the breeder achieve systematic genetic improvement of sorghum for traits of interest (Reddy et al., 2008) (Figure 2.1).

Bicolor	Grains elongate, sometimes slightly obovate, and dorsoventrally symmetrical. Glumes clasping the grain may be completely covered or exposed to about one-fourth of its length at the tip.
Guinea	Grains flattened dorsoventrally, sublenticular in outline, twisting at maturity to about 90° between gaping involute glumes that are nearly as long as or longer than the grain.
Caudatum	Grains are markedly asymmetrical. The side next to the lower glume is flat, or even somewhat concave, with the opposite side being rounded and bulging. Style is persistent, often at the tip of a beak, pointing toward the lower glume. Glumes are half the length of the grain or less.
Kafir	Grains are approximately symmetrical and more or less spherical. Glumes are clasping and variable in length.
Durra	Grains are rounded, obovate, wedge-shaped at the base, and broadest slightly above the middle. Glumes are very wide. The tip has a different texture than the base and often has a transverse crease across the middle.

Source: Reddy et al. (2002).

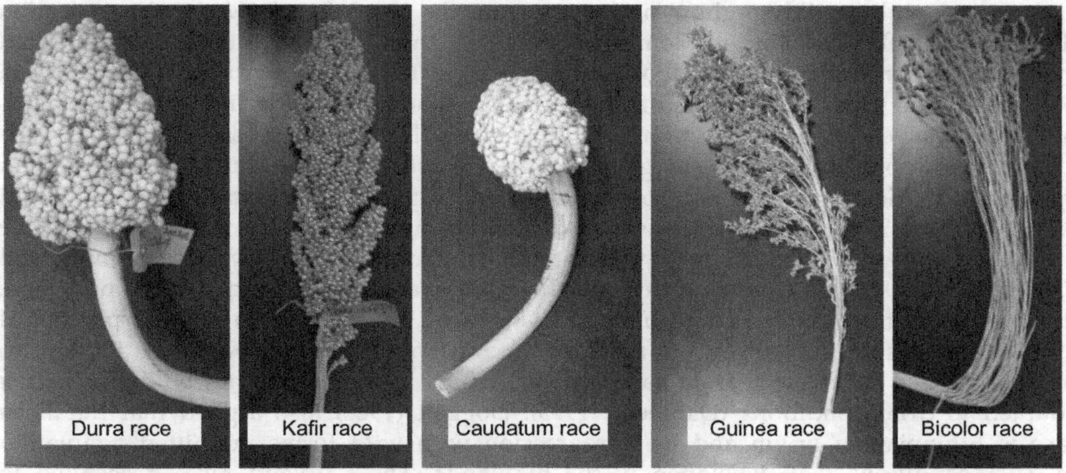

FIGURE 2.1 Races of *S. bicolor*.

2.2 ORIGIN

Mann et al. (1983) hypothesized that the origin and early domestication of sorghum took place approximately 5000 years ago in northeastern Africa. Wendorf et al. (1992) reported new evidence that places the origin and domestication at 8000 years before present (BP) on the Egypt-Sudan border. Thus, there seems to be no argument against the African origin of sorghum (Kimber, 2000), a concept that is also supported by the largest diversity of the cultivated and wild sorghum in Africa (de Wet, 1977; Doggett, 1988). The great diversity of *S. bicolor* has been created through disruptive selection (i.e., selection for extreme types) and by isolation and recombination in the extremely varied habitats of northeast Africa and the movement of people carrying the species throughout the continent (Doggett, 1988). On the Indian subcontinent, evidence for early cereal cultivation was discovered at an archaeological site in the western part of Rojdi (Saurashtra) dating to about 4500 BP (Damania, 2002). The Indian subcontinent is considered to be the secondary center of origin of sorghum (Vavilov, 1992).

Bicolor is widely distributed in Africa and coastwise from India to Indonesia and then to China (de Wet and Price, 1976). It appears that this race arose in east Africa from the subsp. *aethiopicum*, and the great diversity found in Asia occurred after its introduction there. However, according to Dahlberg (1995), early *bicolors* are believed to have arisen from the subspecies *verticilliflorum* in central Africa, and they were thought to have introgressed with wild forms and gave rise to the races *caudatum*, *kafir*, *guinea*, and *durra* (Dahlberg, 1995). As the *bicolors* moved to the west, they came into contact with wild *Sorghum arundinaceum*, from which the race *guinea* evolved (Dahlberg, 2000). Harlan and Stemler (1976) considered *guineas* to be the oldest of the races, mostly found in western and eastern Africa. The morphological affinities and distribution indicate that the *guinea* race was probably derived from selection among wild members of the subsp. *arundinaceum*. The *caudatum* race is thought to have been derived from an introgressed cross of an early *bicolor* with a wild sorghum and arose from the area of early *bicolor* domestication (Dahlberg, 2000). According to Stemler et al. (1975), the *caudatum* race was domesticated later than *bicolor* and *guinea* and thought to be segregated out of *bicolor*. It is dominant in parts of Sudan, Chad, Nigeria, and most parts of Uganda. This is an important race agronomically, especially in combination with other races. The *bicolors* crossed with wild *verticilliflorum* in northern Africa and gave rise to the *kafir* race and carried by the Bantu speakers of Africa to the east and south (Dahlberg, 1995). According to de Wet (1978) and Harlan et al. (1976), the *kafir* race was derived from an early *bicolor* race and may have migrated before the *guinea* race was segregated from the *bicolor* race (Smith and Frederiksen, 2000). It is cultivated from northern Nigeria to west to northern Ghana, where there is a gene flow between the *guinea* and *kafir* races. Its distribution and morphological affinities suggest that it arose from the subsp. *verticilliflorum*. The *durra* race is thought to have originated in Ethiopia from early *bicolors*, which introgressed with wild *aethiopicum* adapted to drier conditions (Dahlberg, 1995). Doggett (1988) also argued that the *durra* race is Ethiopian in origin and its introgression with wild forms permitted adaptation to drier conditions that developed in the highlands. These upland races descended to the lowlands, and the adapted varieties migrated west through Yemen and Saudi Arabia to India.

The *durra* race has a compact panicle, indicating adaptation to low-rainfall environment (Mann et al., 1983). *Durras* are distributed in a belt of 10–15° N latitude from Ethiopia to Mauritania; that is, in the mid-altitude highlands of Ethiopia, the Nile Valley of Sudan and Egypt, and in India and Pakistan (Kimber et al., 2013). There appears to be three centers of morphological diversity: the Ethiopian-Sudan region, the near East, and India (Smith and Frederiksen, 2000).

Most of the winter sorghums grown in India belong to the *durra* race. Harlan and Stemler (1976) felt that *durra* sorghums were selected from early *bicolor*, which had been carried to India before 3000 BP and may have been domesticated in India. Doggett (1988) argues that *durras* originated in Ethiopia since the whole sequence of wild *type-bicolor-durra* is clearly represented there. Doggett (1988) suggested that the early *bicolors* introgressed with wild forms, and they got adapted to drier conditions, which led to the development of the *durra* race. The *durras* moved to the west through Sudan and began to occupy the drier regions below the southern margin of the Sahara. Then they moved through the Horn of Africa and worked their way to India. The compact panicle and predominantly white seeds of the *durra* race are indications that it adapted to low-rainfall environments with a low risk of grain mold (Mann et al., 1983). However, white-grain sorghum is desirable as human food because of its low tannin content and plant improvement programs continue to expand its area of adaptability (Smith and Frederiksen, 2000). The *durras* were widely cultivated by Muslim Africans and Arabic people in Ethiopia. In Ethiopia, the Muslim Oromos (Gallo), who settled in the fertile warm highland almost 500 years ago cultivated the *durra* race, which formed the foundation of their agricultural system (Harlan et al., 1973). The *durras* are presently distributed in the mid-altitude highlands of Ethiopia, the Nile Valley of Sudan and Egypt, and in a belt 10–15° N latitude from Ethiopia to Mauritania. They also are grown in the Islamic and Hindu areas of India and Pakistan (Smith and Frederiksen, 2000). Haaland (1995, 1998) proposed a hypothesis called the *Haaland hypothesis*. The Haaland hypothesis proposes that wild *S. bicolor* was exported to India, where it became domesticated *durra* and was subsequently reintroduced into Africa during Islamic times. In India, *durra* reaches its most extreme forms with creases on both glumes, while in Africa, it is often modified through hybridization with other races.

References

Berenji, J., Dahlberg, J., 2004. Perspectives of Sorghum in Europe. J. Agron. Crop Sci. 1905, 332–338.

Celarier, R.P., 1959. Cytotaxonomy of the Andropogonea. III. Sub-tribe Sorgheae, genus, *Sorghum*. Cytologia. 23, 395–418.

Clayton, W.D., 1961. Proposal to conserve the generic name *Sorghum* Moench (Gramineae) versus *Sorghum* Adans (Gramineae). Taxon. 10, 242–243.

Clayton, W.D., Renvoize, S.A., 1986. Genera Graminum. Grasses of the world. Kew Bulletin Additional Series XIII, Royal Botanic Gardens, London, pp. 338–345.

Dahlberg, J.A., 1995. Dispersal of sorghum and the role of genetic drift. Afr. Crop Sci. J. 3 (2), 143–151.

Dahlberg, J.A., 2000. Classification and characterization of sorghum. In: Smith, C.W., Frederiksen, R.A. (Eds.), Sorghum: Origin, History, Technology, and Production. Wiley, New York, NY, pp. 99–130.

Damania, A.B., 2002. The Hindustan Centre of origin of important plants. Asian Agric. Hist. 6 (4), 333–341.

de Wet, J.M.J., 1977. Domestication of African cereals. Afr. Econ. Hist. 3, 15–32.

de Wet, J.M.J., 1978. Systematics and evolution of sorghum sect. Sorghum (Gramineae). Am. J. Bot. 65, 477–484.

de Wet, J.M.J., Harlan, J.R., 1971. The origin and domestication of *Sorghum bicolor*. Econ. Bot. 25, 128–135.

de Wet, J.M.J., Huckabay, J.P., 1967. The origin of *Sorghum bicolor*. II. Distribution and domestication. Evolution. 21, 787–802.

de Wet, J.M.J., Price, E.G., 1976. Plant domestication and indigenous African agriculture. In: Harlan, J.R., De Wet, J.M.J., Stemler, A. (Eds.), Origins of African Plant Domestication. Mouton, The Hague, pp. 453–464.

Doggett, E., 1988. Sorghum. John Wiley & Sons, New York, NY.

Garber, E.D., 1950. Cytotaxonomic studies in the genus *Sorghum*. Univ. Calif. Publ. Bot. 23, 283–361.

Haaland, R., 1995. Sedentism, cultivation and plant domestication in the Holocene middle Nile region. J. Field Archaeol. 22, 157–174.

Haaland, R., 1998. Theory and evidence of archaeological interpretation of the transition from gathering to domestication. In: Hather, J., Gosden, C. (Eds.), Change in Agrarian Systems. Routledge, London.

Harlan, J.R., de Wet, J.M.J., 1972. A simplified classification of cultivated sorghum. Crop. Sci. 12, 172–176.

Harlan, J.R., Stemler, A., 1976. The races of Sorghum in Africa. In: Harlan, J.R., de Wet, J.M.J., Stemler, A. (Eds.), Origins of African Plant Domestication. Mouton, The Hague, pp. 465–478.

Harlan, J.R., de Wet, J.M.J., Price, E.G., 1973. Comparative evolution of cereals. Evolution. 27, 311–351.

Harlan, J.R., de Wet, J.M.J., Stemler, A.B.L. (Eds.), 1976. Origins of African Plant Domestication. Mouton, The Hague.

House, L.R., 1978. A Guide to Sorghum Breeding. International Crop Research Institute for Semi-Arid Tropics (ICRISAT), Patancheru, India.

IBPGR, ICRISAT, 1993. Descriptors for Sorghum (*Sorghum bicolor* (L.) Moench). International Board for Plant Genetic Resources; International Crops Research Institute for the Semi-Arid Tropics, Rome; Patancheru, India.

Kimber, C., 2000. Origins of domesticated sorghum and its early diffusion to India and China. In: Smith, C.W., Frederickson, R.A. (Eds.), Sorghum: Origin, History, Technology, and Production. Wiley, New York, NY, pp. 3–98.

Kimber, C.T., Dahlberg, J.A., Kresovich, S., 2013. The Gene Pool of Sorghum bicolor and Its Improvement. In: Genomics of the Saccharinae. Plant Genetics and Genomics: Crops and Models, 11. Springer, pp. 23–41. ISBN: 978-1-4419-5946-1 (Print), 978-1-4419-5947-8 (Online).

Mann, J.A., Kimber, C.T., Miller, F.R., 1983. The origin and early cultivation of sorghums in Africa. Texas Agricultural Experiment Bulletin, Texas A&M University, College Station, TX, Report 1454.

Reddy, B.V.S., Ashok Kumar, A., Sanjana Reddy, P., 2008. Designing a sorghum genetic improvement program. In: Reddy, B.V.S., Ramesh, S., Ashok Kumar, A., Gowda, C.L.L. (Eds.), Sorghum Improvement in the New Millennium. International Crops Research Institute for the Semi-Arid Tropics, Patancheru, India, pp. 23–27, 340 pp.

Reddy, V.G., Rao, N.K., Reddy, B.V.S., Prasada Rao, K.E., 2002. Geographic distribution of basic and intermediate races in the world collection of sorghum germplasm. Int. Sorghum Millet Newsl. 43, 15–17.
Smith, C.W., Frederiksen, R.A., 2000. Sorghum: Origin, History, Technology, and Production. John Wiley & Sons, New York, NY.
Stemler, A.B.L., Harlan, J.R., de Wet, J.M.J., 1975. Evolutionary history of cultivated sorghums (*Sorghum bicolor* [Linn.] Moench) of Ethiopia. Bull. Torrey Bot. Club. 102, 325–333.
Vavilov, N.I., 1992. The Origin and Geography of Cultivated Plants. Cambridge University Press, Cambridge.
Wendorf, F., Close, A.E., Schild, R., Wasylikowa, K., Housley, R.A., Harlan, J.R., et al., 1992. Saharan exploitation of plants 8,000 years bp. Nature. 359, 721–724.

CHAPTER

3

Morphology and Breeding Behavior

OUTLINE

This chapter intends to brief novices and first-time sorghum researchers with the basics of the sorghum plant's morphological features, growth stages, and breeding behavior. These aspects are useful for developers of crop improvement programs to know.

3.1 MORPHOLOGY

3.1.1 Root

The root provides water and nutrients to the seedling. Upon germination, an embryonic/primary root appears, and later, several such roots develop that remain unbranched or sparsely branched. The primary roots subsequently die. First, a set of secondary roots develops from the first node. They occupy a 5–15-cm area in the soil around the base of the stem. Another type of secondary roots develops from the second node and above. These roots develop into an extensive root system. These roots branch out laterally (about 1 m^2), interlacing the soil vertically and reaching a depth of 2 m. They are the main suppliers of nutrients to the plant. Brace roots may appear later on the lowermost nodes and may be numerous if the plant is unadapted. They are stunted and thick above ground level and thin in the soil. These roots provide anchorage to the plant (House, 1985).

Genetic Enhancement of rabi sorghum – Adapting the Indian Durras.
DOI: http://dx.doi.org/10.1016/B978-0-12-801926-9.00003-0

3.1.2 Leaf

Sorghum leaves are borne alternately on two rows along the stem. A leaf consists of a sheath and a blade/lamina. The sheath is attached to the node and surrounds the internode, and often the node above it as well. The outer surface of the sheath is covered with bloom. The blades are broad at the base and taper upward to a fine point. Leaf-blade margins are smooth or scarbid. The midrib is prominent, greenish, brown, or white. The blades are thicker at the base than at the tip and along the midrib than along the margins. There is a short (1–3-mm), triangular, membranous ligule at the junction of the leaf blade and the sheath. The ligule deflects the lamina from the stem at an angle. In cultivated sorghum, there are usually 14–17 leaves. The leaves in the middle of the plant are slightly longer than those in the upper part. The topmost leaf, which is short and broad, is called the *boot leaf* or *flag leaf*. The leaves may be as long as 1 m and may vary in width from 10 to 15 cm (House, 1985). The white, ear-shaped structures on both sides of the base of the lamina are called the *auricles*. They act as hinges to facilitate the movement of the lamina. A unique characteristic of sorghum leaves is the rows of motor cells along the midrib on the upper surface of the leaf. These cells can roll up leaves rapidly during moisture stress.

3.1.3 Stem

The stem or culm of sorghum consists of many alternating nodes and internodes. It ranges from slender to very strong, and from solid and dry to succulent and sweet; it is 0.5–5 cm in diameter near the base, and 0.5–4 m in length (House, 1985). A cross section of the stem appears oval or round. The internodes are covered by a waxy layer called *bloom*, which reduces transpiration. Each node appears as a ring at the base of the leaf sheath. This is the point at which the leaf is attached to the stem. A bud is formed at each node except the one bearing the flag leaf. At times, these buds develop tillers. Basal tillers develop from the axillary buds of the lower nodes, and nodal tillers from the axillary buds of the upper nodes. The topmost internode bearing the panicle is called the *peduncle*. The larger the diameter of the peduncle, the larger is the panicle size. The peduncle may be straight or curved.

3.1.4 Inflorescence

The inflorescence of sorghum is the panicle. The vegetative primordium (growing tip) differentiates into the reproductive primordium. The shoot apex elongates into the main axis of the inflorescence, which is called the *rachis*. The rachis tapers off toward the top and is grooved longitudinally. It possesses three types of hairs: (i) the fine types spread in the rachel furrows; (ii) the long hairs on the rachel ridges; and (iii) the scarbid hairs on the ridges. The rachis elongates and forms branches and branchlets with a rapid increase in dimensions. This results in the formation of primary-, secondary-, and tertiary-branch primordia. The tips of the tertiary-branch primordia develop into two paired spikelet primordia, one hermaphrodite and the other staminate. In some cases, one hermaphrodite and two staminate spikelets are seen. The primordial differentiation into floral parts may take about 30 days after sowing (House, 1985). Sorghum panicles vary morphologically, ranging from compact to open. After the complete unfolding of the flag leaf, the peduncle elongation forces the panicle out of the leaf sheath (boot). The part of the peduncle between the base of the lamina of the flag leaf and the base of the peduncle is exertion. The variation in the shape, size, and length of sorghum panicles is due to variation in rachis length, branch length, distance between whorls, and the angle of branching. Sorghum spikelet development is basipetal: those in the upper region of the panicle develop earlier than those in the lower.

A raceme consists of one or several spikelets. One spikelet of a raceme is always sessile and the other pedicellate, except the terminal sessile spikelet, which is accompanied by two terminal pedicellate spikelets. The length of the raceme varies according to the number of nodes and the length of the internodes. Some species have one to four nodes, and others four to eight. Internode length, thickness, and hairiness also vary from genotype to genotype.

Sessile spikelets: The shape of sessile spikelets ranges from lanceolate to almost round or ovate. Sometimes they are depressed in the middle. At flowering, they are green but then change shades, becoming straw- or cream-colored buff, yellow, red, brown, purple, or almost black at maturity. There are two glumes, which vary from hairy to nonhairy. The glumes are hard and tough with nerves, and are obscure except near the tip. In some species, the glumes are thin and brittle, while in others, they are thin and papery. The lower glume is enclosed by the upper glume with its membranous margin. The lower glume is usually flat and conforms more or less to the shape of the spikelet. The upper glume is more convex or boat-shaped. The seed may be enclosed by the glume or may protrude from it either partially or completely. The number of sessile spikelets in a single inflorescence of cultivated sorghum varies from 2000 to 4000 (House, 1985). There are two lemmae, composed of a delicate, white, thin, and papery tissue. The lower lemma is elliptical or oblong and equal in length to the glume. The upper lemma is short and ovate and may be awned. There are two lodicules and a palea. The spikelet has two pistils and three stamens. The stigma is fluffy, attached to a short style extending to the ovary. The anthers are attached to long, threadlike filaments.

Pedicellate spikelets: These are narrower than the sessile spikelets and are lanceolate. They may be smaller or longer than or of the same size as the sessile spikelets. They are male or neutral, or they may rarely have a rudimentary ovary. The lemmae are short, and the upper lemma rarely has an awn. Three stamens and two lodicules are found between the lemma and the palea. The lodicules at the base of the floret are truncate, fleshy, and ciliate.

3.1.5 Flowering

The floral initiation in cultivated sorghum starts 30–40 days after germination. About 6–10 days before flowering, the boot forms a bulge in the sheath of the flag leaf. Sorghum usually flowers in 55–70 days in warm climates (House, 1985), depending on the genotype. Two days after the emergence of the inflorescence from the boot, the flowers begin to open. The flowering starts in the sessile spikelets at the tip of the inflorescence and progresses toward the bottom over 4 or 5 days. It takes 6 days for the whole inflorescence to complete flowering. The maximum flowering takes place on the third or fourth day. At flowering, the glumes open, and the three anthers fall free while the two stigmas protrude, each in a stiff style (House, 1985). As the stamens emerge from the opening glumes, they rotate and spread outward. The filaments elongate rapidly, and the anthers become pendent. When flowering of the sessile spikelets is halfway through inflorescence, the pedicellate spikelets start opening from the tip and proceed downward, completing flowering earlier than the sessile spikelets in the inflorescence. The time taken from the commencement of glume opening to completion of its closing is about 1–2 h, which varies from cultivar to cultivar. Flowering starts at midnight and continues up to 10 AM, depending on the genotype and climate. The maximum anthesis is between 6 AM and 8 AM. Wet and cool weather delays flowering. The anthers dehisce when they are dry and the pollen grains are ejected into the air and onto the stigma. Sorghum is primarily self-pollinated (cross-pollination is only 2–10%). The florets of some of the very long, glumed types do not open for outcrossing to take place, a phenomenon called *cleistogamy.* The discovery of cytoplasmic male sterility in sorghum has made it possible to produce commercial hybrid seeds. A good male-sterile plant does not develop anthers, or the anthers remain shriveled without pollen (House, 1985).

3.1.6 Pollination and Fertilization

Flower-opening is facilitated by the swelling of the lodicules. When stigma becomes visible, the stamen filaments elongate and the anthers become pendent. This process takes about 10 min. The flower remains open for 30–90 min. After the anther dehiscence, the pollen shedding is through the apical pore. The stigma is pollinated before the emergence of the anthers from the spikelets. When pollen grains land on the stigma, they germinate immediately and develop pollen tubes, each with two nuclei, one vegetative nucleus and two sperm nuclei. One sperm nucleus fertilizes the egg to form an embryo (2n) and the other nucleus fuses with the polar nuclei to form the endosperm (3n). Sorghum has a 20-chromosome complement. After pollination, the glumes close, though the empty anthers and stigmas still protrude. The pollen retain their viability for 5 h at room temperature. In the refrigerator, the pollen retain their viability for 3–4 days (Sanchez and Smeltzer, 1965). The pollen require light to germinate (Artschwager and McGuire, 1949). The stigma remains receptive for 10 days. Under normal conditions, stigma fertilization takes place in 2 h. Organ differentiation occurs over the following 12 days, and the embryo continues to grow until the seed is mature (Schertz and Dalton, 1980).

3.1.7 Seed and Seed Development

After fertilization, the endosperm nuclei form a small number of free nuclei near the zygote. These form a cellular tissue, by which time the zygote will undergo a second nuclear division. The development of the embryo is gradual. The deposition of starch grains begin about 10 days after fertilization. Seed development is in three stages: milk stage, early or soft-dough stage, and late or hard-dough stage (Murty et al., 1994). The pericarp develops with the growth of the embryo. The mesocarp and the hypodermis of the pericarp are chlorophyllous in the developing ovary. As the starch grains develop in the endosperm, the green color disappears (Sundararaj and Thulasidas, 1980). The sorghum seed is a free caryopsis, also called *grain*. Seeds are spherical in shape, but somewhat flat on one side, with the embryo at the base. They are red, brown, white, yellow, or cream-colored, with a dull or pearly luster. The endosperm is usually white, but sometimes it can be yellow. The nucellus is hyaline or brown. Brown nucelli develop seeds that have a poor appearance. The testa, when present, is colored and contains tannin (House, 1985). The seed shape varies from ovate, obovate, elliptical, and orbicular. The seed size also can be small, medium, or large (1–6 g/100 seeds). The seed consists of three parts: pericarp (outer coat, 6% by weight), endosperm (storage tissue, 84%), and embryo (germ, 10%). The pericarp is thin, unlike the mesocarp, which has many layers. The grain is composed of the embryonic axis and the scutellum. The embryo is made of 70% fat and 13% protein. The endosperm varies from comprising 100% soft tissue with a little corneous portion, to being a solid corneous seed. It contains a layer of aleurone cells, with the outer corneous endosperm surrounding a central floury endosperm. The grain matures in 30–35 days after fertilization. At physiological maturity, a dark brown callus tissue is formed at the base where the seed is attached to the spikelet. This callus tissue stops the translocation of nutrients from the plant to the seed. At physiological maturity, the seed contains 25–30% moisture and is fully viable. For safe storage, seed moisture should be brought down to 10–12%.

3.1.8 Growth Stages

Vanderlip and Reeves (1972) described the growth stages in temperate sorghums on a 0–9 scale. The growth stages of temperate sorghums are not adequate to characterize the Indian tropical sorghums, where the growing conditions and seasons are different from those of temperate countries. Phenology and growth stages of tropical sorghum were described by Rao et al. (2004). The growth stages of Indian tropical sorghums have been characterized on a 0 (emergence) to 9 (physiological maturity) scale (Table 3.1). The durations of these growth stages may vary based on planting date, genotype, and location (latitude).

Stage 0 (emergence): Emergence is considered to have occurred when the seedlings are seen above the soil surface. This can be identified when the coleoptile is visible at the soil surface, which takes about 4 days. Furthermore, sorghum emergence will vary depending on the depth of planting, seed vigor and soil moisture, temperate, and physicochemical characteristics.

Stage 1 (three-leaf stage): This stage is identified when the seedlings have three fully expanded leaves, and the collar of three leaves is clearly visible, which occur 6 days after emergence (DAE) (Table 3.1) and the seedling grow to a height of 20 cm.

Stage 2 (five-leaf stage): This stage is identified by the appearance of a visible collar in all five leaves, with continuous visibility of the first leaf with a round tip at 16 DAE (Table 3.1). The seedlings enter a grand period of growth at this stage. The seedlings grow to a height of 50 cm.

Stage 3 (panicle initiation stage): This stage is identified at 32 DAE, when the meristem transforms from the vegetative (leaf producing) to the reproductive phase

TABLE 3.1 Identification of Growth Stage Characteristics in Sorghum

Growth stage number	Days from emergence	Duration (days)	Identification characteristics
0	0	0	Emergence: Coleoptile is visible at soil surface (first leaf is seen with a round tip).
1	6	6	Three-leaf stage: Collar of third leaf visible.
2	16	10	Five-leaf stage: Collar of fifth leaf visible.
3	32	16	Growing point differentiation (panicle initiation): Approximately nine-leaf stage by previous criteria.
4	50	18	Flag leaf visible: Tip of flag leaf (final leaf) visible in the whorl.
5	60	10	Boot: Head extend into flag leaf sheath.
6	68	8	50% flowering: Half of the plants has completed 50% pollination.
7	80	12	Soft dough: Squeezing kernel between fingers results in little or no milk.
8	96	16	Hard dough: Seed cannot be compressed between fingers.
9	106	10	Physiological maturity: Black layer (spot) appears on the hilum, at the base of the seed.

N.B. Planting to emergence takes 4 days.

(panicle producing) (Table 3.1). Seedlings grow to a height of 95–100 cm. Panicle initiation can be observed by splitting the stalk with a sharp knife and examining under a compound microscope. During this stage, seedlings develop 9–10 leaves, depending upon maturity group, and the basal 2–3 leaves may become senescenced. Culm growth increases rapidly following this stage.

Stage 4 [flag leaf (final leaf) visible]: The stage is identified as taking place at 50 DAE, and 18 days after stage 3. The stage can be recognized by observing the appearance of the tip of the flag leaf in the whorl (Table 3.1). Plants exhibit rapid leaf and culm elongation during this stage. All the leaves except the top three to four are expanded, and the basal three to five leaves may be dropped due to senescence. The plant grows to a height of 115–120 cm.

Stage 5 (boot stage): The stage can be identified at 60 DAE as a swollen flag leaf sheath enclosing the panicle, which gives the appearance of a boot shape (Table 3.1). The flag leaf is the last leaf to emerge from the growing tip. Plants produce maximum leaf area and panicle development is completed, and they grow to a height of 125–130 cm. The plants experience high water demand, and hence their response to irrigation is greatest at this stage.

Stage 6 (50% flowering): It can be identified when 50% of the plants in the field are in anthesis, and this takes about 68 – 70 DAE and 8 days after stage 5 (Table 3.1). The plants grow to a height of 150 – 160 cm. Flowering typically start 5–7 days after panicle exertion and progresses from the tip to the bottom of the panicle. The crop is said to be at 50%

flowering when anthesis occurs on 50% of the plants in the field. The flowering duration (from starting to end) usually takes 4−9 days.

Stage 7 (soft dough stage): Following flowering, seed development progresses from the milky stage through the soft dough stage, which can be identified when a kernel is squeezed between the fingers to test if little or no milk is present. This stage can be identified at 80 DAE and 12 days after flowering. This stage signals the end of culm elongation. About 8−10 functional leaves are observed and may vary with the cultivar. Plants grow to about 170 cm tall.

Stage 8 (hard dough stage): At this stage, the seed cannot be compressed between the fingers, and it occurs about 96 DAE (Table 3.1). Plants become susceptible to lodging and charcoal rot if the crop suffers from severe moisture stress. Lodging also occurs by defoliation due to insect pests and leaf diseases during flowering through this stage. Also, heavy rain or hail driven by wind may cause lodging.

Stage 9 (physiological maturity): This stage can be identified when a dark spot (black layer) appears at the basal portion of seed, which signals the end of photosynthate supply to the seed. Physiological maturity occurs in about 106 DAE and 10 days after stage 8. Seed moisture content at this stage varies between 25% and 35%, and seeds gain maximum dry weight. The crop can be harvested at 20% seed moisture content. The seeds must be dried to 12−14% moisture content for safe storage. On average, 1000 seeds weigh 25 g, but they may range from 13 to 40 g (Figure 3.1).

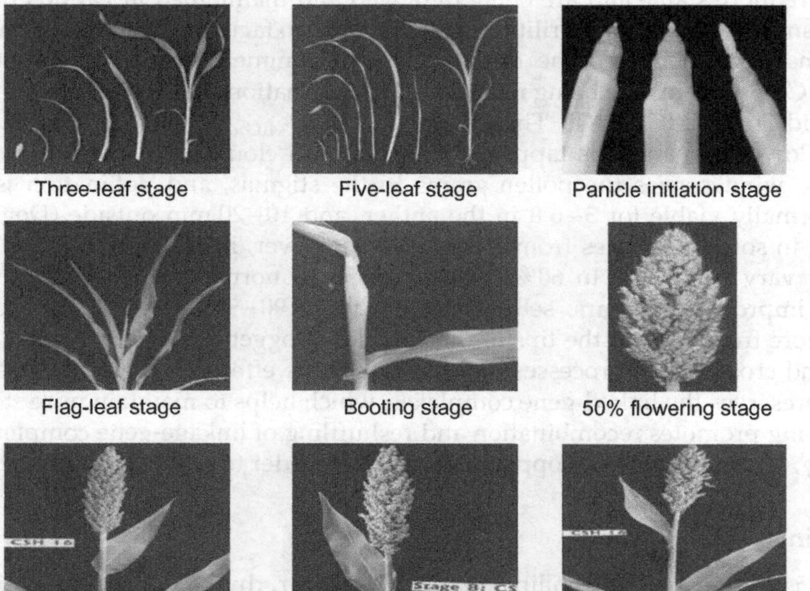

FIGURE 3.1 Growth stages of sorghum.

3.2 BREEDING BEHAVIOR AND POLLINATION CONTROL

Plant breeding is basically a directed evolution. Presence of genetic variability is a prerequisite for a plant breeder to direct and control this variability to develop improved cultivars or hybrids. Selfing and crossing are essential tools in the regulation of variability in plant-breeding programs. An understanding of breeding behavior is essential for systematic and scientific improvement of traits of interest to sorghum researchers and their clientele (i.e., producers), as well as for deciding the breeding methods to be followed (House, 1985).

Breeders should know the breeding behavior of sorghum before launching a breeding program because breeding methods largely depend on the pollination control mechanisms (Reddy et al., 2008). The inflorescence in sorghum is called *panicle*, with racemes on tertiary rachis, each with one or several spikelets. One spikelet is sessile and the other pedicillate, except the terminal sessile spikelet, which is accompanied by two pedicillate spikelets. The sessile spikelets have both male (androecium) and female (gynaecium) parts, and the pedicillate are usually male or female in sex. Outcrossing occurs to an extent of 5 – 20%, depending on the weather conditions and genotypes. However, it is usually handled as self-pollinated species in breeding. Outcrossing is mediated by the wind. Anthers mature before stigmas and come out of glume (called *protandry*), but there is variation among landraces. Anthesis (flower opening) begins from florets at the top to those at the base of the panicle, usually in the morning hours after 0800 (House, 1985). Outcrossing in sorghum can be facilitated by the use of genetic male sterility. There are nearly eight different recessive genes in homozygous conditions that contribute to male sterility. Among these, Ms_3 and Ms_7 genes are more stable and are being deployed and maintained in various populations. The cytoplasmic-nuclear male sterility (CMS) system also facilitates outcrossing in sorghum. CMS systems have facilitated the development of commercial hybrids in sorghum. As many as six CMS systems are being maintained at International Crops Research Institute for the Semi-Arid Tropics (ICRISAT). These are A_1, A_2, A_3, $A_{4(G)}$, $A_{4(VZM)}$, and $A_{4(M)}$.

When a flowering panicle is tapped with a finger, a cloud of yellow pollen grains can be seen. The wind carries the pollen grains to the stigmas, and pollination is achieved. Pollen is normally viable for 3–6 h in the anther, and 10–20 min outside (Doggett, 1988). Outcrossing in sorghum varies from 1% to 10%. However, in types with loose/open panicles, it may vary from 30% to 60% (House, 1985). In normal compact or semi-compact panicles in improved cultivars, selfing can be up to 90–95%, with 5–10% outcrossing occurring more frequently at the tips of the panicles (Doggett, 1988).

Selfing and crossing are processes that have opposite effects. Selfing promotes homozygosity and preserves the linked-gene complexes, which helps to maintain pure stocks of cultivars. Crossing promotes recombination and reshuffling of linkage-gene complexes leading to variability, which provides an opportunity to the breeder to select upon (Allard, 1960).

3.2.1 Selfing

Sorghum is primarily a self-pollinated crop. However, due to occurrence of 5–10% outcrossing (by wind), it is essential that the breeder is careful while selfing. Complete selfing in breeding programs is accomplished by using kraft paper bags. Before covering the panicle (up to 5–8 cm of the peduncle) with kraft paper bags, it is essential to cut the tip of the panicle and flowered florets to prevent chance cross-pollination. Bags can be removed after 10–15 days and the panicles suitably labeled.

3.2.2 Crossing

Crossing is a process of transferring pollen grains from a floret of one panicle to the stigma of a floret of another panicle. In nature, it is usually affected by the wind, as stigmas remain receptive up to a week or more after anthesis depending upon temperature and humidity (Doggett, 1988). Stigmas are most receptive during the first 3–5 days after their emergence.

The gain or the efficiency from breeding is directly proportional to the amount of useful variability created by crossing two or more parents. Hybridization or crossing of sorghum on a field scale is made feasible through the use of genetic (Ayyangar and Ponnaiya, 1939; Stephens, 1937), cytoplasmic, and cytoplasmic-genetic male sterility systems (Stephens and Holland, 1954) (450). In addition, limited scale crossing can be carried out through (i) emasculation with hot water and the plastic bag technique or (ii) hand-emasculation and pollination techniques (House, 1985). Emasculation with hot water and the plastic bag technique is a cumbersome method that requires a lot of preparation. It leaves some selfs in the F_1 that need to be thoroughly checked and rouged out. It is always safer to follow the hand-emasculation method, which can be easily done by unskilled staff with some training (House, 1985).

References

Allard, R.W., 1960. Principles of Plant Breeding. Wiley, New York, NY.

Artschwager, E., McGuire, R.C., 1949. Cytology of reproduction in Sorghum vulgare. J. Agric. Res. 78, 659–673.

Ayyangar, G.N.R., Ponnaiya, B.W.X., 1939. The occurrence and inheritance of panicle tip sterility in sorghum. Curr. Sci. 8, 116.

Doggett, E., 1988. Sorghum. John Wiley & Sons, New York, NY.

House, L.R., 1985. A Guide to Sorghum Breeding, second ed. International Crops Research Institute for the Semi-Arid Tropics, Patancheru, India.

Murty, D.S., Tabo, R., Ajayi, O., 1994. Sorghum Hybrid Seed Production and Management. Information Bulletin no. 41. International Crops Research Institute for the Semi-Arid Tropics, Patancheru, India.

Rao, S.S., Seetharama, N., Kiran Kumar, K.A., Vanderlip, R.L., 2004. Characterization of Sorghum Growth Stages. NRCS Bulletin Series no. 14. National Research Centre for Sorghum, Rajendranagar, India, 20 pp.

Reddy, B.V.S., Ashok Kumar, A., Sanjana Reddy, P., 2008. Designing a sorghum genetic improvement program. In: Reddy, B.V.S., Ramesh, S., Ashok Kumar, A., Gowda, C.L.L. (Eds.), Sorghum Improvement in the New Millennium. International Crops Research Institute for the Semi-Arid Tropics, Patancheru, India, pp. 23–27, 340 pp.

Sanchez, R.L., Smeltzer, D.G., 1965. Sorghum pollen viability. Crop Sci. 111–113.

Schertz, K.F., Dalton, L.G., 1980. Sorghum. Hybridizat Ion of Crop Plants. American Society of Agronomy—Crop Science Society of America, Madison, WI, pp. 577–588.

Stephens, J.C., 1937. Male-sterility in sorghum: its possible utilization in production of hybrid seed. J. Am. Soc. Agron. 29, 690–696.

Stephens, J.C., Holland, R.F., 1954. Cytoplasmic male-sterility for hybrid sorghum seed production. Agron. J. 46, 20–23.

Sundararaj, D.D., Thulasidas, G., 1980. Botany of Field Crops. The MacMillan Company India Ltd, New Delhi, India, 508 pp.

Vanderlip, R.L., Reeves, H.E., 1972. Growth stages of sorghum. Agron. J. 64, 13–16.

CHAPTER

4

Genetic Variability for Qualitative and Quantitative Traits

OUTLINE

The success of crop genetic diversification programs depends on the magnitude of the variability that exists in the genetic resources in hand. Genetic variability in *Sorghum bicolor* has been assessed by several researchers at three levels: (i) morphological/phenotyping, (ii) biochemical, and (iii) DNA/nucleotide.

4.1 MORPHOLOGICAL/PHENOTYPIC LEVEL

At the phenotypic level, genetic variability can be assessed by employing both univariate (range, variance, etc.) and multivariate statistical tools (Mahalanobi's D^2 statistic, principal component, and canonical and factor analysis).

1. *Univariate analysis*: Voluminous literature is available on this subject, and it is evident from the literature that a vast reservoir of variability exists for several traits of interest in *sorghum*. To quote a few, Mahajan et al. (2011), Kamatar et al. (2011), and Shinde et al. (2010) have attempted to assess genetic variability in *rabi* sorghum using univariate statistical tools (Table 4.1). This serves as a raw material for the genetic enhancement/prebreeding of winter sorghum and the development of varieties and hybrids. One of the constraints for rapid genetic enhancement has been thought to be the lack of sufficient genetic variability in *rabi* sorghum. However, many workers have reported high variability, heritability, and genetic advance for several of the traits (Table 4.2). A few studies have reported low genetic variability (Lonc, 1969;

Genetic Enhancement of rabi *sorghum – Adapting the Indian Durras.*
DOI: http://dx.doi.org/10.1016/B978-0-12-801926-9.00004-2

TABLE 4.1 Variability for Agronomic and Yield Traits in *Rabi* Sorghum

Trait	Range	GCV	PCV	Heritability	Genetic advance	Reference
Days to 50% flowering	53–86	3.7–8.8	4.3–8.8	73–98	4.3–11.1	Mahajan et al. (2011), Kamatar et al. (2011), Shinde et al. (2010)
Days to maturity	105–140	4.7–5.2	4.7–6.0	77–97	11–11.7	Mahajan et al. (2011), Kamatar et al. (2011)
Panicle length (cm)	8–32.5	11.6–18.3	14.4–40.9	11–88	1.8–8.5	Mahajan et al. (2011), Kamatar et al. (2011), Shinde et al. (2010)
Panicle width (cm)	3.4–8.3	15.4–27.2	17.4–28.18	78–93	1.7–3.14	Mahajan et al. (2011), Shinde et al. (2010)
Plant height (cm)	138–278	13.3–16.1	13.4–16.8	73–98	45.3–67.1	Mahajan et al. (2011), Kamatar et al. (2011), Shinde et al. (2010)
Number of leaves per plant	7–12	10	12.6	64	1.6	Shinde et al. (2010)
Number of internodes per plant	7–12	10	12.6	64	1.6	Shinde et al. (2010)
Number of primary branches per panicle	35–87	18.3–20.6	19.7–21.4	87–93	20.6–24.7	Mahajan et al. (2011), Shinde et al. (2010)
Number of grains per panicle	567–3625	20.9–27.1	23.7–28.2	78–92	473.4–1286.8	Mahajan et al. (2011), Shinde et al. (2010)
Test weight (g)	14.8–36	14.2–15.47	15.2–17.21	81–87	6.8–7.1	Mahajan et al. (2011), Shinde et al. (2010)
Harvest index (%)	14.4–55.1	35.49	36.72	93	23.72	Mahajan et al. (2011)
Grain yield/panicle (g)	12.8–94.3	21.9–33.9	23.4–38.4	56–96	13.4–39.6	Mahajan et al. (2011), Kamatar et al. (2011), Shinde et al. (2010)
Fodder yield/panicle (g)	32.6–84.5	18	19.4	87	20.9	Shinde et al. (2010)
Seed protein content	7.1–12.4%					Deshpande et al. (2003), Chavan et al. (2009)
Seed starch content	58.1–76%					Deshpande et al. (2003), Chavan et al. (2009)
Soluble/free sugars in grain	0.6%–2.5%					Deshpande et al. (2003), Chavan et al. (2009)
Phenolic % in grain	0.1–0.5					Chavan et al. (2009)

GCV: Genotypic coefficient of variation; PCV: Phenotypic coefficient of variation.

TABLE 4.2 Variability, Heritability and Genetic Advance for Quantitative Traits in Sorghum

	References		
Character	**Variability**	**Heritability**	**Genetic advance**
Days to 50% flowering	Sindagi and Singh (1970), Basu (1971), Khanure (1993), Reddy et al. (1996)	Liang et al. (1969), Chung and Liang (1970), Basu (1971), Phul et al. (1972), Naphade (1973), Singh and Singh (1973), Bhat (1975), Eckebil et al. (1977), Dhimer and Desai (1978), Pauli (1980), Singh and Makne (1980), Nagabasaih (1981), Kukadia et al. (1983), Bello and Obilana (1985), Phul and Allah Rang (1986), Worthman et al. (1987), Khanure (1993), Reddy et al. (1996)	Basu (1971), Phul et al. (1972), Bhat (1975), Reddy et al. (1996)
Plant height	Krantikumar et al. (1970), Basu (1971), Naphade and Ailwar (1977), Singh and Makne (1980), Patel et al. (1980b), Berenji (1990)	Naphade (1973), Singh and Singh (1973), Shinde and Nayeem (1979), Singh and Makne (1980), Salilkumar and Singhania (1984), Kumar and Singh (1986), Worthman et al. (1987), Rao and Patil (1996), Biradar et al. (1996a,b), Nguyen et al. (1999)	Basu (1971), Naphade (1973), Shinde and Nayeem (1979), Singh and Makne (1980), Patel et al. (1980a), Salilkumar and Singhania (1984), Khanure (1993), Sankarapandian et al. (1996), Biradar et al. (1996a,b), Nguyen et al. (1999)
Panicle length	Swarup and Chaugale (1962), Khanure (1993), Biradar et al. (1996a,b)	Swarup and Chaugale (1962), Fanous et al. (1971), Singh and Singh (1973), Bhat (1975), Patel et al. (1980a), Khanure (1993), Biradar et al. (1996a,b)	Bhat (1975), Shinde (1981), Patil and Thombre (1986), Biradar et al. (1996a,b)
Number of primaries per panicle	Berenji (1990), Khanure (1993), Biradar et al. (1996a,b)	Giriraj and Goud (1981), Desai et al. (1983), Kukadia et al. (1983), Dinakar (1985), Patil and Thombre (1986), Khanure (1993), Biradar et al. (1996a,b)	Desai et al. (1983), Patil and Thombre (1985), Khanure (1993), Biradar et al. (1996a,b)
Panicle weight	Asthana et al. (1996), Biradar et al. (1996a,b), Rao and Patil (1996)	Shinde and Nayeem (1979), Kumar and Singh (1986), Khanure (1993), Biradar et al. (1996a,b)	Khanure (1993), Biradar et al. (1996a,b)
1000-grain weight	Sindagi and Singh (1970), Abu EL-Gasim (1975), Shinde and Nayeem (1979), Singh and Makne (1980), Prabhakar (2001)	Shinde and Nayeem (1979), Obilana and Okoh (1984), Kumar and Singh (1986), Phul and Allah Rang (1986), Nguyen et al. (1999), Prabhakar (2001)	Singh and Makne (1980), Shinde (1981), Nguyen et al. (1999)
Grain yield	Abu EL-Gasim (1975), Shinde et al. (1978), Kumar and Singh (1986), Rao and Patil (1996), Reddy et al. (1996), Biradar et al. (1996a,b) Can and Yoshida (1999), Prabhakar (2001)	Shinde and Nayeem (1979), Desai et al. (1983), Salilkumar and Singhania (1984), Kumar and Singh (1986), Phul and Allah Rang (1986), Rao and Patil (1996), Prabhakar (2001)	Singh and Makne (1980), Phul and Allah Rang (1986)

Naphade and Ailwar, 1977; Patel et al., 1980; Raja and Parikh, 1980; Berenji, 1990; Wenzell et al., 1998), low heritability (Ciobanu, 1968; Liang and Walter, 1968; Liang et al., 1969; Fanous et al., 1971; Singh and Singh, 1973; Jan-Orn, 1974; Miller, 1975; Rao and Goud, 1979; Kanaka and Goud, 1982; Obilana and Okoh, 1984; Kulkarni and Shinde, 1987), and low level of genetic advance (Fanous et al., 1971; Naphade, 1973; Singh and Singh, 1973; Patel et al., 1980b; Khanure, 1993).

2. *Multivariate analysis*: Crop improvement efforts embark on the concurrent upgradation of a number of component traits like duration, disease, and pest resistance, drought resistance, and of course, grain and fodder yield and quality. Therefore, genotypes are better differentiated simultaneously on a set of traits of agronomic importance. Multivariate methods are useful for the evaluation of a large number of genotypes that are to be assessed for several characters of agronomic and physiological importance (Peeters and Martinelli, 1989). In sorghum, multivariate statistical tools, such as Mahalanobi's D^2 statistic (1936), principal component analysis (Pearson, 1901), and factor and canonical analysis (Spearman, 1904), have been used to differentiate the genotypes/genetic resources (Reddy et al., 2006). The multivariate methods were used for the first time as a measure of genetic divergence to classify germplasm resources in crop plants in the Biometrical Genetics Unit of Indian Agricultural Research Institute during the 1960s (Arunachalam et al., 1998). The potential of these methods was demonstrated in a reclassification of the species variability in the genus, with *Eu-sorghum* replacing the classification made by Snowden (1936) based on herbarium specimen measurements (Chandrasekhariah et al., 1969). Earlier, Murty et al. (1967) employed the combination of Mahalonobi's D^2 statistic and canonical and factor analysis to catalogue and classify a collection of genetic stocks of *sorghum* from all over the world. Murty and Arunachalam (1967) demonstrated the use of factor analysis in assessing genetic diversity in the genus *Sorghum*. Ayana and Bekele (1999) have used multivariate methods, including principal component, cluster, and discriminant analyses, to assess the patterns of morphological variation and to group 415 sorghum accessions for 15 quantitative characters. Thus, multivariate statistical tools were used to group the genotypes into different clusters. In general, divergence analysis have been used in attempts to identify suitable parents for realizing heterotic F_1s, which in turn can be exploited commercially or can be used to derive superior recombinant inbred lines for further selection. As demonstrated theoretically by Cress (1966), the higher the genetic divergence between the parents, the higher is the heterosis of the F_1s resulting from them. Vast literature is available to endorse this assumption, although a few contradictory results are also reported. Nevertheless, divergence and clustering analysis of sorghum has been carried out by several researchers. To cite only a few, Ayana and Bekele (1999), Barthate et al. (2000), Narkhede et al. (2001), Kadam et al. (2001), Singh et al. (2001), and Umakanth et al. (2002) are some of the recent researchers who have attempted to assess diversity in sorghum germplasm using multivariate methods.

Days to flowering, panicle weight, panicle length, number of primaries per panicle, dead heart percentage, and grain yield contributed to diversity and the clustering pattern and indicated no relationship between the geographical diversity and genetic diversity

(Chittapur et al., 2013). Genetic diversity in postrainy sorghum was studied by many researchers (including Umakanth et al., 2003; Arunkumar et al., 2004), and these studies found that days to 50% flowering contributed most to genetic divergence, followed by plant height, panicle length, number of primary branches, and panicle weight (Biradar et al., 1996a,b). However, one of the studies reported that the seed yield contributed the most (94.20%) to the genetic divergence of the genotypes, followed by plant height (6.49%) and days to 50% flowering (2.10%), while panicle length and test weight indicated a narrow range of diversity among the genotypes under study (Kumar et al., 2010).

4.2 BIOCHEMICAL LEVEL

Biochemical markers can be grouped into low-molecular-weight markers (secondary metabolites), protein markers, and DNA markers (Bretting and Widrlechner, 1995). Low-molecular-weight markers include flavanoids and anthocyanin pigments, nonprotein amino acids, cyanogens, polyacetylenes, alkaloids, and peptides. Proteins are direct products of messenger RNA (mRNA). Mutations of the DNA may result in slightly different protein changes due to amino acid substitutions. This results in differences in mobility when separated by gel electrophoresis. These mobilities have been used for diversity studies since they can directly reveal genetic polymorphism through demonstrating multiple forms of a specific enzyme. Basically, there are three different types of protein markers: seed proteins, isozymes, and allozymes.

Genotyping differentiation can also be achieved by assessing the variation in isozymes (i.e., enzymes with same catalytic activity but with different molecular weight and different mobility in an electric field). Different forms of an enzyme sharing the same catalytic activity but coded by different alleles at the same locus are called *allozymes*, and those coded by more than one gene locus are called *isozymes* (Ford-Lloyd and Painting, 1996). However, the term *isozyme* is generally used to mean both classes.

The difference in enzyme mobility is caused by point mutations that result in amino acid substitution, such that isozymes reflect the products of different alleles rather than genes (Chahal and Gosal, 2002). The banding pattern observed on electrophoresis of enzymes from a genotype is characteristic of its alleles. The enzyme bands are representative of alleles, and the banding pattern of different genotypes results in polymorphism. This principle has been exploited in plant breeding for quantifying genetic variability, characterizing germplasm, and identifying varieties or hybrids. Of the several isozyme systems, esterase and peroxidases that are easily assessable have been commonly employed in crop plants to characterize the germplasm or to assess the variability in genotypes or in the identification of varieties/hybrids (Chahal and Gosal, 2002).

While Arti (1993) and Reddy and Jacobs (2000) used soluble proteins and isozymes markers, respectively, Schertz et al. (1990) and Aldrich et al. (1992) used allozymes to assess the genetic variability of sorghum germplasm. However, these biochemical markers are not widely used in quantifying the genetic divergence in sorghum, as the number of useful and easily assessable isozymes is a limiting factor (Figure 4.1).

FIGURE 4.1 Diversity in sorghum panicles.

4.3 DNA LEVEL

Differentiation among the sorghum genotypes was higher using molecular markers than using morphological markers. No two individuals are similar with respect to the nucleotide sequence of their DNA. This is because it is estimated that crop plants have a total of about 10^8-10^{10} nucleotides of DNA (Paterson et al., 1991). Wide differences exist between any two individuals chosen at random from a population of plants than between individuals that are clonally or apomictically propagated. Genetic variation consists of sequence variation and structure alteration. Sequence variation normally is manifested by single-nucleotide polymorphisms (SNPs), microsatellites or simple sequence repeats (SSRs), short sequence insertions and deletions (indels), and transposable elements. Structural alteration is generally described as presence/absence variations and copy number variations, which include large-scale deletions, insertions, duplications, inversions, and translocations (Zheng et al., 2011).

The variation in nucleotide sequence can be exploited to assess the genetic diversity in the germplasm or in breeding material or genetic stocks. However, assessing the individual differences in the nucleotide sequences would be laborious, tedious and cost ineffective. Hence, a large number of DNA markers, such as restriction fraction length polymorphism (RFLP), random amplified polymorphic DNA (RAPD), sequence characterized amplified region markers, and SSRs, have been developed based on the methodology of assaying and the type of DNA sequences to characterize DNA polymorphism (Chahal and Gosal, 2002). These DNA markers have been used to assess and characterize genetic variability in sorghum genetic resources as well.

Some of the studies in which DNA markers have been used to quantify sorghum germplasm diversity include Cui et al. (1995), Yang et al. (1996), Ahnert et al. (1996), Menkir et al. (1997), Dean et al. (1999), Ayana et al. (2000), Thimmaraju et al. (2000), Jeya Prakash et al. (2006), Vittal et al. (2010), and several studies looked at winter sorghum germplasm diversity, including Sameer Kumar et al. (2010), Madhusudhana et al. (2012), and

Ganapathy et al. (2012). However, this list is incomplete, and the few references listed here are just to illustrate the use of DNA markers in estimating sorghum germplasm diversity. These studies have indicated that DNA markers are effective in detecting the genetic variability, and in most of them, RFLP and RAPD markers were extensively used. Thudi and Fakrudin (2011) studied the genetic diversity among 42 *rabi* sorghum accessions representing landraces (19), advanced breeding lines (16), local cultivars (2), and release varieties (5), and 30 SSR markers revealed about 7.6 mean number of alleles per locus, showing 93.3% polymorphism, and an average polymorphism information content of 0.78 was observed. The average heterozygosity and effective number of alleles per locus were 0.8 and 6.65, respectively.

Upon comparison of SSR data sets across genotyping technologies and laboratories, discrepancies in terms of genotyped data are frequently observed. This technical concern introduces biases that hamper any synthetic studies or comparison of genetic diversity between collections. To prevent this for *S. bicolor*, a kit for diversity analysis (http://sat.cirad.fr/sat/sorghum_SSR_kit/) was developed by two laboratories [Centre de Cooperation Internationale en Recherche Agronomique pour le Developpement (CIRAD), in France, and International Crops Research Institute for the Semi-Arid Tropics (ICRISAT), in India] using two commonly used genotyping technologies (polyacrylamide gel-based technology with LI-COR sequencing machines and capillary systems with ABI sequencing apparatus). It contains information on 48 technically robust sorghum microsatellite markers and 10 DNA controls. It can be used further to calibrate sorghum SSR genotyping data acquired with different technologies and compare those to genetic diversity references (Billot et al., 2012).

References

Abu EL-Gasim, E.H., 1975. Variety and interrelations among characters in indigenous grain sorghums of the Sudan. East Afr. Agric. Forest J. 41 (2), 25–133.

Ahnert, D., Lee, M., Austin, D.F., Woodman, W.L., Openshaw, S.J., Smith, J.S.C., et al., 1996. Genetic diversity among the elite *Sorghum* inbred lines assessed with DNA markers and pedigree information. Crop Sci. 36, 1385–1392.

Aldrich, P.R., Doebley, J., Schertz, K.F., Stec, A., 1992. Patterns of allozyme variation in cultivated and wild *Sorghum bicolor*. Theor. Appl. Genet. 85, 451–460.

Arti, M., 1993. Water-soluble proteins of Indian sorghum cultivars. Sorghum News. 34, 53.

Arunachalam, V., Prabhu, K.V., Sujata, V., 1998. Multivariate methods of quantitative evaluation for crop improvement: conventional and molecular approaches revisited. In: Chopra, V.L., Singh, R.B., Varma, A. (Eds.), Crop Productivity and Sustainability—Shaping the Future. Proc. 2nd International Crop Science Congress. Oxford & IBH Publishing Co., New Delhi, pp. 793–807.

Arunkumar, B., Biradar, B.D., Salimath, P.M., 2004. Genetic variability and character association studies in *rabi* sorghum. Karnataka J. Agric. Sci. 17 (3), 471–475.

Asthana, O.P., Asthana, N., Sharma, R.L., Shukla, K.C., 1996. Path analysis for immediate components of grain yield in exotic sorghum II 100-grain weight. Adv. Plant Sci. 9 (2), 29–32.

Ayana, A., Bekele, E., 1999. Multivariate analysis of morphological variation in sorghum [*Sorghum bicolor* (L.) Moench] germplasm from Ethiopia and Eritrea. Genet. Resour. Crop Evol. 273–284.

Ayana, A., Bekele, E., Bryngelsson, T., 2000. Genetic variation of Ethiopian and Eritrean sorghum [*Sorghum bicolor* (L.) Moench] germplasm assessed by random amplified polymorphism DNA (RAPD). Genet. Resour. Crop Evol. 47, 471–482.

Barthate, K.K., Patil, J.V., Thete, R.V., 2000. Genetic divergence in sorghum under different environments. Indian J. Agric. Res. 34, 85–90.

Basu, A.K., 1971. Note on a variability and heritability on winter season sorghum crops. Indian J. Agric. Sci. 41, 1116–1117.

Bello, S.A., Obilana, A.T., 1985. Inheritance study in grain sorghum. Sorghum Newsl. 28, 76.

Berenji, J., 1990. Variability and interrelation of characters in different genotypes of broom corn [*Sorghum bicolor* (L.) Moench]. Biltenza Hamelj Sivalai: Lekovito Bilju. 22, 69.

Bhat, G.M., 1975. A study of genetic variability and formulation of selection indices in five F2 populations of sorghum (*Sorghum bicolor* (L) Moench). Mysore J. Agric. Sci. 9, 198–199.

Billot, C., Rivallan, R., Sall, M.N., Fonceka, D., Deu, M., Glaszmann, J.C., et al., 2012. A reference microsatellite kit to assess for genetic diversity of *Sorghum bicolor* (Poaceae). Am. J. Bot. 99 (6), 245–250.

Biradar, B.D., Gowda, P.P., Hunaje, R., Sajjan, A.S., 1996a. Variability studies among restorer and maintainer genotypes of *rabi* sorghum [*Sorghum bicolor* (L.) Moench]. J. Res., Andhra Pradesh Agric. Univ. (India). 24, 13–16.

Biradar, B.D., Parameshwarappa, R., Patil, S.S., Goud, J.V., 1996b. Heterosis studies involving diverse sources of cytoplasmic genetic male sterility system in sorghum. Karnataka J. Agric. Sci. 9, 627–634.

Bretting, P.K., Widrlechner, M.P., 1995. Genetic markers and plant genetic resource management. In: Janick, J. (Ed.), Plant Breeding Reviews, vol. 13. John Wiley & Sons, Canada, pp. 11–86.

Can, N.D., Yoshida, T., 1999. Genotypic and phenotypic variances and covariances in early maturing grain sorghum in a double cropping. Plant Prod. Sci. 2 (1), 67–70.

Chahal, G.S., Gosal, S.S., 2002. Principles and Procedures of Plant Breeding—Biotechnological and Conventional Approaches, Narosa Publishing House, New Delhi, India, 486.

Chandrasekharaiah, S.R., Murty, B.R., Arunachalam, V., 1969. Multivariate analysis of divergence in the genus Eu-sorghum. Proc. Nat. Inst Sci. India B. 35, 172–195.

Chavan, U.D., Patil, J.V., Shinde, M.S., 2009. Nutritional and roti quality of sorghum genotypes Indonesian. J. Agric. Sci. 10 (2), 80–87.

Chittapur, R., Biradar, B.D., Salimath, P.M., Sajjanar, G.M., 2013. Genetic divergence for grain quality and productivity traits in *rabi* sorghum. Int. J. Agric. Innov. Res. 1 (5), 132–135.

Chung, J.H., Liang, G.H.L., 1970. Some biometrical studies on nine agronomic traits in grain sorghum (*Sorghum bicolor* L. Moench) I. Variance components and heritability estimates. Can. J. Genet. Cytogenet. 12, 288–296.

Ciobanu, U., 1968. The heritability coefficient, a genetic test of quantitative characteristics of hybrid grain sorghum. Bullet. Saint Univ. Caralova. 10, 419–426.

Cress, C.E., 1966. Heterosis of the hybrid related to gene frequency differences between two populations. Genetics. 53, 269–274.

Cui, Y.X., Xu, G.W., Magill, C.W., Schertz, K.F., Hart, G.E., 1995. RFLP based assay of *Sorghum bicolor* (L.) Moench genetic diversity. Theor. Appl. Genet. 90, 787–796.

Dean, R.E., Dahlberg, J.A., Hopkins, M.S., Mitchell, S.E., Kresovich, S., 1999. Genetic redundancy and diversity among orange accessions in the U.S. National *Sorghum* collection as assessed with simple sequence repeat (SSR) markers. Crop Sci. 39, 1215–1221.

Desai, M.S., Desai, K.B., Kukadia, M.V., 1983. Heterobeltiosis for grain and its components in sorghum. Sorghum Newsl. 26, 88.

Deshpande, S.P., Borikar, S.T., Ismail, S., Ambekar, S.S., 2003. Genetic studies for improvement of quality characters in *rabi* sorghum using Landraces. Int. Sorghum Millets Newsl. 44, 6–8.

Dhimer, I.R., Desai, K.B., 1978. Genetic variability correlation and path coefficient analysis of grain yield in some winter sorghum. Sorghum Newsl. 21, 23.

Dinakar, B.L., 1985. Genetic Analysis of Some Quantitative Characters in Sorghum (*Sorghum bicolor* L. Moench) (M. Sc. (Agri.) Thesis). University of Agricultural Science, Bangalore, Karnataka, India.

Eckebil, J.P., Ross, W.M., Gardner, C.O., Marnille, J.W., 1977. Heritability estimates, genetic correlation and predicted gains from 51 progeny tests in three sorghum random mating populations. Crop Sci. 17, 373–377.

Fanous, M.A., Weibel, D.C., Morrison, R.D., 1971. Quantitative inheritance of some head and seed characteristics in sorghum [*Sorghum bicolor* (L.) Moench]. Crop Sci. 11, 787–789.

Ford-Lloyd, B., Painting, K., 1996. Measuring Genetic Variation Using Molecular Markers. IPGRI, Rome, Italy, pp. 39–43.

Ganapathy, K.N., Gomashe, S.S., Rakshit, S., Prabhakar, B., Ambekar, S.S., Ghorade, R.B., et al., 2012. Genetic diversity revealed utility of SSR markers in classifying parental lines and elite genotypes of sorghum (*Sorghum bicolor* L. Moench). Aust. J. Crop Sci. 6 (11), 1486–1493.

Giriraj, K., Goud, J.V., 1981. Heterosis for vegetative characters in grain sorghum. Indian J. Hered. 13, 9–13.

Jan-Orn, J., 1974. Estimates of genetic and environmental components of variance in some quantitative genetic traits from families derived from NP3R random mating sorghum population and their application in breeding system. Diss. Abstr. Int. 35, 885–886.

Jeya Prakash, S.P., Biji, K.R., Michael Gomez, S., Ganesa Murthy, K., Chandra Babu, R., 2006. Genetic diversity analysis of sorghum (*Sorghum bicolor* L. Moench) accessions using RAPD markers. Indian J. Crop Sci. 1 (1–2), 109–112.

Kadam, D.E., Patil, F.B., Bhor, T.J., Harer, P.N., 2001. Genetic diversity studies in sweet sorghum. J. Maharashtra Agric. Univ. 26, 140–143.

Kamatar, M.Y., Kotragouda, M., Shinde, D.G., Salimath, 2011. Studies on variability, heritability and genetic advance in F3 progenies of kharif x *rabi* and *rabi* x *rabi* crosses of sorghum (*Sorghum bicolor* L. Moench). Plant Arch. 11 (2), 899–901.

Kanaka, S.K., Goud, J.V., 1982. Inheritance of quantitative characters in sorghum. Mysore J. Agric. Sci. 15, 36–39.

Khanure, S.K., 1993. Variability, Correlation, Path Analysis and Stability Analysis for Quantitative Characters in *Rabi* Sorghum (*Sorghum bicolor* L. Moench) (M.Sc. (Agri.) thesis). University Agricultural Science, Dharwad, Karnataka (India).

Krantikumar, Chandola, R.P., Buouta, D.S., 1970. Sorghums of Rajasthan—variability studies. Poona Agric. Coll. Mag. 60, 33–41.

Kukadia, M.U., Desai, K.V., Desai, M.S., Patel, R.H., Rajan, K.R.V., 1983. Estimates of heritability and other genetic parameters in sorghum. Sorghum Newsl. 26, 31.

Kulkarni, N., Shinde, V.K., 1987. Genetic analysis of yield components in *rabi* sorghum. J. Maharashtra Agric. Univ. 12, 378–379.

Kumar, C.V.S., Sreelakshmi, Ch., Shivan, D., 2010. Genetic diversity analysis in *rabi* sorghum (*Sorghum bicolor* L. Moench) local genotypes. Electron. J. Plant Breed. 1 (4), 527–529.

Kumar, R., Singh, K.P., 1986. Genetic variability, heritability and genetic advance in grain sorghum [*Sorghum bicolor* (L.) Moench]. Farm Sci. J. 1, 1–2.

Liang, G.H.L., Walter, T.L., 1968. Heritability estimates and gene effects for agronomic traits in grain sorghum (*Sorghum vulgare* Pers). Crop Sci. 8, 77–81.

Liang, G.H.L., Overlx, C.B., Casady, A.J., 1969. Interrelations among agronomic characters in grain sorghum (*Sorghum bicolor* L. Moench). Crop Sci. 9, 299–302.

Lonc, W., 1969. Heritability of a few morphological characters in sorghum. Genet. Polon. 10, 113.

Madhusudhana, R., Balakrishna, D., Rajendrakumar, P., Seetharama, N., Patil, J.V., 2012. Molecular characterization and assessment of genetic diversity of sorghum inbred lines. Afr. J. Biotechnol. 11 (90), 15626–15635.

Mahajan, R.C., Wadikar, P.B., Pole, S.P., Dhuppe, M.V., 2011. Variability, correlation and path analysis studies in sorghum. Res. J. Agric. Sci. 2 (1), 101–103.

Menkir, A., Goldsbrough, P., Ejeta, G., 1997. RAPD based assessment of genetic diversity in cultivated races of sorghum. Crop Sci. 37, 564–569.

Miller, F.R., 1975. Characterization of seed size in sorghum [*Sorghum bicolor* (L.) Moench]. Diss. Abstr. Int. Biol. 36, 19b.

Murty, B.R., Arunachalam, V., 1967. Factor analysis of genetic diversity in the genus *Sorghum*. Indian J. Genet. 27, 123–135.

Murty, B.R., Arunachalam, V., Saxena, M.B.L., 1967. Cataloguing and classifying a world collection of genetic stocks of sorghum. Indian J. Genet. 27A, 1–312.

Nagabasaih, K.H.M., 1981. Genetic Analysis of Ten Quantitative Characters in F2 Generation of a Seven Parent Diallel Set in Sorghum [*Sorghum bicolor* (L.) Moench] (M.Sc. (Agri.) thesis). University of Agricultural Science, Dharwad, Karnataka, India.

Naphade, D.S., 1973. Heritability and genetic advance for grain yield, flowering and plant height following a sorghum cross. PMKV Res. J. 1, 153–155.

Naphade, D.S., Ailwar, V.L., 1977. Variability, heritability and path analysis in jowar [*Sorghum bicolor* (L.) Moench]. Coll. Agric. Nagpur Mag. 49, 17–23.

Narkhede, B.N., Akade, J.H., Awari, V.R., 2001. Genetic diversity in *rabi* sorghum local types [*Sorghum bicolor* (L.) Monech]. J. Maharashtra Agric. Univ. 25, 245–248.
Nguyen, D., Harydnio, T.A., Yoshida, T., 1999. Genetic variability and character association analysis in grain sorghum. J. Fac. Agric. Krishi Univ. 43, 25–30.
Obilana, A.T., Okoh, P.N., 1984. Relationship between some agronomic and physicochemical traits in long season sorghum. Zeitschrift fur pflanzezii chtung. 92, 239–248.
Patel, R.H., Desai, K.S., Kukadia, M.V., Desai, D.T., 1980. Component analysis in sorghum. Sorghum Newsl. 23, 23–24.
Patel, R.H., Desai, K.B., Kabadia, M.U., 1980a. Combining ability analysis in sorghum. Sorghum Newsl. 23, 23–24.
Patel, R.H., Desai, K.B., Raj, K.R.V., Parikh, R.K., 1980b. Estimates of heritability and genetic advance and other genetic parameters in an F2 populations of sorghum. Sorghum Newsl. 23, 22–23.
Paterson, A.H., Tanksley, S.D., Sorells, M.E., 1991. DNA markers in plant improvement. Adv. Agron. 46, 39–90.
Patil, R.C., Thombre, M.V., 1985. Inheritance of shoot fly and earhead midge resistance in sorghum. Curr. Res. Rep. 1, 44–48.
Patil, R.C., Thombre, M.V., 1986. Genetic parameters, correlation coefficient and path analysis in F1 and F2 generations of a 9 × 9 diallel cross of sorghum. J. Maharashtra Agric. Univ. 8 (2), 162–165.
Pauli, M.H., 1980. Variance components for different characters in grain sorghum. Sorghum Newsl. 29, 18–19.
Pearson, K., 1901. On the lines and planes of closet fit to a system of points in a space. Philos. Mag. 2, 557–572.
Peeters, J.P., Martinelli, J.A., 1989. Hierarchical cluster analysis as a tool to manage variation in germplasm collections. Theor. Appl. Genet. 78, 42–48.
Phul, P.S., Allah Rang, 1986. Genetic evaluation of sorghum genotypes for grain yield and quality traits. Res. Dev. Rep. 3, 44–49.
Phul, P.S., Arora, N.D., Mahndiratta, 1972. Genetic variability and correlations of fodder yield and its components in sorghum. Punjab Univ. J. Res. 9, 422–427.
Prabhakar, 2001. Heterosis in *rabi* sorghum [*Sorghum bicolor* (L.) Moench]. Indian J. Genet. Plant Breed. 61, 364–365.
Raja, K.R.V., Parikh, R.K., 1980. Estimates of heritability and other genetic parameters in sorghum. Sorghum Newsl. 23, 24–25.
Rao, M.J.V., Goud, J.V., 1979. Genetic analysis of per cent protein in grains of five sorghum inbreds. Curr. Res. Commun. 4, 441–448.
Rao, M.R.G., Patil, S.J., 1996. Variability and correlation studies in F2 population of *kharif* and *rabi* sorghum. Karnataka J. Agric. Sci. 9 (1), 78–84.
Reddy, B.V.S., Ramesh, S., Reddy, P.S., 2006. Sorghum genetic resources, cytogenetics, and improvement. In: Singh, R.J., Jauhar, P.P. (Eds.), Genetic Resources Chromosome Engineering and Crop Improvement, Cereals, vol. 2. CRC Press, Taylor & Francis Group, Boca Raton, FL, pp. 309–363.
Reddy, N.P.E., Jacobs, M., 2000. Polymorphism among kakirins and esterases in normal and lysine-rich cultivars of *Sorghum*. Indian J. Genet. 60, 159–170.
Reddy, P.R.R., Das, N.D., Sankar, G.R.M., Girija, A., 1996. Genetic parameters in winter sorghum (*Sorghim bicolor*) genotypes associated with yield and maturity under moisture stress and normal conditions. Indian J. Agric. Sci. 66 (11), 661–664.
Salilkumar, Singhania, D.L., 1984. Genetic advance and heritability estimates for grain yield and its components. Sorghum Newsl. 27, 15.
Sameer Kumar, C.V., Shreelakshmi, Ch., Shivani, D., 2010. Genetic diversity analysis in *rabi* sorghum (*Sorghum bicolor* L. Moench) local genotypes. Electron. J. Plant Breed. 1 (4), 527–529.
Sankarapandian, R., Rajarathinam, S., Mupidathi, 1996. Genetic variability, correlation and path coefficient analysis of jaggery yield and related attributes in sweet sorghum. Madras Agric. J. 83, 628–631.
Schertz, K.F., Stec, A., Deobley, J.F., 1990. Isozyme genotypes of sorghum lines and hybrids in the United States. Miscellaneous publication, Texas Agricultural Experimental Station 1719: 15.
Shinde, D.G., Biradar, B.D., Salimath, P.M., Kamatar, M.Y., Hundekar, A.R., Deshpande, S.K., 2010. Studies on genetic variability among the derived lines of B × B, B × R and R × R crosses for yield attributing traits in *rabi* sorghum (*Sorghum bicolor* (L.) Moench). Electron. J. Plant Breed. 1 (4), 695–705.
Shinde, V.K., 1981. Genetic variability, inter relationship and path analysis of yield and its components in sorghum [*Sorghum bicolor* (L.) Moench]. J. Maharashtra Agric. Univ. 6, 30–32.
Shinde, V.K., Nayeem, K.A., 1979. Evaluation of advanced generation material for protein and lysine content in sorghum. Sorghum Newsl. 22, 19.

Shinde, V.K., Nayeem, K.A., Ramesh, C.R., 1978. Study of heritability and genetic advances in grain sorghum (*Sorghum bicolor* (L.) Moench). Sorghum Newsl. 44, 12.

Sindagi, S.V., Singh, D., 1970. Variation and heritability of some quantitative characters in F2 progenies of inter varietal crosses of sorghum. Indian J. Genet. 30, 660–664.

Singh, A.R., Makne, V.G., 1980. Estimates of variability parameters in sorghum [*Sorghum bicolor* (L.) Moench]. J. Maharashtra Agric. Univ. 5, 80–81.

Singh, D., Singh, V., 1973. Study of heritability and genetic advance in *Sorghum vulgare*. Sci. Cult. 39, 455–456.

Singh, G., Singh, H.C., Ramakrishna, Singh, S., 2001. Genetic divergence in *Sorghum bicolor* (L.) Moench. Ann. Agric. Res. 22, 229–231.

Snowden, J.D., 1936. The Cultivated Races of Sorghum. Allard & Co., London, UK.

Spearman, C., 1904. General intelligence objectively determined and measures. Am. J. Phychol. 25, 201–293.

Swarup, V., Chaugale, D.S., 1962. Studies on genetic variation I (phenotypic variations and its heritable components in some important quantitative characters contributing towards yield). Indian J. Genet. 22, 31–36.

Thimmaraju, R., Krishna, T.G., Kuruvinashetti, M.S., Ravikumar, R.L., Shenoy, V.V., 2000. Genetic diversity among sorghum genotypes assessed with RAPD markers. Karnataka J. Agric. Sci. 13, 564–569.

Thudi, M., Fakrudin, B., 2011. Identification of unique alleles and assessment of genetic diversity of *rabi* sorghum accessions using simple sequence repeat markers. J. Plant Biochem. Biotechnol. 20 (1), 74–83, ISSN 0971-7811.

Umakanth, A.V., Madhusudhana, R., Madhavi Latha, K., Hema Kumar, P., Swarnalata Kaul, 2002. Genetic architecture of yield and its contributing characters in post rainy-season sorghum. Int. Sorghum Millet Newsl. 43, 37–40.

Umakanth, A.V., Madhusudhana, R., Madhavi Latha, K., Swaranalatha, K., Rana, B.S., 2003. Heterosis studies for yield and its components in *rabi* sorghum [*Sorghum bicolor* (L.) Moench]. Indian J. Genet. Plant Breed. 63, 159–160.

Vittal, R., Ghosh, N., Weng, Y., Stewart, B.A., 2010. Genetic diversity among *Sorghum bicolor* L. Moench genotypes as revealed by prolamines and SSR markers. J. Biotech Res. 2, 101–111.

Wenzell, W.G., van den Berg, J., Pretorius, A.J., 1998. Sources for combining ability, malt quality, and resistance to the aphid *Melanaphis sacchari* and stem borer *Chilo partellus* in sorghum inbred lines. Appl. Pl. Sci. 12, 53–56.

Worthman, C.S., Eastin, J.D., Andrews, A., 1987. The quantitative genetics of yield components in grain sorghum population. Sorghum Newsl. 30, 18–19.

Yang, W., de Oliveira, A.C., Godwin, L., Schertz, K.C., Bennetgen, J.L., 1996. Comparison of DNA marker technologies in characterizing plant genome diversity: variability in Chinese *Sorghums*. Crop Sci. 36, 1669–1676.

Zheng, L.Y., Guo, X.S., He, B., Sun, L.J., Peng, Y., Dong, S.S., et al., 2011. Genome-wide patterns of genetic variation in sweet and grain sorghum (*Sorghum bicolor*). Genome biology. 12 (11), R114. Available from: http://dx.doi.org/10.1186/gb-2011-12-11-r114.

CHAPTER

5

Genetics and Cytogenetics

OUTLINE

5.1 GENETICS

Economically important traits of crop plants can be grouped into two classes: (i) qualitative traits, which include the presence or absence of awns, male sterility, etc., which are controlled by one or two major genes and whose expression is not significantly influenced by the environment; and (ii) quantitative traits, whose inheritance depends on the gene differences at many loci, the effect of which are not individually distinguishable. However, knowledge about the nature of genetic control of both qualitative and quantitative characters of agronomic importance is fundamental in systematic and rapid improvement of these traits (House, 1985). In their review, Schertz and Stephens (1966) compiled and recommended standardized gene symbols for the sorghum species as recommended by the Sorghum Improvement Committee of North America (Rooney, 2000). There is a vast amount of literature on genetics of both qualitative and quantitative traits. An excellent review on genetics of qualitative traits is found in Doggett (1988), Murty and Rao (1997) and Rooney (2000). It can be easily realized from the literature that characteristics such as color of various plant parts, endosperm, and awn are mostly controlled by single genes (simply inherited) with multiple alleles at single loci. Genetics of resistance to diseases such as anthracnose, charcoal rot, rust, and downy mildew are also simply inherited with one to three genes. Thus, these traits are easily amenable for manipulation through breeding programs based on the needs. Genetics of some of the important traits are given later in this chapter.

Maturity: The initiation of flowering in *rabi* sorghum is sensitive to day length or photoperiod. Four maturity loci have been identified, and multiple alleles have been characterized at each locus. Of the four loci, Ma_1 conditions a photoperiod response, and under the long days of temperate environments, ma_1ma_1 genotypes flower 25–30 days earlier than plants

Genetic Enhancement of rabi sorghum – Adapting the Indian Durras.
DOI: http://dx.doi.org/10.1016/B978-0-12-801926-9.00005-4

carrying at least one Ma_1 allele, regardless of the allelic condition of the other maturity loci. Under short days, all Ma_1 genotypes flower at the same duration (Miller et al., 1968). Expression at the remaining loci, Ma_2, Ma_3, or Ma_4, is variable. The presence of the ma_3^R allele causes extreme earliness regardless of the genotypes present at any other locus. Childs et al. (1997) used the ma_3^R variants to clone the Ma_3 locus and found that Ma_3 codes a 123-kD phytochrome B and a frame shift mutation causes the absence of this protein in ma_3^R genotypes. The effect of temperature on maturity may be observed at the Ma_4 locus. Recessive ma_4 genotypes normally flower 20 days earlier than Ma_4 genotypes, but under high temperatures, ma_4 genotypes behave as dominant Ma_4 genotypes. Quinby (1967) reported that 13, 13, 16, and 12 different alleles had been identified at Ma_1, Ma_2, Ma_3, and Ma_4, respectively. In some photoperiod-sensitive landrace types of sorghum, the photoperiod sensitivity response is not under the control of Ma_1, Ma_2, Ma_3, or Ma_4. Molecular analysis of maturity in sorghum has identified at least six quantitative trait loci (Lin et al., 1995). Rooney and Aydin (1999) identified two dominant loci, Ma_5 and Ma_6, controlling photoperiod-sensitive response.

Plant height: Plant height in sorghum is conditioned by four major genes: Dw_1, Dw_2, Dw_3, and Dw_4 (Quinby and Martin, 1954; Hadley, 1957). The four genes are unlinked, and tall is incompletely dominant to short. The height of the plant depends on the number of loci at which the recessive alleles are present. The specific height reduction associated with each dwarfing allele is variable and depends on the number of recessive loci already present in the genotype. For example, a height reduction of 50 cm is associated with a single dwarfing gene, but this reduction is reduced if dwarfing alleles are present at the other height loci (Quinby, 1974). The plant height is not related to the days taken to maturity. Though growing conditions affect plant height, the general response of height is relatively consistent across environments.

Fertility: Genetic factors affecting fertility in sorghum can be divided into nuclear genetic sterility and cytoplasmic-nuclear sterility.

Genetic sterility: Male sterility may be due to the absence of pollen production or pollen abortion in sorghum. Three independent genes were identified and designated as ms_1-ms_3 by Ayyangar and Ponnaiya (1937b), Stephens (1937), and Webster (1965). For these three loci, anthers are normal, but the pollen is nonfunctional. Ayyangar (1942) described a genetic male sterile that results from the absence of pollen. Karper and Stephens (1936) described a male sterility mutant that resulted in the absence of anthers. Barabas (1962) used mutagens to induce male sterility mutants, ms_6, in which anther development is slow; anthers are small and without pollen.

Casady et al. (1960) first reported female sterility caused by a two-gene complementary dominant interaction (Fs_1, Fs_2) that leads to rudimentary stigma, style, and ovary. The effect of Fs genes seemed to be dosage-dependent, double heterozygous ($Fs_1fs_1Fs_2fs_2$) caused female sterility, and the presence of three dominant alleles ($Fs_1Fs_1Fs_2fs_2$) resulted in plants that were dwarfed and never headed. The homozygous dominant would have been possibly lethal.

Cytoplasmic-nuclear sterility: The discovery of male sterility resulting from the interaction of cytoplasmic and nuclear genes by Stephens and Holland (1954) laid the foundation and revolutionized the development of hybrid cultivar and hybrid seed production technology. Maunder and Pickett (1959) determined that a single gene, msc1, caused pollen sterility in the presence of a milo cytoplasm. Erichsen and Ross (1963) described a second locus, msc_2, that interacts in a complementary way with msc_1, in which sterility in milo cytoplasm was caused by a homozygous recessive genotype at either msc_1 or msc_2. Brengman (1995) identified a single dominant fertility restoration gene that fully restores fertility without the production of partially fertile plants. The locus was designated as Rf_1, but the genetic relationship between Rf_1 and msc_1 has not been determined (Brengman, 1995).

The original source of the cytoplasm was the milo race, which induced male sterility in the nuclear background of the *kafir* race and is designated as A_1 cytoplasm. Since then, several sources and types of male-sterile-inducing cytoplasms (A_1–A_6) have been discovered and are reported in Table 5.1. In all these cytoplasms, a single/oligo recessive gene in the

TABLE 5.1 Male-Sterility Inducing Cytoplasms of Sorghum

Cytoplasm fertility group	Identity	Source line	
		Race	**Origin**
A_1	Milo	D	–
	IS 6771C	G-C	India
	IS 2266C	D	Sudan
	IS 6705C	G	Burkina Faso
	IS 7502C	G	Nigeria
	IS 3579C	C	Sudan
	IS 8232C	(K-C)-C	India
	IS 1116C	G	India
	IS 7007C	G	Sudan
A_2	IS 1262C	G	Nigeria
	IS 2573C	C	Sudan
	IS 2816C	C	Zimbabwe
A_3	IS 1112C	D-(DB)	India
	IS 12565C	C	Sudan
	IS 6882C	K-C	USA
A_4	IS 7920C	G	Nigeria
9E	IS 7218	G	Nigeria
	IS 112603C		Nigeria
A_5	IS 7506C	B	Nigeria
A_6	IS 1056C	D	India
	IS 2801C	D	Zimbabwe
	IS 3063C	D	Ethiopia

Source: Schertz (1994).

nucleus and sterile cytoplasm induces male sterility. These male sterile cytoplasms are differentiated based on the inheritance patterns of their fertility restoration, which is unclear but depends on the specific cytoplasm and nuclear combinations. Fertility restoration is controlled by a single gene in some combinations (e.g., A_1) but is controlled by two oligo genes when the same nuclear genotype interacts with a different cytoplasm (Schertz, 1994) (Figure 5.1).

FIGURE 5.1 Isonuclear allocytoplasmic A/B-lines of sorghum.

Although diverse male-sterile cytoplasms have been identified, only the milo cytoplasm (A_1) male-sterility system is widely used because the hybrids based on this cytoplasm produce sufficient heterosis (20–30%) over the best available pure lines in sorghum. Although A_2 cytoplasm is as good as A_1 cytoplasm for mean performance as well as heterosis for economic traits such as grain yield, days to 50% flowering, and plant height, it is less popular than the A1 cytoplasm. The anthers of A_2 male-steriles, unlike the A_1 male-steriles, mimic the fertile or maintainer lines and lead to difficulties in monitoring the purity of hybrid seed production. Other alternate sources are not useful, primarily because restorer frequencies are low and male steriles cannot be readily distinguished from male fertiles. As such, there is a need to search for more useful forms of male sterility different from milo (A_1). Milo restorers need to be diversified in a guinea background to further enhance the yield advantage in hybrid development. Therefore, there is a need to identify and breed for high-yielding nonmilo cytoplasm restorers. Based on A_2 CMS systems, only one hybrid Zinza No. 2 has been developed and released in China for commercial cultivation (Shan et al., 2000). The genetics of other qualitative traits are given in Table 5.2.

TABLE 5.2 Genetics for Qualitative Traits in Sorghum

S. No.	Trait	Number of genes	Gene action	Reference
1	Seedling growth habit	Single gene (*So*)	Dominant	Ayyangar and Ponnaiya (1939a)
2	Coleoptile color	One to two genes	Dominant	Karper and Conner (1931), Ayyangar and Reddy (1942), Woodworth (1936)
3	Leaf color	Two genes (C_1, C_2)	Additive	Ayyangar and Nambiar (1941b)
4	Midrib and stem juiciness	Single gene (*D*)	Dominant for white midrib and recessive for juicy stem	Porter et al. (1978)
5	Liguleless	Single gene (lg_m)	Recessive	Singh and Drolsom (1973)
6	Waxy bloom on leaf sheath	1–2 *bm* genes		Ayyangar and Nambiar (1941a), Peterson et al. (1982)
7	Time of tiller development	Single "*tu*" gene		Ayyangar and Ponnaiya (1939a)
8	Tillering	Single "*Tx*" gene		Ayyangar and Ponnaiya (1939a)
9	Stem versus nonsweet stem	Single "*X*" gene	Dominant	Ayyangar et al. (1936)
10	Plant color	Two genes (*PQ*)	Tan color recessive	Ayyangar et al. (1933)
11	Panicle shape	Pa_1	Loose dominant to compact	Ayyangar and Ayyar (1938)
12	Spikelet shedding	Single gene Sh_1	Persistence dominant to shedding	Ayyangar et al. (1937)

(*Continued*)

TABLE 5.2 (Continued)

S. No.	Trait	Number of genes	Gene action	Reference
13	Pedicellate spikelets	Single gene "*hps*"	Hermaphrodite recessive to normal	Casady and Miller (1970)
14	Glume size	Single gene "*Sg*"	Short glumes dominant to long	Graham (1916)
15	Glume shape		Broad truncate dominant to narrow ovate	Vinall and Cron (1921)
16	Presence of hairs on glumes and hair color	Three independent genes		Ayyangar and Ponnaiya (1941a)
17	Glume texture	Single gene "*Py*"	Coriaceous dominant to papery	Ayyangar and Ponnaiya (1939b)
18	Glume color	*Gep* locus, two genes governing straw or faded glumes	Dominant	Ayyangar and Ponnaiya (1937a), Ayyangar and Ponnaiya (1941b)
19	Awn	Single gene "*A*"	Awnless dominant to awned	Vinall and Cron (1921)
20	Epicarp color	Two genes "*RY*"	Red dominant, recessive white	Graham (1916), Vinall and Cron (1921)
21	Purple splotching in pericarp	Single gene "*Pb*"	Dominant	Ayyangar et al. (1939)
22	Mesocarp thickness	Single gene "*Z*"	Thin mesocarp dominant over thick	Ayyangar et al. (1934)
23	Presence of testa	Two genes "B_1, B_2"	Complementary dominant	Laubscher (1945), Stephens (1946)
24	Waxy endosperm	Single gene "*wx*"	Recessive	Karper (1933)
25	Sugary endosperm, shrunken kernals	Single gene "*su*"	Recessive	Karper and Quinby (1963)
26	Lysine	Single gene "*hl*"	Recessive	Singh and Axtell (1973)
27	Scented grain	Single gene "*sc*"	Recessive	Ayyangar (1939)

Source: Rooney (2000).

It would be a huge task to present the entire literature on genetics of quantitative traits. It is evident that the nature of gene action varies among and within characteristics. This differential nature of gene action within a characteristic was due to the use of different analytical approaches, such as generation mean analysis, diallel analysis, line × tester analysis, and triple-test cross-analysis to estimate these parameters of genetic components. Further, the nature and magnitude of genetic parameters include not only the property of genes per se, but also the specific population from which estimates are being obtained. A population sample for a given number of loci will yield different estimates of genetic parameters at different stages of its improvement, depending upon the changes in the allele frequency (Chahal and Gosal, 2002). Nevertheless, these estimates provide a relative contribution of additive, dominant, and epistatic components toward genotypic variance, on which breeding methods to be followed are decided. A perusal of the literature clearly demonstrates that the component characteristics of grain and fodder yield, such as stem girth (Sankarapandian et al., 1996), number of tillers (Manickam and Das, 1994), resistance to diseases such as anthracnose (Sifuentes et al., 1992) and grain mold (Ghorade et al., 1997), and resistance to insect pests such as midge (Ratnadass et al., 2002) are predominantly governed by additive gene action. Inheritance of resistance to sorghum shoot fly (*Atherigona soccata*) depends on the resistance mechanism (Singh and Rana, 1986). The resistance was a quantitative trait with both additive and nonadditive effects (Singh and Rana, 1986; Khush and Brar, 1991). Sharma and Rana reported that the two resistance mechanisms, nonpreference for oviposition and dead heart formation, were both controlled by two independent, complementary recessive genes.

Hence, it follows that simple recurrent selection to improve populations or recurrent selection for high general combining ability (GCA) to develop hybrid parents would be effective for sorghum. For the rest of the traits, including grain and fodder yield, both additive and nonadditive with a predominance of nonadditive gene action were important. These traits are best improved through heterosis exploitation after developing hybrid parents through recombination breeding.

Resistance to charcoal rot (*Macrophomina phaseolina*) was polygenic, moderate to lowly heritable, and partially dominant (Rana et al., 1982). Bramel-Cox et al. (1988) indicated that GCA effects were most important for charcoal rot resistance and selection was effective at improving charcoal rot resistance. Tenkouano et al. (1993) reported that charcoal rot resistance was controlled by two dominant genes that are modified by a third locus. They also noted that charcoal rot and nonsenescent phenotypes were controlled by independent loci, even though these traits are often linked. Rosenow and Clark (1982) described drought stress in two phases. The first occurs if the plants are stressed prior to flowering, and the second occurs if the plants are stressed after flowering. In both cases, distinct and independent sources of resistance were identified. Sorghum genotypes with the "stay green" trait are consistently more drought-tolerant than non–stay green genotypes (Duncan et al., 1981). Walulu et al. (1994) reported that a single gene influences stay green in the genotype B 35. The gene shows varying levels of dominant action depending on the evaluation environment. Using molecular markers, Crasta et al. (1999) reported that three major quantitative trait loci were detected for drought tolerance; and Tuinstra et al. (1996) identified six quantitative trait loci for preflowering drought tolerance.

5.2 CYTOGENETICS

Sorghum bicolor has a haploid chromosome number of 10, and it is classified as a diploid ($2n = 2x = 20$). Most species within the genus *Sorghum* are diploid ($2n = 20$), but several species, most notably *Sorghum halapense*, are tetraploid ($2n = 4x = 40$). As the basic chromosome number in the *Sorghastrae* is 5, it has been hypothesized that *Sorghum* may be of tetraploid origin (Rooney, 2000). Earlier studies on the meiotic chromosome pairing analysis do not provide any evidence for the tetraploid origin of *S. bicolor* (Brown, 1943; Endrizzi and Morgan, 1955), and information on the existence of homologous segments in the chromosomes of *S. bicolor* is very meager and the chromosomes were therefore regarded as distinct. Subsequently, a few recent studies provide limited evidence that *Sorghum* is of tetraploid origin (Gomez et al., 1997, 1998). The molecular marker mapping studies of the genome by Chittenden et al. (1994), Pereira et al. (1994), Gomez et al. (1997), and Dufour et al. (1997) demonstrated duplicated loci on the map, suggesting that sorghum has a tetraploid origin. However, Subudhi and Nguyen (2000) contended that this evidence of tetraploidy is unsatisfactory. They argued that in the analysis by both Chittenden et al. (1994) and Dufour et al. (1996), the duplicated loci found on the mapped genome is only to the extent of 8% and 11%, respectively. In a more recent study, Peng et al. (1999) concluded that there is not enough evidence for tetraploid origin of sorghum. Therefore, the cultivated sorghum could be considered as diploid from the perspective of genome organization (Subudhi and Nguyen, 2000).

Cytogenetic maps of sorghum chromosomes 3–7, 9, and 10 were constructed on the basis of the fluorescence in situ hybridization (FISH) by Kim et al. (2005a). In these chromosomes, euchromatic DNA spans approximately 50% of the sorghum genome, ranging from approximately 60% of chromosome 1 (SBI-01) to approximately 33% of chromosome 7 (SBI-07), and this portion of the sorghum genome is predicted to encode approximately 70% of the sorghum genes (approximately 1 gene model/12.3 kbp). While the heterochromatin spans approximately 411 Mbp of the sorghum genome, a region characterized by an approximately 34-fold lower rate of recombination and approximately threefold lower gene density compared to euchromatic DNA (Kim et al., 2005a). FISH of bacterial artificial chromosomes (BACs) that contain the most proximal linkage markers enabled the localization of *Rf1* to an approximately 0.4-Mbp euchromatic region of LG-08 (Kim et al., 2005b).

1. *Karyotype*: Karyotype analysis of sorghum chromosome has been difficult due to similarities in chromosome size and structure (Huskins and Smith, 1932; Doggett, 1988). Nevertheless, several researchers like Garber (1950), Venkateswarlu and Reddi (1956), Celarier (1959), Gu et al. (1984), Mohanty et al. (1986), and Yu et al. (1991) have attempted to describe the karyotype of sorghum. Yu et al. (1991) was able to identify all 10 chromosomes based on chromosome size, arm ratio, and C-banding patterns, while Kim et al. (2002) used FISH and BAC libraries to identify the 10 chromosomes. The genome size of *S. bicolor* was reported to be 735 Mb, while that for *S. halapense* was reported to be 1617 Mb by Laurie and Bennett (1985). In later studies, Arumunganathan and Earle (1991) estimated the genome size of *S. bicolor* to be 750 Mb and Peterson et al. (2002) estimated the size as 692 Mb.
2. *Euploid variation*: It has been estimated that one-third of all domesticated species of plants (and over 70% of forage grasses) are euploids, with multiples of either the basic or the genome number of chromosomes (House, 1985).
   a. Haploids: Sorghum haploids are usually shorter, with slender stalks, narrow leaves, and smaller panicles. The panicles are sterile, and this may account for the extensive tillering noted in some sorghum haploids (Murty and Rao, 1974).

b. Triploids: In most cases, sorghum triploids were not morphologically different from tetraploids except for high levels of sterility. Meiotic pairing of chromosomes was irregular, with trivalents being the most common meiotic pairing configuration (Murty and Rao, 1974).

c. Tetraploids: Tetraploid sorghums are generally slower to flower and produce larger grains. In most cases, sterility was observed, along with a lower seed set. The reduced fertility is associated with abnormalities in meiotic chromosome pairing, but selection for improved fertility also improved meiotic pairing in autotetraploid sorghums (Murty and Rao, 1974; Doggett, 1988; Luo et al., 1992).

Because of the increased seed size, there was interest in developing autotetraploid sorghums for grain production (Doggett, 1962). The initial limitation was the high level of sterility observed in autotetraploid, but selection for improved fertility was successful and fertility levels were near that of diploids (Doggett, 1962; Luo et al., 1992). However, little research in this direction thereafter resulted in a lack of realization of tetraploid sorghum for grain production.

3. *Aneuploid variation*: The most common form of aneuploidy observed in sorghum is trisomy, followed by translocation. Schertz (1966) produced and characterized five primary trisomics based on plant morphology and chromosome behavior. Later, he identified another set of five trisomics from different genetic backgrounds (Schertz, 1974). Hanna and Schertz (1970) used a series of translocation tester stocks to label these 10 primary trisomics. These trisomics are useful in localizing molecular genetic markers from the genetic linkage groups to specific chromosomes using FISH (Gomez et al., 1997).

4. *Apomixis*: Hanna et al. (1970) reported apomixis in sorghum, in which the embryo is formed by apospory from a somatic cell in the nucellus, and up to 25% of the progeny developed apomictically. Rao and Murty (1972) presented evidence of apomixis in the line R 473 at a significantly higher frequency. They hypothesized that this line was an obligate apomict when self-pollinated, but further research indicated that it was a facultative apomict. In most cases, apospory is the primary mechanism of apomixis in sorghum, although diplospory may occur (Murty et al., 1979).

Apomixis provides a mechanism to perpetuate a high-performing hybrid through self-pollination. Obligate apomixis is necessary to develop such a system, but all the reports of apomixis in sorghum involve only facultative apomixes, and efforts to increase the frequency of apomicts have not been successful (Reddy et al., 1980). To utilize facultative apomixis, the use of "hybrids" has been proposed (Murty et al., 1984; Murty, 1986). The vybrids are the partial apomicts derived from crosses involving facultative apomicts. They were compared with pure-line and hybrid varieties (Murty and Kirti, 1983) but the degree of apomixis in such progenies is highly variable (Murty et al., 1985).

References

Arumunganathan, K., Earle, E.D., 1991. Nuclear DNA content of some important plant species. Plant Mol. Biol. Rep. 9, 208–219.

Ayyangar, G.N.R., 1939. Studies in sorghum. J. Madras Univ. 11, 131–143.

Ayyangar, G.N.R., 1942. The description of crop plant characters and their ranges of variation. IV. Variability of Indian sorghum. Indian J. Agric. Sci. 12, 528–563.

Ayyangar, G.N.R., Ayyar, M.A.S., 1938. Linkage between a panicle factor and the pearly-chalky mesocarp factor Zz in sorghum. Proc. Indian Acad. Sci. 8, 100–107.

Ayyangar, G.N.R., Nambiar, A.K., 1941a. The inheritance of the manifestation of purple colour at the pulvinar regions in the panicle of sorghum. Curr. Sci. 10, 80.

Ayyangar, G.N.R., Nambiar, A.K., 1941b. Inheritance of depth of green colour in the leaves of sorghum. Curr. Sci. 10, 492.

Ayyangar, G.N.R., Ponnaiya, B.W.X., 1937a. The occurrence and inheritance of purple pigment on the glumes of sorghum close on emergence from the boot. Curr. Sci. 5, 590.

Ayyangar, G.N.R., Ponnaiya, B.W.X., 1937b. The occurrence and inheritance of earheads with empty anther sacs in sorghum. Curr. Sci. 5, 390.

Ayyangar, G.N.R., Ponnaiya, B.W.X., 1939a. Studies in *Sorghum sudanense*, Stapf: the sudan grass. Proc. Indian Acad. Sci. 10, 237–254.

Ayyangar, G.N.R., Ponnaiya, B.W.X., 1939b. Cleistogamy and its inheritance in sorghum. Curr. Sci. 8, 419–420.

Ayyangar, G.N.R., Ponnaiya, B.W.X., 1941a. Sorghums with felty glumes. Curr. Sci. 10, 533–534.

Ayyangar, G.N.R., Ponnaiya, B.W.X., 1941b. Two new genes conditioning the tint of the colour on the glumes of sorghum. Curr. Sci. 10, 410–411.

Ayyangar, G.N.R., Reddy, T.V., 1942. Seedling-adult colour relationships and inheritance in sorghum. Indian J. Agric. Sci. 12, 341.

Ayyangar, G.N.R., Vijiaraghavan, C., Pillai, V.G., Ayyar, M.A.S., 1933. Inheritance of characters in sorghum—the great millet. II. Purple pigmentation on leaf-sheath and glume.. Indian J. Agric. Sci. 3, 589–604.

Ayyangar, G.N.R., Vijiaraghavan, C., Ayyar, M.A.S., Rao, V.P., 1934. Inheritance of characters in sorghum—the great millet. VI. Pearly and chalky grains. Indian J. Agric. Sci. 4, 96–99.

Ayyangar, G.N.R., Ayyar, M.A.S., Rao, V.P., Nambiar, A.K., 1936. Mendelian segregations for juiciness and sweetness in sorghum stalks. Madras Agric. J. 24, 247–248.

Ayyangar, G.N.R., Rao, V.P., Reddy, T.V., 1937. The inheritance of deciduousness of the pedicelled spikelets of sorghum. Curr. Sci. 5, 538–539.

Ayyangar, G.N.R., Rao, V.P., Nambiar, A.K., 1939. The occurrence and inheritance of purple blotched grains in sorghum. Curr. Sci. 8, 213–214.

Barabas, Z., 1962. Observation of sex differentiation in sorghum by use of induced male-sterile mutants. Nature. 195, 257–259.

Bramel-Cox, P.J., Stein, I.S., Rodgers, D.M., Claflin, L.E., 1988. Inheritance of resistance to *Macrophomina phaseolina* (Tassi) Gold and *Fusarium moniliform* Sheldon in sorghum. Crop Sci. 28, 37–40.

Brengman, R.L., 1995. The Rf1 gene in grain sorghum and its potential use in alternative breeding methods. Queensland Department of Primary Industries Monograph Series Information series No QI95003. p. 16.

Brown, M.S., 1943. Haploid plants in sorghum. J. Hered. 34, 163–166.

Casady, A.J., Miller, F.R., 1970. Inheritance of hermaphrodite pedicelled spikelets of sorghum. Crop Sci. 10, 612.

Casady, A.J., Heyne, E.G., Weibel, D.E., 1960. Inheritance of female sterility in sorghum. J. Hered. 51, 35–38.

Celarier, R.P., 1959. Cytotaxonomy of the Andropogonea. III. Sub-tribe Sorgheae, genus, *Sorghum*. Cytologia. 23, 395–418.

Chahal, G.S., Gosal, S.S., 2002. Principles and Procedures of Plant Breeding—Biotechnological and Conventional Approaches. Narosa Publishing House, New Delhi, India, p. 486.

Childs, K.L., Frederick, R.M., Pratt, M.M.C., Pratt, L.H., Morgan, P.W., Mullet, J.E., 1997. The sorghum photoperiod sensitivity gene, Ma3 encodes a phytochrome B. Plant. Physiol. 113, 611–619.

Chittenden, L.M., Schertz, K.F., Lin, V.R., Wing, R.A., Paterson, A.H., 1994. A detailed RFLP map of *Sorghum bicolor* x *S. propinquum*, suitable for high density mapping, suggests ancestral duplication of sorghum chromosomes or chromosomal segments. Theor. Appl. Genet. 87, 925–933.

Crasta, O.R., Xu, W.W., Rosenow, D.T., Mullet, J.E., Nguyen, H.T., 1999. Mapping of post-flowering drought resistance traits in grain sorghum: association between QTLs influencing premature senescence and maturity. Mol. Gen. Genet. 262, 579–588.

Doggett, E., 1988. Sorghum. John Wiley & Sons, New York, NY.

Doggett, H., 1962. Tetraploid hybrid sorghum. Nature. 196, 755–756.

Dufour, P., Grivet, L., D'Hont, A., Deu, M., Trouche, G., Glaszmann, J.C., et al., 1996. Comparative genetic mapping between duplicated segments on maize chromosomes 3 and 8 and homeologous regions in sorghum and sugarcane. Theor. Appl. Genet. 92, 1024–1030.

Dufour, P., Deu, M., Grivet, L., D'Hont, A., Paulet, F., Bouet, A., et al., 1997. Construction of a composite sorghum genome map and comparison with sugarcane, a related complex polyploid. Theor. Appl. Genet. 94 (3–4), 409–418.

Duncan, R.R., Bockholt, A.J., Miller, F.R., 1981. Descriptive comparison of senescent and non-senescent sorghum genotypes. Crop Sci. 21, 849.

Endrizzi, J.E., Morgan Jr., D.T., 1955. Chromosomal interchanges and evidence for duplication in haploid *Sorghum vulgare*. J. Hered. 46, 201–208.

Erichsen, A.W., Ross, J.G., 1963. Irregularities at microsporogenesis in colchicines-induced male sterile mutants in *Sorghum vulgare* Pers. Crop Sci. 3, 481–483.

Garber, E.D., 1950. Cytotaxonomic studies in the genus *Sorghum*. Univ. Calif. Publ. Bot. 23, 283–361.

Ghorade, R.B., Gite, B.D., Sakhare, B.A., Archana Thorat, 1997. Analysis of heterosis and heterobeltiosis for commercial exploitation of sorghum hybrids. J. Soils Crops. 7, 185–189.

Gomez, M.I., Islam-Faridi, M.N., Woo, S.S., Schertz, K.F., 1997. FISH of a maize sh2-selected sorghum BAC to chromosomes of *Sorghum bicolor*. Genome. 40, 475–478.

Gomez, M.I., Islam-Faridi, M.N., Zwick, M.S., Czeschin Jr., D.G., Wing, R.A., Stelly, D.M., et al., 1998. Tetraploid nature of *Sorghum bicolor* (L.) Moench. J. Hered. 89, 188–190.

Graham, R.J.D., 1916. Pollination and cross-fertilization in the juar plant (Andropogon sorghum, Brot.). India Dept. Agric. Mem. Bot. Ser. 8, 201–216.

Gu, M.H., Ma, H.T., Liang, G.H., 1984. Karyotype analysis of seven species in the genus *Sorghum*. J. Hered. 75, 196–202.

Hadley, H.H., 1957. An analysis of variation in height in sorghum. Agron. J. 49, 144–147.

Hanna, W.W., Schertz, K.F., 1970. Inheritance and trisome linkage of seedling characters in *Sorghum bicolor* (L.) Moench. Crop Science. 10, 441–443.

Hanna, W.W., Schertz, K.F., Bashaw, E.C., 1970. Apospory in *Sorghum bicolor* (L.) Moench. Science. 170, 338–339.

House, L.R., 1985. A Guide to Sorghum Breeding, second ed. International Crops Research Institute for the Semi-Arid Tropics, Patancheru, India.

Huskins, C.L., Smith, S.G., 1932. A cytological study of genus *Sorghum*. J. Genet. 25, 241.

Karper, R.E., 1933. Inheritance of waxy endosperm in sorghum. J. Hered. 24, 257–262.

Karper, R.E., Conner, A.B., 1931. Inheritance of chlorophyll characters in sorghum. Genetics. 16, 291–308.

Karper, R.E., Quinby, J.R., 1963. Sugary endosperm in sorghum. J. Hered. 54, 121–126.

Karper, R.E., Stephens, J.C., 1936. Floral abnormalities in sorghum. J. Hered. 27, 183–194.

Khush, G.S., Brar, D.S., 1991. Genetics of resistance to insects in crop plants. Adv. Agron. 45, 223–274.

Kim, J.S., Childs, K.L., Islam-Faridi, M.N., Menz, M.A., Klein, R.R., 2002. Integrated karyotyping of sorghum by *in situ* hybridization of landed BACs. Genome. 45, 402–412.

Kim, J.S., Faridi, M.N.I., Klein, P.E., Stelly, D.M., Price, H.J., Klein, R.R., et al., 2005a. Comprehensive molecular cytogenetic analysis of sorghum genome architecture: distribution of euchromatin, heterochromatin, genes and recombination in comparison to rice. Genetics. 171 (4), 1963–1976.

Kim, J.S., Klein, P.E., Klein, R.R., Price, H.J., Mullet, J.E., Stelly, D.M., 2005b. Molecular cytogenetic maps of sorghum linkage groups 2 and 8. Genetics. 169 (2), 955–965.

Laubscher, F.X., 1945. A genetic study of sorghum relationships. Union A. Afr. Sci. Bull. 242, 1–22.

Laurie, D.A., Bennett, M.D., 1985. Nuclear DNA content in the genera *Zea* and *Sorghum*, intergeneric, interspecific and intraspecific variation. Heredity. 55, 307–313.

Lin, Y.R., Schertz, K.F., Paterson, A.H., 1995. Comparative analysis of QTLs affecting plant height and maturity across the Poaceae, in reference to an interspecific sorghum population. Genetics. 141, 391–411.

Luo, Y.W., Yen, X.C., Zhang, G.Y., Liang, G.H., 1992. Agronomic traits and chromosomal behavior of autotetraploid sorghums. Plant Breed. 109, 46–53.

Manickam, S., Das, L.D.V., 1994. Line x tester analysis in forage sorghum. Int. Sorghum Millets Newsl. 35, 79–80.

Maunder, A.B., Pickett, R.C., 1959. The genetic inheritance of cytoplasmic-genetic male sterility in grain sorghum. Agron. J. 51, 47–49.

Miller, F.R., Barnes, D.K., Cruzado, H.J., 1968. Effect of tropical photoperiods on the growth of *Sorghum bicolor* (L.) Moench, when grown in 12 monthly plantings. Crop Sci. 8, 499–502.

Mohanty, B.D., Maiti, S., Ghosh, P.D., 1986. Establishment of karyotype in *Sorghum bicolor* through somatic metaphase and pachytene analysis. In: Manna, G.K., Sinha, U. (Eds.), Perspectives in Cytology and Genetics, vol. 5. Proceedings of the Fifth All India Congress of Cytology and Genetics, Bhubaneshwar, 7–10 October 1984, pp. 559–563.

Murty, U.R., 1986. Apomixis: achievements, problems and future prospects. In: Advanced Methods in Plant Breeding. Oxford and IBH, New Delhi, India.

Murty, U.R., Rao, N.G.P., 1997. Sorghum. In: Bahl, P.N., Salimath, P.M., Mandal, A.K. (Eds.), Genetics, Cytogenetics and Breeding of Crop Plants, vol. 2, Cereal and Commercial Crops. Oxford & IBH Publishing Co. Pvt. Ltd., New Delhi, pp. 197–239.

Murty, U.R., Schertz, K.F., Bashaw, E.C., 1979. Apomictic and sexual reproduction in sorghum. Indian J. Genet. Plant Breed. 39, 271–278.

Murty, U.R., Kirti, P.B., Bharathi, M., Rao, N.G.P., 1984. The nature of apomixes and its utilization in the production of hybrids (vybrids) in *Sorghum bicolor* (L.) Moench. Z. Pflanzenzuch. 92, 30–39.

Murty, U.R., Kirti, P.B., 1983. The concept of vybrids. I. Comparative performance of pureline, hybrid and vybrid varieties. Cereal Res. Commun. 11, 229.

Murty, U.R., Kirti, P.B., Bharathi, M., 1985. The concept of vybrids in sorghum. II. Mechanism and frequency of apomixes under cross pollination. Z. Pflanzenzuecht. 95, 113.

Peng, Y., Schertz, K.F., Cartinhour, S., Hart, G.E., 1999. Comparative genome mapping of *Sorghum bicolor* (L.) Moench using an RFLP map constructed in a population of recombinant inbred lines. Plant Breed. 118, 225–235.

Pereira, M.G., Lee, M., Bramel-Cox, P., Wordman, W., Doebley, J., Whitkus, R., 1994. Construction of an RFLP map in sorghum and comparative mapping in maize. Genome. 37, 236–243.

Peterson, D.G., Schulze, S.R., Sciara, E.B., Lee, S.A., Bowers, J.E., Nagel, A., et al., 2002. Integration of cot analysis, DNA cloning, and high-throughput sequencing facilitates genome characterization and gene discovery. Genome Res. 12 (5), 795–807.

Peterson, G.C., Suksayretrup, K., Weibel, D.E., 1982. Inheritance of some bloomless and sparse-bloom mutants in sorghum. Crop Sci. 22, 63–67.

Porter, K.S., Axtell, J.D., Lechtenberg, V.L., Colenbrander, V.F., 1978. Phenotype, fiber composition, and *in vitro* dry matter disappearance of chemically induced brown midrib (bmr) mutants of sorghum. Crop Sci. 18, 205–208.

Quinby, J.R., 1967. The maturity genes of sorghum. In: Norman, A.G. (Ed.), Advances in Agronomy, vol. 19. Academic Press, New York, NY, pp. 267–305.

Quinby, J.R., 1974. Sorghum Improvement and the Genetics of Growth. Texas A&M University Press, College Station, TX.

Quinby, J.R., Martin, J.H., 1954. Sorghum improvement. Adv. Agron. 6, 305–359.

Rana, B.S., Anahosur, K.H., Jaya Mohan Rao, V., Parameshwarappa, R., Rao, N.G.P., 1982. Inheritance of field resistance to sorghum charcoal rot and selection for multiple disease resistance. Indian J. Genet. 42, 302–310.

Rao, N.G.P., Murty, U.R., 1972. Further studies on obligate apomixis in grain sorghum, *Sorghum bicolor* (L.) Moench. Indian J. Genet. 32, 379–383.

Ratnadass, A., Chantereau, J., Coulibaly, M.F., Cilas, C., 2002. Inheritance of resistance to the panicle-feeding bug (*Eurystylus oldi*) and the sorghum midge (*Stenodiplosis sorghicola*) in sorghum. Euphytica. 123, 131–138.

Reddy, C.S., Schertz, K.F., Bashaw, E.C., 1980. Apomictic frequency in sorghum R 473. Euphytica. 29, 223–226.

Rooney, W.L., 2000. Genetics and cytogenetics. In: Smith, C.W., Frederiksen, R.A. (Eds.), Sorghum: Origin, History, Technology and Production, John Wiley & Sons, New York, NY, pp. 261–307.

Rooney, W.L., Aydin, S., 1999. The genetic control of a photoperiod sensitive response in *Sorghum bicolor* (L.) Moench. Crop Sci. 39, 397–400.

Rosenow D.T., Clark L.E., 1982. Drought tolerance in sorghum. In: Proceeding Corn and Sorghum Research Conference. American Seed Trade Association, Washington, DC, 37, pp. 18–30.

Sankarapandian, R., Rajarathinam, S., Mupidathi, 1996. Genetic variability, correlation and path coefficient analysis of jaggery yield and related attributes in sweet sorghum. Madras Agric. J. 83, 628–631.

Schertz, K., Stephens J.C., 1966. Compilation of gene symbols, recommended revision and summary of linkages for inherited characters of *Sorghum vulgare* Pers. Texas Agric. Exp. Stn. Tech. Monograph 3, Texas A&M University, College Station, TX.

Schertz, K.F., 1966. Morphological and cytological characteristics of five trisomics of *Sorghum vulgare* Pers. Crop Sci. 6, 519–523.

Schertz, K.F., 1974. Morphological and cytological characteristics of five additional trisomics of *Sorghum bicolor* (L.) Moench. Crop Sci. 14, 106–109.

Schertz, K.F., 1994. Male sterility in sorghum: its characteristics and importance. In: Witcombe, J.R., Duncan, R.R. (Eds.), Use of Molecular Markers in Sorghum and Pearl Millet Breeding for Developing Countries. Proceedings of an ODA Plant Sciences Research Conference. 29th March–1st April, 1993, Overseas Development Administration (ODA), Norwich, UK, pp. 35–37.

Shan, L.Q., Ai, P.J., Yin, L.T., Yao, Z.F., 2000. New grain sorghum cytoplasmic male-sterile A2V4A and F1 hybrid Jinza No.12 for northwest China. Int. Sorghum Millets Newsl. 41, 31–32.

Sifuentes, J.A., Mughogho, L.K., Thakur, R.P., 1992. Inheritance of resistance to sorghum leaf blight. Pages 16-17 in Cereals Program Annual Report. International Crops Research Institute for the Semi-Arid Tropics, Patancheru, India.

Singh, B.U., Rana, B.S., 1986. Resistance in sorghum to the shootfly, *Atherigona soccata* Rondani. Insect Sci. Appl. 7, 577–587.

Singh, R., Axtell, J.D., 1973. High lysine mutant gene (hl) that improves protein quality and biological value of grain sorghum. Crop Sci. 13, 535–539.

Singh, S.P., Drolsom, P.N., 1973. Induced recessive mutations affecting leaf angle in *Sorghum bicolor*. J. Hered. 64, 65–68.

Stephens, J.C., 1937. Male-sterility in sorghum: its possible utilization in production of hybrid seed. J. Am. Soc. Agron. 29, 690–696.

Stephens, J.C., 1946. A second factor for subcoat in sorghum seed. J. Am. Soc. Agron. 38, 340–342.

Stephens, J.C., Holland, R.F., 1954. Cytoplasmic male-sterility for hybrid sorghum seed production. Agron. J. 46, 20–23.

Subudhi, P.K., Nguyen, H.T., 2000. New horizons in biotechnology. In: Smith, C.W., Frederiksen, R.A. (Eds.), Sorghum: Origin, History, Technology and Production. John Wiley & Sons, New York, NY, pp. 349–397.

Tenkouano, A., Miller, F.R., Frederiksen, R.A., Rosenow, D.T., 1993. Genetics of non-senescence and charcoal rot resistance in sorghum. Theor. Appl. Genet. 85, 644–648.

Tuinstra, M.R., Grote, E.M., Goldsbrough, P.B., Ejeta, G., 1996. Identification of quantitative trait loci associated with pre-flowering drought tolerance in sorghum. Crop Sci. 36, 1337–1344.

Venkateswarlu, J., Reddi, V.R., 1956. Morphology of pachytene chromosomes and meiosis in *Sorghum subglabrescens*, a eu-Sorghum. J. Indian Bot. Soc. 35, 344–356.

Vinall, H.N., Cron, A.B., 1921. Improvement of sorghums by hybridization. J. Hered. 12, 435–443.

Walulu, R.S., Rosenow, D.T., Wester, D.B., Nguyen, H.T., 1994. Inheritance of stay green trait in sorghum. Crop Sci. 34, 970–972.

Webster, O.J., 1965. Genetic studies in *Sorghum vulgare* (pers.). Crop Sci. 5, 207–210.

Woodworth, C.M., 1936. Inheritance of seedling stem colour in a broomcorn-sorghum cross. J. Am. Soc. Agron. 28, 325–327.

Yu, H., Liang, G.H., Kofoid, K.D., 1991. Analysis of C-banding chromosome patterns of sorghum. Crop Sci. 31, 1524–1527.

CHAPTER

6

History of Winter Sorghum Improvement in India

OUTLINE

Most grain sorghums of India belong to *Sorghum durra*, *S. cernuum*, and *S. subglabrascens* according to the classification given by Snowden (1936). Postrainy season–adapted sorghums grow well in short day length (photoperiod sensitivity), flower and mature irrespective of temperature fluctuations and sowing dates (thermoinsensitivity within the postrainy season varieties), are tolerant of terminal moisture stress and resistant to stalk rot/charcoal rot, produce high biomass (grain and stover), and have large lustrous grain with semicorneous endosperm. They are tolerant to shoot fly, lodging (mechanical) and rust. Rapid progress has not been achieved in the postrainy sorghum improvement program, unlike the rainy sorghum improvement program. More than 80% of the postrainy sorghum area is still dominated by two important cultivars: *Maldandi*, a local landrace; and M 35-1, a selection from *Maldandi* released in 1937 (Patil et al., 2014). The age-old association and the consequent local preferences for the taste of the respective local varieties are the dominant reasons for their continued cultivation in spite of their erratic behavior and low yields (Rao and Murthy, 1972). As the postrainy sorghum is an important source of food and fodder, there is a pressing need to develop high-yielding sorghum cultivars with superior grain and fodder quality. Hence, there is a need to study the history of these selections and the progress made in strategic research, as this information may help to formulate suitable breeding strategies for rapid genetic enhancement, as in rainy season–adapted sorghum, and to break the existing yield plateau.

Genetic Enhancement of rabi sorghum – Adapting the Indian Durras.
DOI: http://dx.doi.org/10.1016/B978-0-12-801926-9.00006-6

6.1 THE ORIGIN OF M 35-1

The M 35-1 variety has dominated most of the postrainy tracts for more than seven decades. The reason for lack of significant genetic improvement calls for the detailed study of the release process of this variety. Maldandi, the originator for M 35-1, is the popular landrace that has been grown by farmers mostly in the regions of Maharashtra during postrainy season for decades. The Maldandi variety was observed to contain mainly two kinds of variants: one with large and one with short internodes. Plants with shorter internodes were shorter, had leaf sheaths overlapping each other and fully covering the stem, grew slowly, and were generally late in flowering and did not thrive well in shallow soils. Those with long internodes grew rapidly and were early in flowering. However, the rate of growth in both types was the same in the first month. Later, the plants with long internodes recorded a rapid growth utilizing the available moisture and also expressed good earhead emergence percentage. Hence, the short internode types of Maldandi sorghum were removed, as 75% of the soils in the place of its cultivation were shallow in nature. The 35-1 variety evolved from this was a long internode type (Table 6.1).

During 1927–1928, at Mohol farm in Maharashtra, India, 58 diverse earheads selected from Maldandi were evaluated in three rows, each being 30 m in length and with a spacing of 45 cm × 30 cm, with Maldandi used as a check. They were further tested in subsequent years, and the final 14 selections (from Maldandi), together with 30 others brought from Maldandi growing regions of Solapur district of Maharashtra state, were grown along with Maldandi as a check. Based on grain yield and grain size, six selections (namely, 47-3, 35-1, N4, N11, N9, and 31-2) were made and further tested against Maldandi local in 100-m^2 plots replicated seven times. Only 31-2 was significant over local Maldandi. The 35-1, N4, and N11 selections gave higher yields, to an extent of 5%. The genotypes 47-3 and N9 did not fare well due to Chikta disease.

During 1935–1936, the 35-1, 47-3, M-9, and 31-2 varieties were tested against local Maldandi in station trials in 75-m^2 plots, and the trial was replicated five times. In this

TABLE 6.1 Comparison of Short and Long Internode Variants of Maldandi

	Block I (deep black soils)			Block II (fairly deep soils)			Block III (shallow soils)		
Genotype No.	No. of plants	No. of plants bearing earheads	% of plants with ears	No. of plants	No. of plants bearing earheads	% of plants with ears	No. of plants	No. of plants bearing earheads	% of plants with ears
Maldandi check	91	60	65.8%	90	50	55.5%	100	40	40%
30-3 (short internode type)	66	38	57.7%	82	35	42.6%	78	6	7.7%
35-1 (long internode type)	90	72	80.0%	96	69	72.0%	101	59	58.4%

FIGURE 6.1 M 35-1, *rabi* variety.

study, 31-2 was highly significant over local than 35-1 and 47-3. Though 31-2 showed good yielding capacity, it flowered late and had poor grain size. The genotypes 35-1, M-9, and 47-3 are early, medium, and late, respectively, and are fairly good in yield and grain size. The period of 1936–1937 suffered from scanty rainfall. The Maldandi strains 35-1, 47-3, M-9, and 31-2 were tested against local, graded, and rejected seed lots from the market in seven replications in 100-m^2 plots. The 35-1 variety recorded the highest yields compared to the other varieties. Also, random grain samples were sent to the Mohol market for valuation. The Maldandi strains 35-1, 47-3, and M-9 were grown on a large scale for yield tests on cultivators' fields in most of the talukas of Sholapur, parts of Bijapur, Poona, Nagar, Satara, and Khandwah districts of the Maharashtra state. The 35-1 variety was appreciated for its high yield, large grain, and earliness. The 47-3 variety, being late, performed well in irrigated tracts.

Of the four Maldandi strains, strain 31-2, being late and small-grained, was dropped from the trial, and the strains 35-1, 47-3, and M-9 were tested against local, graded, and rejected seed lots in 100-m^2 plots with four replications for two consecutive years (i.e., the postrainy seasons of 1937–1938 and 1938–1939) in cultivators' fields of Sholapur, Nagar, Poona, and Satara districts. The three selections were significantly superior to local, while M 35-1 proved its merit by earliness, high yield, and larger grains. These were also valued in the local market of Mohol. The grain of M 35-1 fetched a premium price of 11.5–15.5% over local, while the other two varieties fetched 4.4–8.9% over local for two consecutive years of testing. From then on, the variety spread rapidly in the farmers' fields and has also become the check variety in the All India Coordinated Sorghum Improvement Project trials on postrainy sorghum improvement. The M 35-1 variety was released in 1969 from Mohol station in Maharashtra for cultivation in post-rainy tracts across the country (Figure 6.1). Prominent researchers of the early period included G. N. Rangaswamy Ayyangar and his colleagues, of the then-state of Madras, and Rao Sahib Kottur and his colleagues of the then-state of Bombay. Prominent varieties developed during the early period were the *Co* series in Tamilnadu; the Nandyal, Guntur, and Anakapalle series of Andhra Pradesh; the *PJ kharif* and *rabi* selections,

Saonar, Ramkel, Aispuri, the *Maldandi* and *Dagadi* (compact-headed) selections of Maharashtra; the *Bilichigan, Fulgar white, Fulgar yellow, Kanvi, Nandyal, Hagari,* and *Yanigar* varieties of the Mysore state; the *Budh Perio, Sundhia,* and *Chasatio* of Gujarat; the selections of Gwalior and Indore from Madhya Pradesh; the *RS* selections of Rajasthan and a few others (Rao and Murthy, 1972).

In the early 1960s, the Indian Council of Agricultural Research (ICAR), with Rockefeller Foundation assistance, initiated research on hybrid sorghum. Under this direction, the All India Coordinated Sorghum Improvement Project (AICSIP) was formed in 1969 to oversee the sorghum research activities at the national level. These programs initiated public research and conducted multilocation testing for improved characteristics of sorghum hybrid varieties, with support from state agricultural universities and other research stations in India.

Focused breeding of *rabi* sorghum was initiated in the early 1970s, which over the years led to the release of several state and central release varieties. At the national level, the CSV 7R variety was released in 1974, CSV 8R in 1979, Swati in 1984, CSV 14R in 1992, Sel 3 in 1995, Phule Yashoda in 2000, CSV 18 in 2005, CSV 22 in 2007, CSV 26 in 2012, and CSV 29 in 2013 (Table 6.2). The released postrainy sorghum varieties, CSV 8R, CSV 14R, CSV 18, and Swati, were better received than the postrainy hybrids such as CSH 7R and CSH 8R. Several varieties were released at the state level, which included Mukti, Parbhani Moti/SPV 1411 from Parbhani center, and NTJ 2 and NTJ 3 from the Nandyala station of Andhra Pradesh state (Lakshmaiah et al., 2004). The sorghum program of Karnataka state released the DSV 4 and DSV 5 varieties. The Maharashtra state sorghum improvement program released varieties like Swati, Selection 3, CSV 216/Phule Yashoda/SPV 1359, RSLG 262/Phule Maulee, Phule Anuradha, Phule Revati, Phule Vasudha, Phule Chitra, and Phule Suchitra (Table 6.3) (Patil et al., 2014).

Most of the postrainy sorghum varieties are only of the *Durra* type, whereas rainy sorghum cultivars belonged to *Caudatum* and *Kafir* races (Reddy et al., 2003). Genetic improvement of postrainy sorghum, therefore, is hindered by the lack of phenotypic variability among breeding lines. On the other hand, postrainy sorghum landraces form an important source of genetic variation. The fodder obtained during postrainy season is more important than that of rainy season–grown sorghum. Therefore, postrainy sorghum grain productivity has to be accompanied by normal or better fodder productivity. Unlike the rainy sorghum cultivars, higher levels of resistance against major pest (shoot fly) and disease (charcoal rot), stringent maturity duration to suit different receding soil moisture regimes and certain levels of thermoinsensitivity are essential in postrainy sorghum cultivars for better adaptability. Grain quality is also as important as the grain yield. The quality benchmark is that of the popular land race, Maldandi 35-1 (M 35-1). In adaptability criteria such as shoot fly resistance, as well as grain quality, the varieties are superior to hybrids. Unlike in the case of rainy sorghum, postrainy sorghum varieties have better preference over hybrids for reasons of adaptability and grain quality. The released postrainy sorghum varieties, CSV 8R, CSV 14R, CSV 18, and Swati, were better received than the postrainy hybrids such as CSH 7R and CSH 8R.

Several varieties, such as CSV 7R, CSV 8R, and CSV 14R, were developed using selections from segregating populations derived from the crosses among Indian locals (namely, M 35-1 and IS 2644 with American germplasm lines). Marginal improvement

TABLE 6.2 Pedigree and Origin of Postrainy Season Adapted Sorghum Varieties Released at National Level

S. No	Variety	Year of release	Pedigree of variety	Pedigree of parental lines of variety	Center which developed the variety
1	M 35-1	1969	Landrace sel. from local maldandi bulk	–	Mohol
2	CSV 7R	1974	IS 2950 × M35-1	IS 2950 = *Guinea durra* (USA)	NRCS
3	CSV 8R	1979	R24 × R16	R 24 = IS 3687 (*kafir*—USA) × M 35-1 R 16 = IS 2950 (*Guinea-durra*, USA) × M 35-1	NRCS
4	Swati	1985	SPV86 × M 35-1	SPV 86 = R 24 × R 16 R 24 = IS 3687 (*kafir*—USA) × M 35-1 R 16 = IS 2950 (*Guinea-durra*, USA) × M 35-1	Rahuri
5	CSV 14R	1992	(M 35-1 × (CS 2947 × CS 2644) × M 35-1	CS 2947 = IS 2947 (*kafir*—USA) CS 2644 = IS 2644 (*durra*—India)	NRCS
6	Sel 3	1995	Sel. from Bidar Postrainy local	It is a reselection from Maldandi local	Rahuri
7	CSV 216	2000	Landrace Sel. from postrainy germplasm Dhulia	Pureline Sel. from Tapi river valley; RSLG 112-1-6	Rahuri
8	CSV 18	2005	A selection from cross (CR 4 × IS 18370)	A selection from cross (CR 4 × IS 18370)	Parbhani
9	CSV 22	2007	SPV 1359 × RSP 2	SPV 1359 = CSV 216	Rahuri
10	CSV 26	2012	SPV 655 × SPV 1538	(SPV-655: Selection from cross M35-1 × M-148 SPV-1538: Selection from Phul Mallige, Chandakpur village, Aland Tq, Gulbarga dist, Karnataka)	Directorate of Sorghum Research
11	CSV 29 (SPV 2033)	2013	(GRS 1 × CSV 216R) × CSV 216R	GRS-1: Selection from Natte maldandi of Gulbarga CSV-216R: Landrace Sel. from postrainy germplasm Dhulia, Pureline Sel. from Tapi river valley; RSLG 112-1-6	Regional Agricultural Research Station, Bijapur

Source: Reproduced from Patil et al. (2014).

was achieved for grain yield over the most popular landrace variety, M 35-1, until 2000. Most of these varieties could not become popular, as they did not match shoot fly resistance level and grain quality of M 35-1. Among the yield component traits, long panicles, bold grains, and number of grains per panicle, 100-seed weight contributed for grain yield, and most of these traits have high heritability, which enables the plant breeder to improve these traits through simple selection. The gap between flag leaf sheath and panicle base should be minimal to have good grain filling and the glume coverage on grains must be less for higher threshability. Grain size can be visually judged, and grain color can be

TABLE 6.3 Pedigree and Origin of Postrainy Season Adapted Sorghum Varieties Released at State Level

S. No	Variety	Year of release	Pedigree of variety	Area of adaptation	Center which developed the variety
1	Phule Maulee	1999	RSLG 262	Maharashtra	Rahuri
2	Phule Vasudha	2007	RSLG 206 × SPV 1047	Maharashtra	Rahuri
3	Phule Chitra	2006	SPV 655 × RSLG 112	Maharashtra	Rahuri
4	Phule Revati	2010	CSV 216 × SPV 1502	Maharashtra	Rahuri
5	Phule Anuradha	2010	RSLG 559 • RSLG 1175	Maharashtra	Rahuri
6	Phule Suchitra	2012	SPV 1359 × SPV 1502	Maharashtra	Rahuri
7	Parbhani Moti	2004	Sel. from ICRISAT population GD 31-4-2-3	Maharashtra	Parbhani
8	GJ 9	1979	Pur line selection from local broonch district	Gujarat	Surat
9	DSV 4	1998	E 36-1 × SPV 86	Karnataka	Dharwad
10	DSV 5	1996	Pureline selection from the local variety "Natte Maldandi" of Gulbarga region	Karnataka	Dharwad
11	NTJ 2	1990	Pureline sel. from E-1966	Andhra Pradesh	Nandyal
12	NTJ 3	1995	NJ2092 × POD 24	Andhra Pradesh	Nandyal
13	N 14	1990	Pureline sel. from Prodatur local	Andhra Pradesh	Nandyal

Source: Reproduced from Patil et al. (2014).

selected as per the consumer/market preference in the given adaptation (House, 1980; Reddy et al., 2009).

Productivity depends not only on the moisture availability, but also on the soil types under which it is grown and the genotypes. Much of the postrainy season sorghum is grown on residual and receding soil moisture on shallow and medium-deep soils. An exhaustive work of landrace exploration was taken up at MPKV, Rahuri in an ICAR-funded project called "Exploration of postrainy sorghum landraces for superior substitute to M 35-1," which ran from 1998 to 2002. A total of 307 landraces were collected from post-rainy sorghum–growing areas of Satpura ranges and Tapi, Purna, and Godavari river basins. A total of 1175 local types collected earlier to this period were also evaluated. Wide variability was recorded among the germplasm accessions (Table 6.4). The characteristics of the germplasm collected from a district resembled each other in terms of certain characteristics (Table 6.5).

The elite landraces identified from the evaluations were tested for their nutritional quality (protein) to isolate superior types. The average protein content ranged from 4.5%

TABLE 6.4 Range Among the Postrainy Sorghum Germplasm Lines for Agronomic and Quality Traits in Sorghum

S. No.	Character	Range
1	Plant height (cm)	40–220
2	Days to 50% flowering	71–94
3	Number of internodes	6–14
4	Internode length (cm)	9.5–25.9
5	Number of leaves	5–15
6	Leaf width (cm)	4.5–9.5
7	Grain yield/plant (g)	5.5–115
8	Fodder yield/plant (g)	24–172
9	Earhead width (cm)	2.5–11.5
10	Earhead length (cm)	6.5–25.5
11	1000 grain weight (g)	24.0–49.5
12	Dead heart (%)	00–85
13	Protein content (%)	5.6–13.5
14	Total sugar (%)	0.45–1.65
15	Starch (%)	49.5–75.0

to 12.5%. The landraces were screened for shoot fly and charcoal rot resistance and for drought tolerance in shallow and medium soils. Genotypes selected through pureline selection were evaluated in coordinated trials. As a result, Phule Yashoda (pureline selection from RSLG 112) was released at the national level in 2000. It had a grain yield superiority of 18.2% and fodder yield superiority of 8.3% over M 35-1. It had less shoot fly incidence, was more tolerant of charcoal rot, and had more adventitious roots than M 35-1. The quality of roti (Bhakari) was superior to M 35-1 in sweetness and softness. The variety contained 10% protein, as opposed to 9.73% in M 35-1. Germplasm lines suitable to both the soil conditions and to specific soils were identified by Pawar et al. (2005). Under shallow soils, the genotypes were shorter, flowered and matured early, and in medium-deep soils, mean leaf area, grain number, and 1000 grain mass, grain and fodder yields were higher (Rafiq et al., 2003). Breeding varieties suitable for varying soil depths was emphasized at Mahatma Phule Krishi Vidyapeeth, Rahuri, Maharashtra, and several varieties were released. The RSLG 262/Phule Maulee variety was released for shallow to medium soils of Maharashtra in 1999. Its grain is medium-sized and creamy white. It is tolerant of shoot fly and charcoal rot with excellent roti quality. It has high water use efficiency and photosynthetic efficiency. Later, Phule Chitra (SPV 655 × RSLG 112) was released for medium soils in 2006, Phule Vasudha (RSLG 206 × SPV 1247) was released for deep soils in 2007, and Phule Anuradha (RSLG 539 × RSLG 1175) was released for shallow

TABLE 6.5 Characteristics of Postrainy Sorghum Landraces Collected from Different Tracts of Maharashtra

Name of district and total no. of collections	Name of taluka	No. of collections	Special features of landraces collected
Ahmednagar (166)	Ahmednagar	68	Semicompact elliptic panicles, charcoal rot tolerant, wide variability for grain yield, fodder yield, number of internodes, leaf length and leaf width
	Parner	14	
	Newasa	12	
	Karjat	30	
	Rahuri	28	
	Pathardi	2	
	Jamkhed	12	
Solapur (133)	Solapur	2	Pearly white bold grains (36–48 g per 1000 grain), compact panicles, early flowering, drought tolerant
	Mangalvedha	55	
	Karmala	34	
	Pandharpur	3	
	Madha	5	
	Barshi	20	
	Mohol	13	
	Sangola	1	
Pune (34)	Pune	9	Landraces tall (1.4–2.3 m) and late flowering, suitable for medium to heavy soils
	Haveli	3	
	Shirur	5	
	Bhor	7	
	Indapur	10	
Aurangabad (36)	Aurangabad	32	Landraces of M 35-1 type and not much diversity observed
	Gangapur	4	
Satara (69)	Mann	1	Compact panicles variable for grain yield, fodder yield and 1000 grain weight
	Satara	8	
	Karad	60	
Sangali (21)	Sangali	21	Compact panicle, grain with more flour recovery
Nasik (20)	Nasik	11	Not much variability but variable for plant height, number of internodes and peduncle length
	Yawala	9	
Dhule (265)	Dhule	26	Wide variability existed in qualitative traits such as color and coverage of glumes, grain color, internode length, leaf orientation, etc.
	Shirpur	42	
	Shindkheda	192	
	Nardana	5	

(*Continued*)

TABLE 6.5 (Continued)

Name of district and total no. of collections	Name of taluka	No. of collections	Special features of landraces collected
Nandurbar (30)	Nandurbar	2	Wide variability as in Dhule region. Landraces with high protein content, starch and sugar
	Akkalkuwa	15	
	Shahada	9	
	Navapur	4	
Jalgaon (254)	Jalgaon	169	Diverse from M 35-1, medium in height, resistant to charcoal rot and shoot fly
	Nashirabad	22	
	Bhusawal	11	
	Chopda	20	
	Yarandol	1	
	Yawal	33	
Osmanabad (28)	Osmanabad	6	Compact panicles, early flowering, drought tolerant
	Tuljapur	15	
	Bhum	7	
Latur (5)	Latur	5	Compact panicles, early flowering, drought tolerant
Beed (19)	Beed	15	Compact panicles, early flowering, drought tolerant
	Ashti	4	
Buldhana (3)	Buldhana	3	Compact panicles, early flowering, drought tolerant
Amravati (2)	Amravati	2	Compact panicles, early flowering, drought tolerant
Nagpur (5)	Nagpur	5	Compact panicles, early flowering, drought tolerant
Bijapur (14)	Bijapur	14	Compact panicles, early flowering, drought tolerant

Source: Reproduced from Patil et al. (2014).

soils in 2008. These were developed by pedigree breeding from the crosses made among the landraces. The Phule Revati variety was released for medium-to-deep soils in 2010 and Phule Suchitra was released for medium soils in 2012. These varieties have been performing well in specific soil situations. Therefore, there is a need for the development of varieties adapted to specific soil situations in postrainy season to enhance production and productivity levels.

On the comparison of yield attributes and physiological traits of Phule Chitra with M 35-1, a widely cultivated postrainy season variety with broad adaptation, and Phule Maulee, a variety suitable for medium soils, the ideotype was reflected in the higher per-day grain and fodder productivity, higher harvest index, greater earhead exertion, higher relative water content, and slow leaf senescence (Sanjana Reddy et al., 2012). Hence, while breeding for new varieties adaptable to medium soils, these traits have to be focused (Figures 6.2–6.4).

FIGURE 6.2 Phule Anuradha variety, suitable for cultivation in shallow soils. *Courtesy: Gawali, IIMR.*

FIGURE 6.3 Phule Chitra variety, suitable for cultivation in medium to deep soils. *Courtesy: Gawali, IIMR.*

6.1.1 Grain Quality

Postrainy season sorghum is known for its quality, due to which it is mostly preferred for human consumption, which is characterized as being lustrous, pearly white, and attractive. Developing genotypes with high yield potential coupled with nutritionally superior-quality grain is the prime objective of the breeding program. Studies on F_2

FIGURE 6.4 Phule Vasudha variety, suitable for cultivation in deep soils. *Courtesy: Gawali, IIMR.*

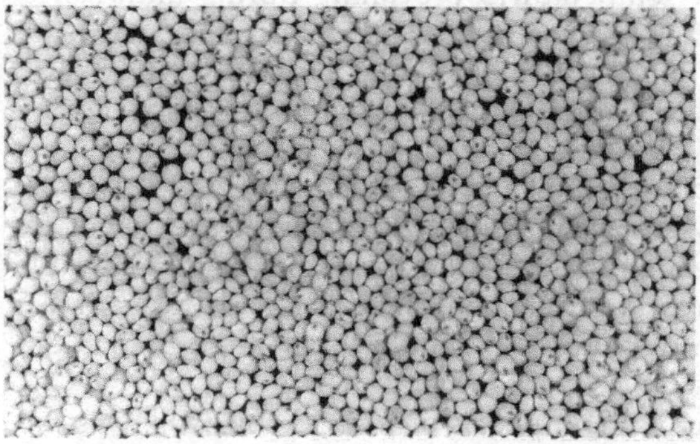

FIGURE 6.5 Rabi variety with good grain quality.

segregation pattern in the postrainy × rainy cross CSV 216R × 401 B indicated that seed luster was under the influence of recessive epistasis, and the intensity of seed luster depends on the recessive homozygous alleles at either both loci or one locus, while the F_2 segregation pattern in the postrainy × postrainy cross CSV216R × 104B, polymeric gene interaction was noted. Further, based on both crosses, it was interpreted that seed luster was controlled by two pairs of genes (Rekha, 2006). For the mesocarp thickness, the dominant gene was attributed to the expression of thin mesocarp and recessive gene to the thick mesocarp (Reddy, 2007). M 35-1 (Shoba et al., 2008), Phule Vasudha (RSV 423), CSV 22, and Phule Chitra (SPV 1546) were found to be most promising varieties for roti quality (Chavan et al., 2009). Proximate composition (i.e., moisture, protein, fat, ash, crude fiber, and total carbohydrates) of postrainy sorghum genotypes differed significantly ($P < 0.01$) and ranged from 6.64–8.58%, 8.73–12.81%, 1.22–2.36%, 1.14–1.72%, 1.21–2.48%, and 81.82–87.58%, respectively (Shoba et al., 2008) (Figure 6.5).

6.1.2 Hybrids

Although efforts were made to introgress farmer-preferred traits such as bold, lustrous, and semicorneous grain type and juicy stalks into the hybrids targeted for postrainy season cultivation by crossing improved Indian landraces as pollinators with the established exotic and elite female parent, CK 60A (*milo* or A_1 cytoplasm), the resulting hybrids lacked significant heterosis, had threshing difficulties, and were too tall, which is not amenable for increasing plant population per unit area. Induced dwarf mutants of an Indian variety, Gidda maldandi, have been used as pollinator parents by workers at Dharwad. Several studies in the past have indicated only modest levels of heterosis for economic traits in postrainy season sorghum, as most of the parents utilized in postrainy hybrid programs were related by descent. The hybrids, CSH 7R and CSH 8R, developed from the improved parents and released in 1977, were not acceptable to farmers even though they had high heterosis, for they lacked grain luster, resistance to shoot fly, and lodging. Later, CSH 12R, released in 1986, also could not progress well compared to other varieties and failed to make any impact in farmers' fields. However, Rana et al. (1997) performed a study that indicated appreciable levels of heterosis for grain yield and other agronomic traits. The hybrids developed and released did not attract farmers, as these hybrids lacked matching grain quality and shoot fly resistance compared to the most popular landrace variety, M 35-1. Also, poor seed setting (mainly caused due to low temperatures prevailing during anthesis) is observed in hybrids. Greater yield heterosis was observed in derivative × tropical (African) varietal crosses due to the diversity of genes (Rana et al., 1985). Rana and Murty (1978) also reported that increase in number of seeds per panicle branch in short compact headed varieties (tropical) and increase in the panicle branches in the long panicle type (temperate) by introgression of genes from African germplasm result in yield heterosis. Large heterotic response for grain yield and harvest index were accompanied by stalk rot and shoot fly susceptibility (Rao, 1982). For increasing the grain yield within the limits of the available water supply, the choice of female parent for hybrid production should be made for both leaf area and photosynthetic rate and the selection of pollinators should be made for maximum seed numbers per panicle (Krieg, 1988).

The second phase of *rabi* sorghum breeding, with emphasis on hybrid cultivars, was initiated in the late 1980s. During this period, 250 experimental hybrids were evaluated in the AICSIP Initial Hybrid Trials, which resulted in the identification of two hybrids, SPH 504 and SPH 677, for central release as CSH-13R and CSH-15R (Table 6.6). CSH 13R has significant yield superiority over M 35-1 but is highly vulnerable to shoot fly and low temperatures and had inferior grain quality. CSH 15R based a *rabi* MS line (104A) developed at Mohol center had a marginal yield advantage over M 35-1. It was felt that *rabi* hybrids will have a tangible impact only when the parental lines have *rabi* adaptability and desired combining ability (Rao et al., 1986) (Figures 6.6 and 6.7).

A mission mode project on development of hybrid crops under the National Agricultural Technology Project (NATP) funded by the World Bank was operating specifically to develop *rabi* sorghum hybrids at NRCS and 7 AICSIP centers working on *rabi* sorghum at the national level from 1999 to 2005 (Rana and Kaul, 2005). The objective of the project was to develop high yielding, shoot fly–resistant *rabi* sorghum hybrids of early to medium duration suitable for variable soil depths, and to develop suitable parental lines

TABLE 6.6 Pedigree and Origin of Postrainy Season Adapted Sorghum Hybrids Released at National Level

S. No.	Hybrid	Year of release	Pedigree of the hybrid	Pedigree of the parental lines of the hybrid	Center which developed the hybrid
1	CSH 7R	1977	36A × 168	36A = CK 60B × PJ8K CK 60B-*kafir* (USA) PJ8K-Parbhani jowar (rainy)	NRCS
2	CSH 8R	1977	36A × PD 3-1-11	36A—CK 60B × PJ8K PD 3-1-11 = temperate US dwarf × BP 53 (Gujarat rainy local)	Parbhani
3	CSH 12R	1986	296A × M 148-138	296A—IS 3922 × Karad local IS 3922-*kafir durra* Karad local—rainy local from Maharashtra M 148-138 = mutant of Maldandi	Dharwad
4	CSH 13R	1991	296A × RS 29	RS 29-SC 108 × SPV 126 SC 108-Purdue (USA) SPV 126-tall mutant of CS 3541	NRCS
5	CSH 15R	1995	104A × RS 585	104A = 296B × Swati 296B = IS 3922 × Karad local IS 3922-*kafir durra*—yellow endosperm. Karad local—rainy local from Maharashtra Swati = SPV86 × M 35-1 RS 585 (CS 3541 × M 35-1) × Nandyal postrainy local	NRCS
6	CSH 19R	2000	104A × R354	104A- = 296B × Swati R 354-((SPV-504 (20 KR) × (SPV 504 × R 263)) × R-67-4	Akola

Source: Reproduced from Patil et al. (2014).

with resistance to shoot fly, charcoal rot, drought, and low temperatures by utilizing *milo* and non-*milo* cytoplasms to diversify the hybrid base. As part of the project, more than 4900 new hybrids were developed and tested during 1998–2005 across seven centers. SPH 1010, bred at Akola center, was released as CSH 19R in 2000 (Table 6.6). While CSH 15R was suitable for general cultivation, CSH 19R was ideal for favorable locations (Madhusudhana et al., 2003). A new drought-tolerant, early *rabi* sorghum hybrid called DSH 4R was developed and released for Karnataka in 2001. New cytoplasmic-nuclear male sterility (CMS) lines developed across Rahuri, Dharwad, Bijapur, Parbhani, NRCS, and Akola were 104A, 116A, 117A, 1409A early, 1409A medium, PMS 20A (tan type), 41A, 49A, 59A, 67A, 89A, 95A, 109A, 127A, 133A, 147A, 163A, 169A, 185A, 187A, 203A, 215A, 237A, DNA1, DNA2, DNA4, DNA5, BJMS 1A, BJMS 2A, BJMS 3A, BRJ 204 A, RS 20 A, RS 70A, RS 132A, RS 137A, RS 159A, RS 412A, and AKRMS nos. 14A, 43A, 46A, 47A, 63A, 67A, 68A, 72A, and 82A. Several lines on A_2 CMS were developed that included RS 530 A2, ICSV 8603 A2, SPV 932 A2, RS 71 A_2 at Rahuri, and PMS 12 A_2, PMS 13 A_2, PMS 15 A_2,

FIGURE 6.6 CSH 13R, *rabi* sorghum hybrid.

FIGURE 6.7 CSH 15R, *rabi* sorghum hybrid.

PMS 23 A_2, PMS 14 A (M 35-1), and PMS 18 A (M 35-1) at Parbhani. At NRCS Hyderabad, 19 pairs of A_2-based CMS lines were developed. Lines with bold-seed (MBMS 7B and SB 101 B), charcoal rot resistance (DCCR 1) and rust resistance (IS 3443) were utilized on M 31-2B at Dharwad to develop new B lines. *Rabi* locals (namely, Barshi Prakash, Muddihalijola, Honnutagi Local, RS 585, and SPV 570) that are maintainers on

M 31-2A were advanced to BC_2 at Bijapur. New MS lines [namely, PMS 47 A (A_2) and PMS 49 A (A_2)] were developed at Parbhani center.

For developing restorers on alternate cytoplasms, 11 A_2 restorers have been identified at Hyderabad. At Parbhani MR 11, MR 13, MR 15, MR 27, and MR 29 were identified as capable of restoring on Maldandi cytoplasm and KR 198 and PVR 350 to be restoring on A_2 cytoplasm. At Bijapur, the lines (namely, BRJ 62, BRJ 67, SPV 1491 and SPV 1556, CR 9, Jewaragi Local, and SPV 1452) have been identified as restorers on Maldandi cytoplasm. At Dharwad, Barshizoot local, Dagdi local, Hagari local, SPV 570, and IS 4582 were identified as restorers on M 31-2A. During 2001–2002, about 33 restorers on non-*milo* cytoplasm were identified at NRCS, Rahuri, and Bijapur. During 2003–2004, five stable restorers on Maldandi cytoplasm were identified—namely, IS 29411, IS 29406, IS 4587, Raichur Local, and Ramke at Bijapur center.

Strategic research in rabi hybrids: Though the released *rabi* hybrids could not progress well compared to other varieties, Rana et al. (1997) indicated appreciable levels of heterosis for grain yield and other agronomic traits. Large heterotic response for grain yield and harvest index were accompanied by stalk rot and shoot fly susceptibility (Rao, 1982). For increasing the grain yield within the limits of the available water supply, the choice of female parent for hybrid production should be made for both leaf area and photosynthetic rate, and the selection of pollinators should be made for maximum seed number per panicle (Krieg, 1988).

Greater yield heterosis was observed in derivative × tropical (African) varietal crosses due to diversity of genes (Rana et al., 1985). Rana and Murty (1978) have also reported that increase in number of seeds per panicle branch in short compact-headed varieties (tropical) and increase in the panicle branches in the long panicle type (temperate) by the introgression of genes from African germplasm result in yield heterosis.

Landrace pollinator-based hybrids, where many of the desirable attributes of landraces are inherited favorably in their hybrids possess moderate levels of shoot fly resistance and desirable grain quality traits. However, they lack lodging resistance and have moderate yielding ability (Reddy et al., 1983). Most of the landraces, including M 35-1, showed segregation for fertility restoration/sterility maintenance ability, indicating the need to select for restoration ability within the landraces. This also explains the partial restoration observed when bulk pollen of M 35-1 was used by many workers.

Rao and his associates converted M 35-1 and IS 3691, a yellow *hegari*, into steriles. Both of these are restorers on the *milo-kafir* sterility system. Additional sources of cytoplasmic sterility have been reported in *durra* (G2, VZM 1 and VZM 2) by Hussaini and Rao (1964). At Raichur, another indigenous sterile M 31-2A is said to owe its origin to induced mutation. Fertility restoration by landraces was poorer on *durra*-derived A_1 CMS lines than on *caudatum*-derived A_2 CMS lines. When both A_1 and A_2 CMS lines were based on *caudatum*, fertility restoration was higher on A_1 than A_2. This finding has a bearing on developing CMS lines involving *caudatum*-based germplasm lines adapted to postrainy season and testing for fertility restoration in hybrids. The hybrids involving *caudatum*-based female parents and *durra*-based landraces showed high heterosis for grain yield, but grain quality was poor. Up to 100% better-parent heterosis using landraces as pollinators over the A-lines 104A and M 31-2A was obtained (Sajjanar et al., 2011). Fertility restoration by landraces was poorer on *durra*-derived A_1 CMS lines than on *caudatum*-derived A_2 CMS lines.

When both A_1 and A_2 CMS lines were based on *caudatum*, fertility restoration was higher on A_1 than A_2. This finding has a bearing on developing CMS lines involving *caudatum*-based germplasm lines adapted to postrainy season and testing for fertility restoration in hybrids. The hybrids involving *caudatum*-based female parents and *durra*-based landraces showed high heterosis for grain yield, but grain quality was poor. *Milo* hybrids exhibited superiority over *maldandi* hybrids for most of the desired characteristics (Pattanashetti et al., 2005). Jayanthi (1997) showed that shoot fly resistance in both parents (or at least in seed parents) realizes a higher frequency of shoot fly–resistant hybrids in postrainy season.

Significant positive heterosis for grain yield was observed to range from 10.1% to 54.9% over checks CSH-15R and CSV-22 in multienvironment trials. Some promising new combinations included 185A × SLR-59, 104A × SLR-47, 104A × SLR-67, SL9A × SLR-57, 104A × SLR-79 and 104A × SLR-57, BJMS-2A × SLR-13, AKR-45A × SPV-570, AKR-45A × SLR-28, AKR-45A × SLR-10, 1409A × RS-585, 104A × SLR-28, 104A × BRJ-358, 1409A × JP-1 1-5, 1409A × SLR-13, and 1409A × SLR-27 (Prabhakar et al., 2013).

6.1.3 Hybrid Parents

The combining ability of compact-headed Indian landraces was found to be relatively poorer than very long panicle types of exotic origin. It was hypothesized that lack of reinforcement between genes responsible for primary axis length and those contributing to girth resulted in lack of marked heterosis for ultimate grain yield (Rao, 1970). Pollinators and female lines from exotic germplasm contributed to poor grain quality. The derivatives *durra-caudatum* (*zera zera*) crosses as pollinators developed for postrainy season and CMS lines developed for *kharif* season from *kafir-caudatum* (*zera zera*) crosses did not attract the attention of the farmers because they lacked grain luster, size and shape, and fodder yield comparable to M 35-1, despite their superiority under late sowings in postrainy season, with yield heterosis of 45–64%. Landrace pollinator-based hybrids, where many of the desirable attributes of landraces are inherited favorably, possess moderate levels of shoot fly resistance and desirable grain quality traits. However, they lack lodging resistance and have moderate yielding ability (Reddy et al., 1983). Most of the landraces, including M 35-1, showed segregation of fertility restoration/sterility maintenance ability, indicating the need to select for restoration ability within the landraces. This also explains the partial restoration observed when bulk pollen of M 35-1 was used by many workers. Jayanthi (1997) showed that shoot fly resistance in both parents (or at least in seed parents) in order to realize higher frequency of shoot fly resistant hybrids in postrainy season.

References

Chavan, U.D., Patil, J.V., Shinde, M.S., 2009. Nutritional and roti quality of sorghum genotypes. Indonesian J. Agric. Sci. 10, 80–87.

House, L.R., 1980. A Guide to Sorghum Breeding. International Crops Research Institute for the Semi-Arid Tropics, Patancheru, India.

Hussaini, S.H., Rao, P.V., 1964. A note on the spontaneous occurrence of cytoplasmic male sterility in Indian sorghum. Sorghum Newsl. 7, 27–28.

Jayanthi, P.D.K., 1997. Genetics of Shoot Fly Resistance in Sorghum Hybrids of Cytoplasmic Male Sterile Lines (Ph. D. Thesis). Acharya N. G. Ranga Agricultural University, Hyderabad, India.

Krieg, D.R., 1988. Water use efficiency of grain sorghum. In: Wilkinson, D. (Ed.), Proceedings of the 43rd Annual Corn and Sorghum Research Conference. American Seed Trade Association, Washington, DC, pp. 27–41.

Lakshmaiah, K., Jayalakshmi, V., Sreenivasulu, M.R., 2004. Nandyal Tella Jonna 4, a high yielding sorghum variety identified. Agric. Sci. Digest. 24, 289–291.

Madhusudhana, R., Umakanth, A.V., Kaul Swarnlata, Rana, B.S., 2003. Stability analysis for grain yield in *rabi* sorghum [*Sorghum bicolor* (L.) Moench.]. Indian J. Genet. Plant Breed. 63, 255–256.

Patil, J.V., Sanjana Reddy, P., Prabhakar, Umakanth, A.V., Gomashe Sunil, Ganapathy, K.N., 2014. History of post-rainy season sorghum research in India and strategies for breaking the yield plateau. Indian J. Genet. 74 (3), 271–285.

Pattanashetti, S.K., Biradar, B.D., Salimath, P.M., 2005. Comparative performance of *Milo v/s Maldandi* based rabi sorghum hybrids in the Transitional Zone of Karnataka. Karnataka J. Agric. Sci. 18, 655–659.

Pawar, K.N., Biradar, B.D., Shamarao, J., Ravikumar, M.R., 2005. Identification of germplasm sources for adaptation under receding soil moisture situations in post-rainy sorghum. Agric. Sci. Digest. 25.

Prabhakar, Kannababu, N., Samdur M.Y., Bahadure, D.M., 2013. Exploitation and assessment of heterosis in *Rabi* sorghum hybrids across environments. Presented at Global Consultation on Millets Promotion for Health & Nutritional Security, held at Directorate of Sorghum Research. 18–20 December 2013, Hyderabad, India.

Rafiq, S.M., Madhusudhana, R., Umakanth, A.V., 2003. Heterosis in post-rainy season sorghum under shallow and medium-deep soils. Int. Sorghum Millet Newsl. 44, 17–21.

Rana, B.S., Kaul, S.L., 2005. Final report on NATP project on development of Rabi sorghum hybrids. National Research Centre for Sorghum, Hyderabad, pp. 1–44.

Rana, B.S., Murty, B.R., 1978. Role of height and panicle type in yield heterosis in some grain sorghums. Indian J. Genet. Plant Breed. 38, 126–134.

Rana, B.S., Jaya Mohan Rao, V., Reddy, B.B., Rao, N.G.P., 1985. Overcoming present hybrid yield in sorghum. In: Genetics of Heterotic System in Crop Plants, Proceedings of Precongress Symposium XV International Genetics Congress. 7–9 December 1983, Tamil Nadu Agricultural University, Coimbatore, India, pp. 48–59.

Rana, B.S., Swarnalata, K., Rao, M.H., 1997. Impact of genetic improvement on sorghum productivity in India. In: Proceedings of an International Conference on the Genetic Improvement of Sorghum and Pearl Millet, held at Lubbock, Texas. 22–27 September 1996, International Sorghum and Millet Research (INTSORMIL)—International Crops Research Institute for the Semi-Arid Tropics (ICRISAT), pp. 142–165.

Rao, N.G.P., 1970. Genetic analysis of some exotic × Indian crosses in Sorghum I. Heterosis and its interaction with seasons. Indian J. Genet. 30, 347–361.

Rao, N.G.P., 1982. Transforming traditional sorghum in India. Sorghum in the eighties. Proceedings of the International Symposium on Sorghum. 2–7 November 1981, International Crops Research Institute for the Semi-Arid Tropics, Patancheru, India, vol. 1, pp. 39–59.

Rao, N.G.P., Murthy, U.R., 1972. Further studies on obligate apomixis in grain sorghum, *Sorghum bicolor* (L.) Moench. Indian J. Genet. 32, 379–383.

Rao, N.G.P., Jaya Mohan Rao, V., Reddy, B.B., 1986. Progress in genetic improvement in *rabi* sorghum in India. Indian. J. Genet. 46, 348–354.

Reddy, B.V.S., Rudrappa, A.P., Prasada Rao, K.E., Seetharama, N., House L.R., 1983. Sorghum improvement for *rabi* adaptation: approach and results. Presented at the All India Coordinated Sorghum Improvement Project Workshop. 18–22 April 1983, Haryana Agricultural University, Hisar, India.

Reddy, B.V.S., Sanjana, P., Ramaiah, B., 2003. Strategies for improving post-rainy season sorghum: a case study for landrace hybrid breeding approach. Paper Presented in the Workshop on Heterosis in Guinea Sorghum. Sotuba, Mali, pp. 10–14.

Reddy, B.V.S., Ramesh, S., Sanjana Reddy, P., Ashok Kumar, A., 2009. Genetic enhancement for drought tolerance in sorghum. Plant Breed. Rev. 31, 189–222.

Reddy, K.A., 2007. Genetic Analysis of Shoot Fly Resistance, Drought Resistance and Grain Quality Component Traits in *Rabi* Sorghum (*Sorghum bicolor* L. Moench) (M Sc (Agri) thesis). University of Agricultural Science, Dharwad, India, 79 pp.

Rekha, B.C., 2006. Genetic Studies on Grain Quality and Productivity Traits in *Rabi* Sorghum (M Sc (Agri) thesis). University of Agricultural Science, Dharwad, India.

Sajjanar, G.M., Biradar, B.D., Biradar, S.S., 2011. Evaluation of crosses involving rabi landraces of sorghum for productivity traits. Karnataka J. Agric. Sci. 24, 227–229.

Sanjana Reddy, P., Patil, J.V., Nirmal, S.V., Gadakh, S.R., 2012. Improving post-rainy season sorghum productivity in medium soils: does ideotype breeding hold a clue? Curr. Sci. 102, 904–908.

Shoba, V., Kasturiba, B., Rama, K.N., Nirmala, Y., 2008. Nutritive value and quality characteristics of sorghum genotypes. Karnataka J. Agric. Sci. 20, 586–588.

Snowden, J.D., 1936. The Cultivated Races of Sorghum. Allard & Co, London, UK.

CHAPTER

7

Breeding Methods for Winter Sorghum Improvement

OUTLINE

Genetic Enhancement of rabi sorghum – Adapting the Indian Durras.
DOI: http://dx.doi.org/10.1016/B978-0-12-801926-9.00007-8

7.1 YIELD AND ADAPTATION BREEDING FOR GRAIN AND FODDER YIELD

Sorghum (*Sorghum bicolor* (L.) Moench) is the fifth most important cereal crop in the world after rice, wheat, maize, and barley. In India, it is grown during both rainy (*kharif*) and postrainy (*rabi*) seasons for multiple uses as food, feed, fodder, and fuel. Sorghum grown in rainy season is mainly utilized as feed, as the grain is often caught in the rain and the quality is affected due to grain molds. However, postrainy sorghum is primarily used as a food owing to its good grain quality, and it also serves as a main source of fodder, especially during dry seasons. There has been a drastic reduction in sorghum area, especially in rainy seasons, but the area under postrainy sorghum has remained stable and is grown predominantly in six districts of Maharashtra (namely, Solapur, Ahmednagar, Pune, Beed, Osmanabad, and Aurangabad) and three districts of Karnataka (namely, Bijapur, Gulbarga, and Raichur), as well as parts of Andhra Pradesh and Tamil Nadu (Seetharama, 2004).

Adaptation is the measure of a cultivar's ability to survive in and respond to a defined target environment and is manifested in different ways (wide/narrow, broad/specific). *Economic yield* is the production of economically desirable plant parts per unit area (e.g., grain, fodder, oil, and protein). Almost all sorghum cultivars have the genetic/physiologic potential to produce some degree of economic yield, but yield potential varies among cultivars. Expression of potential yield depends to a great extent on the environment and the adaptation of the cultivar that allows it to cope with its environment (yield = genotype × environment). It is assumed that the cultivar should possess high absolute yield and a high degree of yield stability across environments. Hence, a widely adapted cultivar is one that responds positively to varying environmental changes. Postrainy sorghum is grown under receding moisture conditions under different depths of soil. Moisture and nutrient stresses set in very fast in shallow soils (almost 1 m ha area in western Maharashtra) resulting in very low yield levels (400–450 kg/ha). Besides widespread moisture stress (drought), cold, susceptibility to shoot fly, charcoal rot resulting in lodging, the cultivation of low-yielding local cultivars are the major yield constraints, resulting in low productivity. The absence of any alternative to postrainy sorghum has in fact resulted in increased area over the last two decades due to the ever-increasing demand for fodder/stover and the resultant stability achieved for grain prices. The stagnation of sorghum productivity in postrainy season is due to various genetic, edaphic, abiotic, and biotic factors that need to be highlighted in moisture-limiting environments. In fact, many of these limitations can be managed to harness better productivity in postrainy season by better contingency planning and adoption of new high-yielding cultivars released at the national levels.

Trait focus: Postrainy season-adapted sorghums are characterized by response to shorter day length (photoperiod sensitivity), flowering, and maturity (more or less the same time), irrespective of temperature fluctuations and sowing dates (thermoinsensitivity within the postrainy season varieties). In general, *rabi*-adapted sorghums are characterized by tolerance to shoot fly, stalk rot, and terminal stress, and by large, lustrous grain with semicorneous endosperm (Reddy et al., 2006). Tolerance to shoot fly, lodging (mechanical), and rust are also required (Seetharama et al., 1990). Though strong efforts have been made to develop

FIGURE 7.1 *Rabi* sorghum panicle.

hybrids with wider adaptability to varied production environments, the results are not encouraging (Madhusudana et al., 2003). Productivity depends not only on moisture availability, but also on the soil types under which it is grown (Jirali et al., 2007). Among the factors influencing adoption of sorghum varieties, farmers rated grain and fodder quality attributes as their first priority (Nagaraj et al., 2011). Genetic improvement of *rabi* sorghum is therefore hindered by lack of phenotypic variability among breeding lines. On the other hand, *rabi* landraces form an important source of genetic variation. The fodder obtained during postrainy season is more important than that in rainy-season-grown sorghum. Therefore, postrainy sorghum grain productivity has to be accompanied by normal or better fodder productivity. Unlike the rainy sorghum cultivars, higher levels of resistance against major pest (shoot fly) and disease (charcoal rot), stringent maturity duration to suit different receding soil moisture regimes and certain levels of thermoinsensitivity are essential in postrainy sorghum cultivars for better adaptability. The soil moisture was comparatively more at deeper soil depths, thereby implying that a genotype with deep penetrating roots is more suitable for cultivation under receding soil moisture conditions (Chapman et al., 1993; Naidu et al., 2001). As the crop concurs terminal drought, stay-greenness is an important attribute to breed for. Grain quality is also as important as the grain yield. The quality benchmark is that of the popular landrace, Maldandi 35-1 (M 35-1) (Figures 7.1 and 7.2).

7.1.1 Trait-Based Breeding Approaches

Crop yield is mainly dependent on the interplay of various physiological and biochemical functions of the plants, as well as the impact of the environment. The cause-and-effect

FIGURE 7.2 *Kharif* sorghum panicle.

relationship is difficult to understand, mainly because of the complexity in understanding the interplay of several processes and functions under limited water supply conditions. Among the yield component traits, long panicles, bold grains, number of grains per panicle, and 100-seed weight contribute to grain yield, and most of these traits have high heritability, enabling the plant breeder to improve these traits through simple selection. The gap between flag leaf sheath and panicle base should be at a minimum to have good grain filling and the glume coverage on grains is to be less for higher threshability. Grain size can be visually judged, and grain color can be selected as per the consumer/market preference in the given adaptation (Reddy et al., 2009; House, 1980).

Plant height: Plant height is influenced by the interaction of environmental conditions and genetic constitution of the plant. Introduction of dwarfing genes in rice and wheat has resulted in compounded rates of yield enhancement, which was later termed as the *green revolution*. On the contrary, some commercially employed dwarfing genes in sorghum have been associated with negative effects on grain yield (Barbara George-Jaeggli, 2009).

Several studies were made to relate plant height and yield under receding soil moisture conditions and differential responses were observed. In general, the landraces and commercial varieties showed higher plant height than the hybrids. Patil (1987) found significant genotypic differences with respect to plant height at all the growth stages (GSs) in postrainy sorghum. This indicates that plant height is also an important parameter for increasing productivity. Salunke et al. (2003) also reported the significant differences in plant height, with the popular landrace M 35-1 recording maximum plant height in postrainy season.

Further, plant height was positively related to dry matter production and grain yield in different studies (Kulkarni et al., 1981; Dhoble and Kale, 1988; Bakheit, 1989; Donatelli et al., 1992; Shivalli, 2000; Sameer et al., 2012). On the other hand, several researchers reported no

FIGURE 7.3 Taller genotypes are preferred for rabi adaptation, as are shorter ones for kharif adaptation.

correlation between plant height and grain yield (Hiremath and Parvatikar, 1985; Gaosegelwe and Kirkhan, 1990; Patil, 2002; Chand and Kumar Singh, 2003; Reddy et al., 2007a,b). Even negative correlation was observed between these parameters (Moila and Reddy, 1974; Pawar and Jadhav, 1996; Mutava et al., 2011). It was found positive with the length and width of the fourth leaf in sorghum (Moila and Reddy, 1974). Similarly, cvs. SPV 1090 and SPV 1217 recorded higher plant height, which was correlated with grain yield under receding soil moisture (Parameshwarappa and Karikatti, 2002). Stay-green genotypes were 3–4 cm shorter than senescent genotypes (Duncan et al., 1981). Kadam et al. (2002) reported less differences in plant height, with RSPG 37 and RSPG 1 recording the highest (202 cm) and lowest (197 cm) plant heights, while the corresponding values for grain yield were 0.80 and 1.66 t/ha, respectively. Taller genotypes also had larger grains (Reddy et al., 2007a,b).

In contrast to wheat, the dwarfing gene *dw3* led to a significant reduction in plant biomass, which was not associated with an increase in harvest index (HI) to avoid yield reduction. The observed reductions in plant biomass were associated with a reduction in radiation use efficiency (RUE) in the short types. Subsequent experiments suggested that an increase in allocation of biomass to the roots, rather than differences in photosynthetic capacity or respiration, was the main cause for the apparent reduction in RUE (Barbara George-Jaeggli, 2009). As lodging may be controlled by other means than height reduction (e.g., stay-green), we suggest that the yield of standard sorghum types used in industrialized countries may benefit from moderate increases in plant height.

Both additive as well as nonadditive gene action were noticed in the inheritance of plant height (Amsalu and Bapat, 1990). However, additive gene action for plant height was reported by Fayed (1975) and Rajguru et al. (2005). Predominance of nonadditive gene action for plant height was reported by Biradar (1995). Tallness was found to be dominant, and epistatic interaction was 7–10 times greater than additive and dominance gene effects (Shinde and Joshi, 1985) (Figure 7.3).

Days to 50% flowering: Though the flowering is controlled genetically, it is influenced by the environment and associated with grain yield.

Associations: While days to flowering showed significant positive correlation with grain yield (Youngquist et al., 1990; Jeyaprakash et al., 1997; Alam et al., 2001 and Patil et al., 2003), its negative association with grain yield showed that delayed flowering and maturity provide more time for plants to grow and produced more biomass, which contributed to more yield (Choudhary, 1992; Patel et al., 1994; Mahalakshmi and Bidinger, 2002).

Gene action: Nonadditive gene action, high heritability, and high predictability ratio were reported for days to 50% flowering (Salunke et al., 1996). Kanaka (1979) and Patel et al. (1995) found the importance of both additive and nonadditive components of genetic variance for 50% flowering date. Whereas predominance of nonadditive gene action was noted by Biradar (1995) and predominance of additive gene action by Rajguru et al. (2005). According to Chandak and Nandanwanvar (1993), additive × dominance interaction was very important for 50% flowering.

Flowering under moisture stress or receding moistures: Since postrainy sorghums often experience terminal stress, earliness is an important drought escape mechanism, especially in shallow soils. Sorghum has been found to be more sensitive to moisture stress during flowering and early grain-filling stages compared to vegetative GSs (Seetharama, 1986). Several studies reported that there is a differential response with respect to days to 50% flowering under receding soil moisture conditions. Blum (1970) reported that sorghum grown under stored soil moisture condition undergoes increasing water stress, with advancement in crop growth due to depletion of soil moisture. Late planting of most common temperate genotypes causes reduction in number of days to floral initiation (Tauli, 1964; Quinby, 1967; Stickler and Pauli, 1969). Reddy and Rao (1978) reported that with the increasing number of days from 50% flowering to physiological maturity, there was an improvement in the yield of hybrids due to increased number of days available for dry matter accumulation in the grain. Verma et al. (1983) observed that water stress not only delayed flowering, but also reduced grain weight. Norem et al. (1985) found that the days required for anthesis were significantly less for drought-tolerant lines than for medium- and low-tolerant lines. The initial increase in inflorescence development in stressed plants was earlier than in control plants, but plants that had experienced stress just prior to inflorescence initiation had faster development (Hermus et al., 1982). Under water stress, there was a decrease in grain-filling duration of 7 days, from a total duration of 28 days in M 35-1, CSV 14R, and IS 1314 genotypes (Rao and Shivaraj, 1988). Mathews et al. (1990b) found that drought delayed the panicle initiation; but once started, the length of reproductive period increased in resistant lines and decreased in susceptible ones. Craufurd et al. (1993) also observed a delay in panicle initiation and flowering by 2–25 days and 1–59 days, respectively, due to drought. They also reported that water stress increased the period between panicle initiation and flowering by retarding the rate of panicle development. Pawar (1996) reported that of the restorer lines, RS 29 took comparatively more days for 50% flowering. Craufurd and Peacock (1993) concluded that the genotype performance in water-limited environments was strongly related to phenology; they also emphasized the importance of the timing of stress and the growth rate during flowering in determining grain number and grain yield. Kamoshita et al. (1996) observed a delay in phenology with water deficit. They concluded that the variation in phenology was mostly due to the effect of water and time of sowing rather than different genotypic responses to either N or N × W interaction. Genotypes that had good capability in relation to the timing of emergence, floral initiation, antheiss, and maturity to temperature and photoperiods under limiting conditions were reported by Hammer et al. (1989) and Muchow and Carberry (1990). Kadam et al. (2002) revealed significant differences in their phenology, and M 35-1 required more days for 50% flowering, while the same genotype took the maximum duration from 50% flowering to maturity and produced a higher grain yield (Shivalli, 2000). There was a negative correlation between days to 50% flowering and HI (Blum, 1970).

Days to physiological maturity: Blum (1970) documented the advantage of grain yield associated with early maturity for dryland sorghum grown in Mediterranean climate. Quinby (1973) reported that early maturity had greater potential benefit in situations where growth is achieved solely on stored water. Blum et al. (1977) opined that early maturity is associated with reduced water use in sorghum, resulting from increased root density and root/shoot ratio. Pawar (1996) found that RS-29, a high-yielding genotype, took comparatively more days for 50% flowering but attained physiological maturity at almost the same time as that of other high- and moderate-yielding genotypes. Delayed flowering and maturity provide more time for the plant to grow and produce more biomass, which contributes to increased grain yield (Alam et al., 2001). Kadam et al. (2002) revealed that early maturity genotypes escape terminal stress by utilizing the maximum stored water. Salunke et al. (2003) reported that RSP 1 and Sel 3 matured earlier than the rest of the genotypes (124 days).

Grain yield parameters: Yield is a complex polygenic trait that is highly influenced by the environment and season. A sound knowledge of the association of various components of yield with biomass and grain yield is a prerequisite for initiating successful breeding program.

Grain number per panicle: Sriram and Rao (1983), Heinrich et al. (1993), Eastin (1983), Pinjari and Shinde (1995), Rao and Singh (1998) and Shinde and Narkhede (1998) emphasized the importance of a higher grain number for realizing the high yield potential. Henrich (1983) observed that high temperatures during panicle development reduced the seed number per head and grain yield in sorghum cultivars. In nonstress conditions, the senescent hybrids produced larger panicles with more grain per panicle and per-unit leaf area (LA) than nonsenescent hybrids (Harden and Krieg, 1983). Further, the glossy genotypes have recorded fewer unfilled grains than nonglossy ones (Rao and Shivaraj, 1988). However, increase in seed number and grain yield without corresponding increase in LA was considered important in dryland stress conditions (Krieg and Dalton, 1990). Youngquist et al. (1993) indicated that yield stability in terms of seeds per panicle was important for low rainfall environments. While the number of grain per panicle had the highest positive and significant direct effect on gross panicle weight (Asthana et al., 1996). Borrell et al. (1999) found that grain size was correlated with relative rate of leaf senescence during grain filling, such that reducing it from 3% to 1% loss of LA per day resulted in doubling grain size from about 15–30 mg in stay-green reserve more quickly and dried up 2 days before the drought resistance sorghum. They also observed that under well-watered (WW) conditions, both sorghums had similar water potential (WP), osmotic potential (OP) and adaxial stomatal resistance. Thus, they suggested that variations in susceptibility to drought between the two genotypes were due to the difference in the rate of soil moisture extraction. Mathews et al. (1990) revealed that resistant lines had slower shoot and root growth rates, slower soil water extraction rates, but higher root-shoot ratios. However, total length of seminal roots (first-order lateral), rate of root production, root elongation, and viability of the root tips and cortex were reduced in drought-stressed plants (Pardales and Kono, 1990). Kramer and Boyer (1995) reported that the roots were usually the site of the greatest resistance in the pathway for liquid phase movement of water through the soil-plant-atmosphere continuum. The efficiency of soil water uptake by the root system is, therefore, a key factor in determining the rate of transpiration and tolerance to drought. Patil (2005) found that varieties that produce higher dry matter per plant exploited more moisture in the profile, leading to low soil moisture status in the top 60-cm soil profile from 60 days after sowing (DAS) until harvest occurs under rainfed situation in sorghum. Saliah et al. (1999) showed that genotypic differences in the transpiration rate due to soil moisture stress were consistent with canopy density and the anatomical structures of stem and root of the sorghum cultivars. Drought tolerance in Gadambalia cultivar is associated with the highest water extraction efficiency, fewer nodal roots per plant, fewer late metaxylem vessels per nodal root, a smaller LA, and a well-developed sclerenchyma. Several studies reported positive relationships of grain number with grain yield in sorghum (Patel et al., 1980; Craufurd and Peacock, 1993; Pinjari and Shinde, 1995; Rao and Singh, 1998; Borrell et al., 1999; Nouri et al., 2004).

Grain weight: Chowdhary and Wardlaw (1978) observed that the potential large loss of grain weight accumulation was due to high temperatures during the grain-filling stage. Zhao et al. (1983) stated that 1000-grain weight and grain number per panicle were correlated with LA and photosynthetic rate. Moisture stress during the grain-filling period had reduced the 1000-grain weight significantly (Bakheit, 1989). Further, some genotypes are superior in panicle weight, and grain yield also showed moderate stability. Rao (1982) reported that the genotype Hegari 1 produced higher biomass but yielded less than M 35-1 because of lower sink size and test weight. The 1000-seed weight and number of grains per panicle showed positive correlation with grain yield (Kadam et al., 2002) and grain and fodder yield under residual moisture conditions (Thombre et al., 1982; Patil, 2002; Awari et al., 2003). Salunke et al. (2003) stated that the test weight of PBS 2, CR 4, CR 6, and CR 9 as 47.6, 41.8, 39.1, and 38.4 g, respectively, as compared to M 35-1, Sel-3, and CSV 14R. Also, 1000-seed weight and grain number per plant had a significant positive correlation with grain and fodder yield (Thombre et al., 1982; Patil, 2002; and Awari et al., 2003). The gene action for 100-grain weight was observed to be predominantly governed by additive × dominance epistasis gene effects (Kachave and Nandanwankar, 1980; Patidar and Dabholkar, 1981). Analysis of 100-grain weight in complete diallel crosses involving nine sets of male sterility lines showed that 100-grain weight was controlled by four alleles, which fitted in the additive-dominance model. Dominant genes, which were frequent, increased the grain weight and recessive genes decreased the grain weight. Rajguru et al. (2005) reported nonadditive gene action for 1000-seed weight.

Harvest index: HI is one of the major yield components for higher grain yield in crop improvement. It is thus an important aspect of differential partitioning of photosynthates and improved HI represents an increased physiological capacity of the crop to mobilize photosynthates and nutrients and translocate them to organs of economic value (Wallace et al., 1972). Donald (1962) defined the HI as the ratio of the grain dry weight to the total aboveground dry weight at maturity of the crop. Yoshida (1976) found the range in the HI reported among crops in several studies was between 0.15 and 0.40, but in bread wheat, it ranged from 0.5 to 0.6 (Jain and Kulashreshta, 1976; Yoshida, 1976). Though the HI is a genotypic character, it is influenced by environmental factors occurring during ripening. In many crops, in recent years, improvement in yield did not occur appreciably due to the ceiling of HI. Willey and Basiime (1973) reported that tall and late-maturing sorghum genotypes had recorded low HI and reduced assimilate partitioning to the panicles. Kulkarni et al. (1981) and Parvatikar and Hiremath (1985) reported a significant positive correlation between HI and grain yield in *rabi* sorghum. Similarly, in sorghum hybrids, HI has shown a significant positive association with grain yield (Muchow, 1989). HI in high osmotic adjustment (OA) was cultivars ranging from 0.38 to 0.40 under no stress, and 0.30 to 0.36 in stress conditions (Ludlow et al., 1989). Higher yields in sorghum cultivars were due to more and large grains that are associated with higher harvest indices and distribution index (Ludlow et al., 1990). Garrity et al. (1984) reported higher HI in normal situation; however, it decreased with variable moisture deficits. Sometimes this effect was extreme, and it could be even as low as zero (Boyle et al., 1991). Santamaria et al. (1990) also reported that higher grain yield realization was solely due to higher HI (27%), which was also associated with higher distribution index (25%) combining higher grain numbers (19%). Blum (1990, 1991) and Blum et al. (1992) also opined that higher grain yield in hybrids was due to higher HI. On the contrary, Gurumurthy (1982) reported that the biological yield did not have any correlation with HI. Similarly, Chaugle et al. (1982) also indicated that there was no relationship between biological yield and HI.

Wenzel et al. (1999) reported that the traits most severally affected by moisture stress were seed mass, HI, and biomass. Maintenance of higher HI by means of channeling assimilates to the developing ear was an important drought-resistant mechanism in sorghum (Wenzel et al., 2000). Higher HI values were found in nonglossy varieties than glossy ones under water stress (Rao, 1999). Kadam et al. (2002) concluded that the higher HI invariably leads to higher grain yield. Salunke et al. (2003) noticed that the HI of test entries PBS 2, CR 4, CR 6, and CR 9 was higher than cultivars M35-1, Sel-3, and CSV 14R. Kusalkar et al. (2003) explained the cause of high grain yield in RSLG 262 was an appreciably high HI of 23%, which showed the efficiency of converting biological yield into economic yield. HI was positively related to grain yield in several studies of sorghum under postrainy seasons, as reported by Muchow (1989), Omanya et al. (1997), Pawar and Chetti (1997), and Shinde et al. (1998).

Grain yield: Grain yield is the manifestation of various physiological and biochemical processes occurring in plants with relation to external environmental factors. Prolonged water shortage affects virtually all metabolic processes and often results in severe reductions in plant productivity. Sorghum is said to be a relatively drought-tolerant crop, but at certain critical stages such as PI, boot, and anthesis, moisture stress causes a reduction in growth and yield. Several researchers have reported that the reduction in grain yield due to water stress was more severe when it occurs at the reproductive and early grain fill phases than during the vegetative phase (Ravindra and Shivraj, 1983; Gonalez-Hernandez, 1985; Garrity et al., 1984; Ludlow et al., 1990; Bakheit, 1989; Baldy et al., 1993; Sankarpandian et al., 1993). Krieg (1983) reported that moisture stress reduced the yield of sorghum up to an extent of 70%. Erick and Musick (1979) reported that reduction in sorghum grain yield was mainly due to reduced grain size when stress was initiated at the heading stage or later. Kulkarni et al. (1983) reported the relationship between physiological parameters and grain yield in *rabi* season and found that genotypes with higher dry matter accumulation usually flowered early and matured early and had a higher number of primary and secondary panicle branches in addition to high yielding. Garrity et al. (1983) noticed that grain weight and dry matter were more sensitive to water stress during the grain-filling period when sorghum grown under temperate climatic conditions of Nebraska. Sriram and Rao (1983) observed that among the yield components, HI, panicle dry weight, and the number of grains per panicle had showed significant positive association with higher grain yield.

Significant differences in grain yield were observed among the cultivars of sorghum under rainfed conditions rather than irrigated (Wright et al., 1983). Further, sorghum hybrids have shown less superiority in water retention capacity over their parents, but followed similar to one of the parents (Gangadhararao and Sinha, 1988). Bapat and Gujar (1990) also noticed higher yields on drought-resistant lines (Sel-3) among other genotypes in drought stress (DS) conditions. Heinrich et al. (1985) suggested using high-seed-weight genotypes in the breeding program could result in higher grain yields. In general, the seed weight component of grain yield was influenced by water stress if stress occurs during the grain-filling stage (Norem et al., 1985). Under field conditions generally, the grain yield reduction due to moisture stress was mainly through both low seed weight and grain number per panicle (Parvatikar and Hiremath, 1985; Rao and Shivaraj, 1988). However, Verma and Eastin (1985) demonstrated no decrease in seed weight in water stress, and the grain yield reduction occurred due to reduction in seed numbers.

Takzure et al. (1998) stated that water stress at the heading or milk stage showed greater adverse effects than at the panicle initiation. Their results revealed that water stress decreased the number of grains per panicle by 18.8–70.0%, grain yield per plant by 13.7–59.4%, and 1000-grain weight by 4.4–35.6%. Rao and Shivraj (1988) reported that all the glossy varieties showed a significant increase in grain dry weight compared with the nonglossy varieties. Also, they concluded that the water stress decreased the grain yield by 54–73% in nonglossy varieties, while in glossy varieties, the corresponding values were 46–54%.

Sandoval et al. (1989) observed the effect of drought at different panicle developmental stages on gain yield and found that DS during microsporogenesis destroyed the whole panicle. Drought prior to microsporogenesis caused a 25–55% reduction in grains per panicle due to abortion of the pinnacle branch primordial. After microsporogenesis, DS reduced individual grain weight by less than 50%. Thus, they concluded that DS at all stages of panicle development reduced the yield. A similar study was conducted for this trial by Ludlow et al. (1990) and reported that water stress prior to synthesis reduced the grain yield more than the postanthesis stage at the same intensity.

Mastroilli et al. (1992) opined that water stress during the boot leaf stage greatly decreased the final biomass and grain yield, but stress applied later had no significant effect. According to Craufurd and Peacock (1993), grain yield was affected by both timing and severity of stress, and a largest reduction of 87% in grain yield occurred when stress was imposed at the boot and flowering stages. But grain yield was not affected when the stress treatments were given during the vegetative phase (GSI). Blum (1990) noticed a threefold (dryland) to fourfold (irrigated) increase in the yield of improved cultivars due to increase in HI by threefold to fourfold rather than increasing the total biomass. Thus, he concluded that drought susceptibility could be measured by estimating the reduction in yield from irrigated to dryland conditions and reported that landraces showed the greatest variability for this trait than hybrids. Santamaria and Fukai (1990) stated that OA was an important adaptive mechanism under drought conditions and the cultivars with high OA produced a higher grain yield than those with low OA. Similarly, Ludlow et al., (1990) noticed that the increase in yield was about 24% in the cultivars having high OA.

Khizzah and Miller (1992) found that grain yield was positively correlated with plant height, HI and 1000-grain weight, and negatively correlated with days to anthesis and green leaf retention. Similarly, Choudhary (1992) reported a positive relationship of grain yield per plant with growth rate of panicle and number of grains per panicle, while the former was negatively related to flowering. Sankarapandian et al. (1993) reported that grain yield had severely reduced by water stress at the flowering and ripening maturity phases, but less so at the vegetative phase, indicating that the crop is able to withstand early season drought. Sankarapandian and Muppidathi (1995) reported that the mean grain yield was 31.19 q/ha under drought compared to 35.77 q/ha under normal conditions. Bharud et al. (1995) reported that the sorghum variety Selection-3 gave a significantly higher grain yield over Swati, Lakadi, and hybrid CSH-13R in *rabi* situations and the moisture use efficiency for grain was higher in Selection-3. Sharma and Kumari (1996) reported that grain yield increases with increased irrigations. Under residual soil moisture conditions, it was observed that the drought-tolerant *cv.* M 35-1 was found most suitable (Dabholkar et al., 1995). Rao (1997) reported that water stress reduced grain yield by 27% in M 35-1 and 63% in PIS-1860. Out of all the varieties, the least reduction (6.3%) in grain yield was observed in glossy PS-18601-2, and water stress also decreased 100-grain weight by 6.3% in M 35-1 and 11% in ICSV11. Rao (1999) reported that glossy lines of sorghum (light green) produced more grain yield under water stress than nonglossy lines (dark green). There was better grain filling in the glossy type when compared to the nonglossy type. Rao et al. (2000, 2001) reported that the drought tolerance of RSLG 262 was mainly due to higher photosynthetic rate during preanthesis and postanthesis drought periods coupled with higher crop water status and transpiration efficiency (TE), which resulted in higher grain yield in sorghum under moisture stress situations. They also reported that among the yield components, greater photoassimilate partitioning during GS3 stage (HI) and more grain per panicle combining greater grain size will enhance the yielding ability of postrainy season sorghums in diverse environments. Patil and Prabhakar (2001) noticed that 1000-grain weight exhibited a significant positive correlation with grain yield under moisture stress. Yadav et al. (2003) reported that the yield reduction in sorghum was related to reduction in both grain number and grain size when water stress was imposed at the anthesis and early grain-filling stages. Kadam et al. (2002) have showed that the high grain yield of RSPS 1 (17.4 q/ha) was mainly associated with ear head length, breadth, and number of grains per ear head. Salunke et al. (2003) reported that among the 14 genotypes evaluated, CSV 14R recorded a significantly higher grain yield (1.58 t/ha), followed by M 35-1 and CR 4 (1.49 t/ha). Kusalkar et al. (2003) conducted an experiment to identify the key physiological parameters governing the yield potential of *rabi* sorghum on medium soils, and three years of study revealed that RSLG 262 recorded a significantly superior grain yield (1449 kg/ha) over Sel-3 (1133 kg/ha), which was due to higher dry matter production in different plant parts and its higher translocation to ear head at physiological maturity in rainfed situations (Patil, 2005).

Predominance of nonadditive gene action for the number of kernels per panicle was reported by Biradar (1995). Nonadditive gene action, high heritability (NS), and high predictability ratio for the inheritance of number of grains per plant were reported (Salunke et al., 1996). For panicle weight, Nayakar et al. (1989) indicated the importance of nonadditive gene and digenic epistasis was reported by Chandak and Nandanwanvar (1993). Information on combining ability is derived from the data on yield per plant, number of leaves per plant, plant height, and days to 50% flowering. Estimates of general and specific combining ability (sca) effects indicated that the presence of both additive and dominance gene action for these characters (Bhadouriya and Saxena, 1997). The inheritance of grain weight/spike was additive, while dominance effects were comparatively large, with additive and superior effects being smaller (Gao, 1993).

7.1.2 Grain Quality Characteristics

Postrainy season sorghum is known for its quality, characterized by lustrous, pearly white, attractive grains, and for this reason, it is mostly preferred for human consumption. Developing genotypes with high yield potential coupled with nutritionally superior quality grains is the prime objective of the breeding program. Studies of the F_2 segregation pattern in the postrainy × rainy cross CSV 216R × 401 B indicated that seed luster was under the influence of recessive epistasis, and intensity of seed luster depends on the recessive homozygous alleles at either both loci or one locus, while for the F_2 segregation pattern in the postrainy × postrainy cross CSV216R × 104B, polymeric gene interaction was noted. Further, based on both crosses, it was interpreted that seed luster was controlled by two pairs of genes (Rekha, 2006). For the mesocarp thickness, the dominant gene was attributed to the expression of thin mesocarp and the recessive gene to the thick mesocarp (Reddy, 2007). The M 35-1 (Shoba et al., 2008), Phule Vasudha (RSV 423), CSV 22, and Phule Chitra (SPV 1546) varieties were found to be most promising for roti quality (Chavan et al., 2009). Proximate composition (that is, moisture, protein, fat, ash, crude fiber, and total carbohydrates of postrainy sorghum genotypes) differed significantly ($P < 0.01$) and ranged from 6.64–8.58%, 8.73–12.81%, 1.22–2.36%, 1.14–1.72%, 1.21–2.48%, and 81.82–87.58%, respectively (Shoba et al., 2008).

Of the several factors that control seed color in sorghum, R and Y genes have been found to control pericarp color, and S and Z genes control mesocarp color (Yark, 1977). In the F_2 of most of the crosses studied, the segregation ratio for grain color indicates that it is conditioned by three dominant genes (Khusnmetdinova and Elkonin, 1989). The inheritance study on seed color using varieties with different seed colors, black and green (Kps1 and Kps2), indicated that inheritance of black and green seed coat color was controlled by one gene, with black gene B as dominant gene, plant purple petiole gene, and black seed color gene having a very close linkage (Chen-Huiming, 2001). Shinde and Sudewad (1980) reported partial dominance or overdominance of larger seed size over smaller seed size and observed additive gene action for seed size. The inheritance of seed size was studied in two crosses involving the popular *rabi* variety M 35-1 and high-grain-weight genotypes of African origin. Seed weight inherited quantitatively, and small seed size was partially dominant over large seed size. Gene action was predominantly additive (Biradar et al., 1996). Generation mean analysis and frequency distribution studies revealed that grain size is governed by dominant genes that are polygenic in nature. Predominance of dominance and epistatic interactions in both crosses indicates that selection for higher grain size would be more effective if the dominance and epistatic effects are first reduced by a few generations of selfing. Biparental mating, therefore, was suggested for developing homozygous bold grain lines (Audilakshmi and Aruna, 2005). Seed characteristics in sorghum, such as seed size and seed weight, were genetically positively correlated with protein content, and plant height, 100-seed weight, and lysine content were positively correlated with sucrose content. Audilakshmi and Aruna (2005) reported that the round grain shape is governed by a single dominant gene and grain luster by two complementary recessive genes. The study suggested that developing a sorghum hybrid with bold, round grain is feasible, provided that either of the parents has bold grain, round grain, or both. However, for the hybrid to be lustrous, both parents need to be lustrous and homozygous for the alleles conferring grain luster at a common locus. Study on seed luster in mungbeans indicated that seed luster possibly controlled by more than two genes (Chen-Huiming, 2001). Whereas studies of the F_2 segregation pattern in the cross CSV 216R × 401 B indicated that seed luster was under the influence of recessive epistasis and intensity of seed luster depends on the recessive homozygous alleles at both loci (or) at one locus, the F_2 segregation pattern in another cross CSV216R × 104B, saw polymeric gene interaction based on both crosses, which was interpreted to mean that seed luster was controlled by two pairs of genes (Rekha, 2006). Endosperm texture was simply inherited, with corneous texture being partially dominant over floury. Further, variation in endosperm texture may be influenced by several modifying genes (Ellis, 1975). Inheritance of endosperm texture in the sorghum kernel was studied. Heritability estimates obtained from five parental lines ranging in endosperm texture from very hard to soft and F_3 segregating grains from crosses among them were generally low. Endosperm texture appeared not to be simply inherited and was strongly influenced by the environment (Saadan and Miller, 1983).

The selection index based on seed yield, plant height, 1000-seed weight, and fodder yield may be considered as an appropriate selection index for seed yield improvement in *rabi* sorghum genotypes (Sameer et al., 2012).

7.1.3 Physiological Traits

Due to depletion of soil moisture, a decline in photosynthesis and growth occurs as a result of stomatal closure (Mwanamwenge et al., 1999). Water stress at different GSs causes various morphophysiological changes in the plant to acclimatize it to such conditions (Ali et al., 2011). Water stress at the seedling stage might lead to higher dry root weights, longer roots, coleoptiles, and higher root/shoot ratios, which could be exploited as selection criteria for stress tolerance in crop plants at very early stages of growth (Takele, 2000; Dhanda et al., 2004; Kashiwagi et al., 2004). However, at later growth phases like the reproductive stage, flag LA (Karamanos and Papatheohari, 1999; Ali et al., 2010), specific leaf weight (SLW), leaf dry matter (Aggarwal and Sinha, 1984), excised leaf weight loss (McCaig and Romogosa, 1991; Bhutta, 2007), relative dry weight (Jones et al., 1980), relative water content (RWC; Fischer and Wood, 1979; Colom and Vazzana, 2003), residual transpiration (Clarke et al., 1991; Sabour et al., 1997), and cell membrane stability

(Premachandra et al., 1992; Ali et al., 2009b) are the characteristics of interest and had been widely exploited as reliable morphophysiological markers contributing to drought tolerance for various crop plants in addition to sorghum.

Higher stomatal frequency is found to be associated with higher water use efficiency (WUE). Higher stomatal frequency was found on abaxial surfaces compared to adaxial surfaces in *rabi* sorghum varieties. Two *rabi* sorghum varieties, Phule Chitra and Phule Maulee, had significantly lower stomatal frequency on both leaf surfaces compared to the popular landrace M 35-1. However, this trait may not be contributing to yield improvement in medium soil conditions, as seen by nonsignificant correlation with grain yield (Sanjana Reddy et al., 2012). Shawesh et al. (1985) observed higher numbers of stomata on abaxial surfaces than on adaxial surfaces. They also found higher stomatal frequency for drought-tolerant genotypes than susceptible genotypes in sorghum.

The chlorophyll stability index (CSI) is associated with desiccation tolerance under terminal water deficit conditions and can be used as one of the reliable selection criteria in rapid screening for postrainy-adapted genotypes for drought tolerance. Nonsignificant differences in CSI among the *rabi* varieties adapted to various soil depths indicated that the trait may not be influenced by the soil type. CSI has a negative association with grain yield, indicating that less stable chloroplasts had resulted in higher photosynthetic rates under water deficit conditions (Kadam et al., 2002; Awari et al., 2003; Sanjana Reddy et al., 2012).

RWC is the ability of a plant to maintain high water in its leaves in moisture stress conditions, and this has been used as an index to determine drought tolerance in crop plants (Barrs and Weatherly, 1962). Blum et al. (1989a,b) reported that higher-leaf RWC allows plants to maintain turgidity, and thus would exhibit relatively less reduction in biomass and yield. RWC exhibited significant correlation with grain yield ($r = 0.38^*$) (Sanjana Reddy et al., 2012). In a study, Phule Chitra was found to have the highest RWC of 87% and was also the highest grain yielder. The physiological parameters of Phule Chitra, such as leaf area index (LAI), relative leaf water content, leaf temperature, CSI, stomatal frequency, HI, grain size, and earhead exertion, were superior to the other varieties in receding moisture conditions, indicating its suitability for cultivation in medium types of soil during postrainy season (Sanjana Reddy et al., 2012).

An increasing number of reports provide evidence of the association between high rate of OA and sustained yield or biomass under water-limited conditions across different cultivars of crop plants (Blum, 2005). Increased deep-soil moisture extraction has been found to be a major contribution of OA in sorghum (e.g., Wright and Smith, 1983).

Leaf area: Growth and yield depends on the LA, which is governed by the number, rate of expansion, and size of leaves. Pathways for constitutive reduction in plant size and LA are smaller leaves, reduced tillering, and early flowering. Reduced growth duration is associated with reduced leaf numbers (Blum, 2004). Narrow and erect leaves will have lower LA, which helps to reduce the depletion of soil moisture due to transpiration. But low productivity in rainfed sorghum appears to be only partially due to slower LA development and faster leaf senescence. Further, the LA development was more sensitive to water stress than stem elongation (Kannangara et al., 1983; Verma and Eastin, 1985). In a field experiment, it was observed that the rate of dry matter accumulation per unit radiation intercepted by the rainfed sorghum was only 67% of the irrigated crop (Seetharama et al., 1978). Retention of the uppermost green leaves in stay-green hybrids during the latter half of the grain-filling period enables these types in assimilating carbon and complete grain filling, as evidenced by the positive correlation between green LA at maturity and grain yield (Fischer and Wilson, 1971).

Gonalez-Hernandez et al. (1986) observed that the soil moisture depletion to the extent of 30–33% of available moisture did not affect the rate of leaf elongation at any stage during panicle development, but decreased rapidly when the soil moisture level was below the critical level [below the permanent wilting point (PWP)], and opined that elasticity of sorghum guard cells declined with leaf and plant age. Moisture stress caused the cessation of leaf elongation, which was higher in hybrids than in their parents. Under no moisture stress, CSH 6 had a higher rate of leaf elongation than CSH 5 (Parameshwara and Krishnasastry, 1982). Garrity et al. (1984) reported that there was 14–26% reduction in canopy photosynthesis observed in sorghum subjected to DS; this was fully accounted for by LA reduction rather than by stomatal response. LA significantly reduced before decrease in stomatal conductance (Blum and Arkin, 1984).

The senescent sorghum hybrids produced greater LA and translocated more organic material from the bottom leaves than other genotypes (Harden and Krieg, 1983). Water stress has not only inhibited the photosynthesis and transpiration, but also decreased the functional area, leading to low grain yield (Gonalez-Hernandez, 1985). Khanna and Sinha (1988) observed the loss of LA and chlorophyll content due to water stress. During stress, older leaves are killed, while the younger leaves retain turgor, stomatal conductance, and assimilation (Blum and Arkin, 1984). At the grain-filling stage, leaf senescence was more rapid in male fertile lines than in sterile lines of sorghum, which suggests that the reproductive sink accentuates drought-induced leaf senescence.

Gangadhar Rao and Sinha (1988) reported that LA produced has a direct bearing on leaf WP under limited soil moisture status. Sorghum hybrids having high LA depleted the limited soil moisture at a faster rate, resulting in decreased soil WPs. These hybrids did not have superiority in water relations over their parents. Drought resistance has been found to depend on root and leaf characteristics (Rosa and Maiti, 1990) and drought-tolerant germplasm accessions; for instance, IS 564 and IS 8311 had higher values for mesocotyl length, root dry weight, shoot dry weight, and LA than those of drought-susceptible ones, such as IS 5484 and IS 3962.

Mathews et al. (1990a) found more leaf rolling in resistant than in susceptible lines. Leaf rolling reduces the effective area of the uppermost leaves by about 75% and alters the leaf surface and microclimate, so that stomata may remain open and growth may continue without a high rate of water loss. The OA helps in maintaining green LA and usage of water during the drying period (Tangspremsri et al., 1991c). Wenzel (1989) reported that drought-resistant sorghum line BT × 623 had higher growth rate, lower rate of moisture loss, low total LA, and high moisture loss/net LA. Sarwar (1983) assessed 13 sorghum varieties and found that 6 were not affected by severe drought, whereas the local Y-75 variety was found to be doing well, with lush green leaves from top to bottom. The glossy genotypes are more drought-tolerant than the nonglossy ones, even at the vegetative stage indicated by less reduction in flag leaf length and area (Ravindranath and Shivaraj, 1983).

A positive correlation was observed between plant dry matter and leaf width and LA in sorghum (Shang, 1989). Rosenow et al. (1983) and Henzell et al. (1992) found that higher

LA at maturity was used as an indicator of postanthesis drought resistance in sorghum, which was produced higher grain yield under stress conditions. No significant correlation was established between the reductions of LA and the extent of seed numbers. Bharud et al. (1995) reported that after anthesis, the sorghum genotypes have optimum LA, which helps in reducing the moisture losses through transpiration. Similarly, both LA and transpiration rates were reduced by 3% and 11%, respectively, in dry conditions, as reported by Saliah et al. (1999). Water deficit reduced the green LA at maturity due to terminal drought compared with full irrigation (Borrell et al., 2000a,b). However, there was no decline in the total LA with water stress, when applied at the boot stage (Munamava and Riddoch, 2001). Patil et al. (2003) reported the positive association of LA with grain yield, which could be considered as selection criterion under DS situations.

LAI: LAI can be taken as the ability of plants to produce grain yield. Reduced LAI is a major mechanism for moderating water use and reducing injury under DS (e.g., Mitchell et al., 1998). But reduced LAI due to moisture stress during flowering reduces grain yield per plant (Bakheit, 1989). Sorghum hybrids have larger LAI and leaf area duration (LAD) than the local cultivars (Sinclair and Ludlow, 1986), which result in poor plant water status. The glossy genotypes possessed higher LAI values, even under water stress, than nonglossy ones (Rao, 1997). The high LAI and LAD of a genotype that produced high dry matters could be considered as indicators of grain yield (Kadam et al., 2002). However, in the current study, LAI is poorly but positively correlated with grain yield ($r = 0.05$) and the varieties showed nonsignificant differences for LAI that ranged from 3.1 to 3.4.

Watson (1947) reported the role of LAI in dry matter production. Erick and Hanway (1965) suggested that the relationship between the photosynthetic rate at 60 DAS and the increase in dry matter from 60 to 90 DAS was positively associated with an increase in plant population and resulted in higher LAI. The differences in crop growth rate (CGR) at early GSs of sorghum were mainly attributed to LA development, especially to the initial LA, but not the leaf growth rate (Fischer and Wilson, 1975). Further, they studied the relationship between LAI and net assimilation rate (NAR) and concluded that there was no greater improvement on yield due to unit increment in LAI. Kudasomannavar (1974) reported that early planted sorghum was exposed to less water stress, as evidenced by lower diffusive resistance prior to heading and small reduction in LAI after heading compared to late planted sorghum. Growth analysis in sorghum hybrids indicated that LAI increased throughout the growth period, but the increase was less rapid after 50 DAS (Santos et al., 1979). The grain yield was closely related to CGR and LAI at 1 week before heading. But these correlations, as well as those between grain yield and total shoot dry weight, were less close in the sparing than in the autumn sorghum crop (Lin and Yeh, 1990). Similarly, Joshi and Jamadagni (1990) reported that LAI and LAD during the grain-filling stage, high dry matter accumulation, and HI were considered to be the most important physiological characteristics responsible for high yield. Ravindranath and Shivraj (1983) reported that LAI, CGR, NAR, and grain yield decreased due to water stress, and grain yield was mostly controlled by LAI. However, the glossy genotypes had high LAI values even under water stress, and the LAI was 1.9 for M 35-1 at the peak stage of stress. Hiremath and Parvatikar (1985) found that in *rabi* season, the LAI values and differences of LAI among genotypes were smaller in range than in *kharif* season.

There was a positive association between LAI and CGR (Eastin, 1983; Myers et al., 1986; Rao and Singh, 1998) Thus, LAI can be taken as an index to indicate satisfactorily the ability of the plant to produce grain yield. But the moisture stress during flowering reduces the LAI and grain yield per plant (Bakheit, 1989). Sorghum hybrids have larger LAI and LAD than the local cultivars, which maintained better water status (Blum, 1991; Blum et al., 1992). The poor plant water status of hybrids was partially ascribed to their larger LAI. The glossy genotypes possessed higher LAI values, even under water stress, than nonglossy ones (Rao, 1997). The sorghum genotypes RS 615, RS 29, C43, and RS 585 maintained more LAI ($>$2.00) at the dough stage, under water stress conditions (Surwenshi, 1999; Chimmad and Kamatar, 2003). The high LAI and LAD of a genotype that produced high dry matter could be considered as indicators of grain yield (Kadam et al., 2002).

Leaf area duration: Sorghum growth is generally related to high CGR, which in turn depends upon LA and LAD (Veeranna, 1972). Krishnamurthy et al. (1973) observed higher LAI and LAD in hybrids CBE-X and CSH-1 than in the Neerajola and swarna varieties. Similarly, higher grain yields were realized by E 57, under DS, which is due to its ability to maintain green leaves for a longer period during the grain-filling stage (Wright et al., 1983). Duncan et al. (1981) observed that nonsenescent genotypes in sorghum. Further, Kulkarni et al. (1981) reported that the dry matter accumulation was correlated with LAI and LAD in sorghum. Sriram and Rao (1983) concluded that the growth parameters such as leaf area ratio (LAR) and its duration had a strong association with grain yield. A positive correlation was observed among plant dry weight, plant height, leaf width, number of tillers per plant, LA per plant, and HI in sorghum (Shang, 1989). Similarly, grain yield in soybean was strongly associated with maximum LAI, but not with NAR, while the grain yields usually depend on LAD after anthesis (Annandale et al., 1984). Henzell et al. (1992) observed a positive association between green LAD and grain yield in sorghum. Higher LAD in the sorghum genotypes indicated the importance of not only higher LA, but also its persistence for a longer time leads to obtaining potential yields and better grain filling (Joshi and Jamadagni, 1990; Kadam et al., 2002). Green LA retention estimated at maturity was identified as one of the component traits of stay-green, which was highly correlated with green LAD (Tenkouano et al., 1993; Walulu et al., 1994; Osterom et al., 1996).

Specific leaf area (SLA) and SLW: Specific leaf area (SLA) and SLW are important physiological traits in assessing plant growth. Yoshida (1972) reported that the photosynthetic rate of single leaves was associated with leaf characteristics such as thickness or SLW, leaf nitrogen content (LNC), stomatal density, carboxylating enzymes, inter veinal distance, and leaf vein frequency. Irvine (1975) observed that in sugarcane, leaf width and SLW were negatively associated with photosynthetic rate, whereas leaf thickness had a positive relationship with photosynthetic rate. Patterson et al. (1977) reported that leaf thickness and leaf weight were the main causes for the difference in photosynthetic rate in cotton. In another study, the effect of air temperature and solar radiation on LAR, NAR, SLA, leaf relative growth rate (RGR), and leaf weight ratio (LWR) in various crops, including sorghum, was that LAR tended to be negatively affected by air temperature and solar radiation in C3 winter crops. However, C4 crops were not affected, but air temperature and solar radiation in c3 winter crops tended to show an immediate response. Change in LAR was due to changes in SLA and LWR (Vong and Murata, 1978). Santos et al. (1979) reported that values of LAR were higher at 30 days after emergence (DAE) in sorghum. Sriram and Rao (1983) stated that low-yielding cultivars of grain sorghum had smaller LAR at the flowering and early grain-filling stages and LAR at flowering exhibited a strong association with yield. The relationship of SLA with growth and yield in wheat were studied, and it was found that SLA decreased from the tip to base of the leaf. It also declined between leaf appearance and 10–12 days later, and then it rose again. However, SLA in the flag leaf was the most stable with age. They further observed that in canopies, the final rise in SLA with leaf age occurred earlier than in spaced plants. The SLA declined approximately 4 cm^2/g for each mol quanta m-2 day-1 increase in radiation. From this, they concluded that differences among SLA in wheat genotypes and ranking were unchanged by the change in radiation (Rawson et al., 1987). Pawar (1996) noticed that the SLA and LAR were at their maximum at the early stage and then decreased at harvest. Such a decrease in SLA and LAR could be due to the impact of the developing moisture stress on cell elongation. They also reported that SLW decreased in N-13, M 35-1, and RS 29 at later stages due to decreasing soil moisture, which has some effect on the thickness of mesophyll cells in sorghum. Reduction in SLA in sorghum was due to increase in crop age (Rao et al., 2003). The genotype RSLG 283 maintained higher SLW, and leaf thickness was compensated by LAI decrease (Shivalli, 2000). Munamava and Riddoch (2001) reported that SLW and SLA decreased with stress, especially when water stress was applied at the boot stage. Rao et al. (2003) concluded that varieties recorded higher SLW (26.5 mg/cm) over the hybrids (16.5 mg/cm) in sorghum, and they also found that SLW was positively correlated with total chlorophyll content and biomass.

Leaf nitrogen: Rego et al. (1998) and Rao et al. (2003) reported that variations in SLN in sorghum at different GSs. At early GS in sorghum, the stay-green hybrids partitioned more carbon and nitrogen to leaves than their senescent counterparts, resulting in higher SLN. However, after anthesis, higher SLN delays the onset and reduced the rate of leaf senescence, and this is associated with stay-green crops taking up more nitrogen from the soil than senescent crops. These processes led to increased grain yield and lodging resistance in stay-green lines under terminal DS.

Borrell and Hammer (2000) stated that higher SLN initiates a chain of responses, including enhanced radiation use and transpiration efficiencies, enabling the plant to set a high yield potential by anthesis. And this response was observed under both high and low N conditions (Borrell and Hammer, 2000).

The SLN in stay-green sorghum hybrids remained above the threshold senescence level for longer than in senescent hybrids for at least three reasons: (i) the leaf N benchmark at anthesis was higher in stay-green than in senescent hybrids; (ii) N-uptake during grain filling was higher in stay-green than in senescent hybrids; and (iii) the remobilization of N from leaves of stay-green hybrids during grain filling was low compared to senescent hybrids (Borrell and Hammer, 2000).

Photosynthesis in the leaves caused an increase in dry matter production and economic yield. Better growth of sorghum was generally related to higher CGR, which in turn depended upon the LA, LAD, and relative leaf growth rate (Veeranna, 1972; Krishnamurhty et al., 1974; Prabhakar, 1975).

The CGR of a sorghum hybrid was greater than either of the parents from emergence to panicle initiation, while the NAR and RGR did not differ among the genotypes. The yield superiority of Rs. 610 was attributed mainly to the combination of high CGR of female parents during the grain-filling period, with effective conversion of dry matter into the grain of male parents (Gibson and Schertz, 1977). In general, the average CGR of all sorghum hybrids reached a peak before anthesis (Bueno and Atkins, 1982). There is a continuous decline of CGR after 40 days in unirrigated sorghums, in contrast to a steady increase in irrigated sorghums (Seetharama et al., 1978). However, Reddy (1980) did not find any genotypic differences for CGR, RGR, and NAR and reported that these three parameters increased up to 60–70 days and declined thereafter. The stressed plants when irrigated showed a similar growth rate as the irrigated ones (Lasavio et al., 1982).

Patanothai and Atkins (1971) concluded that during the grand growth period in sorghum, the absolute rate of vegetative weight accumulation in hybrid was greater than that of the midparent, but their RGR was similar. In general, the stress restricted the expansion of sorghum leaves and resulted in poor dry matter production. However, the dry matter distribution, mean RGR, and NAR were not affected much due to stress. Ferraris (1981) reported in sweet sorghum cultivars, a positive correlation was observed for RGR, NAR, and CGR with the dry matter production at 30-, 60-, and 90-day-old plants. The lower RGR value at the vegetative and flowering stages increased during the grain-filling period of sorghum (Bueno and Atkins, 1982). There is a gradual increase in RGR, which reaches a peak value of 230 mg/g/day at panicle initiation and subsequently declined rapidly up to anthesis and slowly during the grain-filling and physiological maturity stages (Gonalez-Hernandez, 1985). The NAR was associated positively with SLW, RGR, and dry matter production, and negatively with LAR (Murthy et al., 1986).

The NAR in the entire vegetative period was 0.50 $mg/cm^2/day$ in sorghum (Kim et al., 1987). The values of CGR, RGR, LAI, and NAR in sorghum were found to decrease under stress, and this was more dominant in nonglossy genotypes than in glossy genotypes. High CGR and NAR values suggest the maintenance of high rates of photosynthesis by glossy genotypes under water stress. Premachandra et al. (1992) observed that sorghum cultivars (HS and ZT) maintained higher RGR and NAR under water deficit conditions, which did not exhibit greater stress symptoms and also produced higher plant weight than others, indicating their drought tolerance.

Pawar (1996) noticed that CGR values were relatively more in high-yielding sorghum cultivars, while RGR and NAR values were inconsistent at different GSs. Rao and Singh (1998) and Sri Ram and Rao (1983) observed significant variations in LAR, LWR, and LAD among the genotypes. Further, LAR had a significant positive association with LWR, panicle mass, biomass, and grain yield. Similarly, Pawar and Chetti (1997) also reported the positive association between LAR and RGR with grain yield in sorghum under terminal stress conditions.

Rao (1999) observed that water stress during postanthesis decreased growth parameters such as CGR and NAR in glossy and nonglossy sorghum genotypes. Similarly, water stress applied at the vegetative or reproductive stage was found to reduce the relative growth and net assimilation rates of sorghum (Younis et al. 2000).

Dry matter accumulation and its distribution: Dry matter accumulation and distribution are important factors indicating the partitioning efficiency of photosynthetic assimilates. In general, soil moisture determines the distribution and accumulation of dry matter in different plant parts. Further, the moisture stress occurring at various stages ultimately influences the economic yield. In sorghum, terminal stress is a common phenomenon in *rabi* situations, and the grain-filling stage coincides with terminal stress. The dry matter distribution in different plant parts of sorghum revealed that the preflowering contribution is only 12% of grain weight, and 93% is due to assimilation by the top head and four leaves (Fischer and Wilson, 1971). This may be due to their close proximity to the panicle, greater interception of light, and higher metabolic efficiency due to relatively young age. Rao and Singh (1998) observed that the stem weight of sorghum increased up to the first week after anthesis and then fell to a level much lower than that at anthesis, and thereafter declined further. The decline in both stem weight and sugar content were due to remobilization of stem dry matter to the panicle. Wilson et al. (1980) reported that the dry matter accumulation rate of the sorghum plants was greatly reduced by the water deficit and that it was due to reduced LAI and decreased substrate production rate per unit LA photosynthesis. Wong et al. (1983) found that drought affects the panicle dry weight more than the vegetative parts of the plant, since water stress normally occurs after the vegetative stage. They also found that most of the genotypes had reduced dry matter production during the grain-filling period and later recovered. Hukkeri and Shukla (1983) found significant reduction in fodder yield and dry matter by withholding irrigation during any stage of growth. Garrity et al. (1983) noticed that dry matter and grain weight were much less sensitive to season evapotranspiration deficit treatment (irrigation applied throughout the GSs than the irrigation during the grain-filling period), but seasonal WUE in the treatment where irrigation applied throughout the GSs was substantially higher. They concluded that the magnitude of DS conditioning depends on the genotype, phonological timing of treatment, and irrigation regimes employed. Garrity et al. (1984) reported a reduction of grain and dry matter yield by about 36% and 37%, respectively, due to water stress. Bishonoi (1983) reported that dry matter production in pearl millet was higher than that in sorghum and maize due to higher net photosynthetic rate, LAI, and PAR absorption. Hiremath and Parvatikar (1985) noticed that dry matter accumulation in sorghum enhanced with an increase in LAI. However, certain genotypes (SPV 126, SB 3307, SB 2431 and SB 3304) did accumulate a large amount of dry matter even with low LAI due to efficient utilization of the limited available water by reducing transpiration loss. Rego et al. (1988) noticed a decrease in dry matter yield and expansion in the number of leaves after the imposition of high OP treatment, and concluded that plants in relieved water stress regimes produced more dry matter and expanded more leaves than plants in continuous stress regimes.

Muchcow and Coates (1986) observed that there was a reduction in leaf dry weight and total dry matter, and this decrease might be attributed to reduced LA and leaf abscission, which led to a concomitant decrease in the efficiency with which solar radiation was used to accumulate biomass. McCree et al. (1984) revealed that stressed plants accumulated biomass and carbon through the OA with little additional metabolic cost, and the carbon stored during stress was immediately available for the synthesis of biomass upon rewatering. Muchow (1989) reported that in sorghum hybrids, high biomass both at maturity and during the grain-filling stage were positively associated with grain yield. Subramanian et al. (1989) and Rao and Singh (1998) found a positive correlation of panicle mass with grain number per panicle and 1000-grain weight. Dabholkar et al. (1970), Sriram and Rao (1983), Muchow (1989), Pinjari and Shinde (1995), Omanya et al. (1997), Pawar and Chetti (1997), and Rao and Singh (1998) also reported highly significant positive correlation between biomass at maturity, panicle mass, and grain yield. There was a negative correlation between dry matter accumulation and relative moisture loss (Wenzel et al., 1999; Singh et al., 1990). The dry matter production of both resistant and susceptible genotypes was directly proportional to the amount of light intercepted by the canopy and to the plant water loss divided by saturation deficit of the air (Terry, 1990). Genotypes with high OA had greater root length, soil water extraction capacity, and dry matter production during the preanthesis stress period and also found no significant difference in dry matter yield at physiological maturity between low- and high-OA groups (Santamaria et al., 1990). Sorghum hybrids produced more biomass per day than varieties under stress conditions (Blum, 1991; Blum et al., 1992). Thus, in terms of plant water status and mean daily biomass production, cultivars were more drought-resistant than hybrids. However, the physiological superiority of cultivars under DS did not result in higher grain yield because of their relatively inherent poor HI. Donatelli et al. (1992) observed genotypic variation in biomass reduction under limiting water conditions. The dry matter of leaf, stem, and panicle at harvest decreased with DS, when imposed at all the developmental stages, except at physiological maturity (Gonalez-Hernandez et al., 1992; Rao and Singh, 1998). There was a positive correlation between dry matter production and grain yield (Joshi and Jamadagni, 1990; Choudhary, 1992; Craufurd and Peacock, 1993; Sankarapandian et al., 1993; Shinde et al., 1998). Lamani (1996) observed that the genotypes maintained higher dry matter in leaves, stem, and panicle during the postanthesis period and decreased sharply postanthesis to developing grains is important for higher productivity under receding soil moisture conditions.

Shivalli (2000) reported higher leaf, stem, panicle, and total dry matter content combining a higher grain yield in IS 9244, NR 35, RS 29, and DSH 4 genotypes. Kadam et al. (2002) reported that the low-yielding genotypes produced more total dry matter at the early stage and extracted more soil moisture early in the GS. While the genotypes produce more dry matter during later stages, they extracted more water during later GSs and thus sustained productivity when current photosynthesis was inhibited by terminal stress (Blum et al., 1997). Yadav et al. (2003) noticed a decrease in dry matter in response to short-term water deficit at both the anthesis and grain-filling periods. Rao et al. (2003) reported that panicle biomass had significant positive correlation with total biomass, grain number, grain yield, and HI. Salunke et al. (2003) found that the increase in grain yield in PBS 2, CR4, CR6, and CR9 was mainly due to increases in total biomass; and the actual values were 520, 541, 505, and 543 g/m, respectively.

The nonsenescent/stay-green genotypes retain more of their photosynthate in the leaves (Borrell and Hammer, 2000) and stems, whereas rapid leaf senescence may indicate reserve mobilization to the grain under stress (e.g., Fokar et al., 1998; Yang et al., 2001). The delicate balance between stem reserve mobilization and nonsenescence, which involves carbohydrate and nitrogen metabolism, is not quite clear.

Assessment of recombinational variability for combining ability in early F_2 generation: With respect to early-generation testing, Jenkins (1935) presented data about maize, concluding that inbred lines acquired their individuality as parents of top crosses (male sterile line × open pollinated population) very early in the inbreeding process and remained relatively stable for combining ability thereafter. Further, Jenkins (1940) confirmed that limited segregation with the S1 families paved the way for early testing to determine their relative worth with respect to yield-attributing traits. Early-generation testing was primarily designed to categorize the population of lines into good and poor groups in terms of combining ability.

This enables greater emphasis to be placed on further selection and testing only in the good-combiner group. Lonnquist (1950) has considered this an advantage of early testing. Testing of lines (maintains and restorer lines) at an early F_2 stage and after would reduce the amount of time, land, and labor, and helps to exploit transgressive segregation for combining ability (Falconer, 1986). Dankov (1965) compared S1, S2, and S3 inbreds in test crosses. He opined that the percentage of S3 lines showing stable performance for combining ability was not higher than that of S1 and S2 lines. This infers that S1 evaluation is on par with evaluation practiced at later stages (i.e., S3 evaluation).

In maize, Agrawal and Sharma (1972) showed that crosses involving inbreds (made from different levels of inbreeding) with the testers did not display significant differences in per se performance. This suggests that every time the level of inbreeding might not be strictly required for high per se performance of a cross, it is possible to obtain transgressive segregants for combining ability in the desirable direction at an early stage of inbreeding. Based on studies of combining ability, Baldwa et al. (1979) concluded that F_2 generation can be effectively used for identification of good combiners. Several other studies have also agreed on the assessment of genetic variability generated in the segregating generations of the crosses at an early stage to harness the recombinational variants for combining ability and per se performance (Vahtin, 1958; Fawzi, 1962; Gerasenkov and Goncharova, 1972). The response to selection of any characteristic is dependent on the existence of variability in the genetic material. For improvement purposes, breeders must be concerned with the creation of variability and identify superior recombinants by following appropriate breeding methods.

Sorghum improvement strategies in India primarily lay emphasis on the improvement of hybrids by improving the performance of the parents involved. The improvement in per se performance of hybrids is possible by creating variability in maintainer parents as well as restorer parents, followed by selection. This requires the adoption of systematic breeding approaches. One of the approaches used to create variability is through the recombination of genotypes and evaluation of the recombinants in segregating generations for combining ability. In cross-pollinated crops like maize, different recurrent selection procedures for general combining ability and sca are used to serve this purpose. Such procedures cannot be followed entirely in often cross-pollinated crops like sorghum, whereas, due to considerable self-pollination, there are limitations for intermating among the selected plants, which is an important step in recurrent selection methods. But the principle for which these procedures are adapted can be effectively used in sorghum.

Pederson (1974) and Bos (1977) have also argued against recurrent selection methods in autogamous crops. They concluded that selection and selfing in early segregating generations is just as efficient as that of random mating, in order to increase the frequency of favorable alleles. Bos (1977) demonstrated that F (obtained through laborious intermating among F_2 plants) will contain at most only 25% more plants with two desirable alleles than F (the generation obtained from F_2 selection and selfing). For assessing the F_2 population for variation in combining ability, the truncated selection proposed by Pederson (1974) would be better than intermating. In truncated selection, individuals are selected strictly on the basis of their phenotypic values. This procedure helps for identifying the possible transgressive recombinants for combining ability in the desirable direction produced in early F_2 populations. These favorable segregants can be selected primarily based on their phenotypic performance, and are further screened based on the results of progeny

FIGURE 7.4 Selection of the F2 generation of rabi crosses.

evaluation (i.e., evaluation of a progeny hybrid derived by crossing male sterile line × early F_2 segregants) (Patil and Pandit, 1995). In improving the parental lines of a hybrid, often B × B and R × R crosses are made to extract new B and R lines, respectively (Madhusudhan, 1993). Among the earlier studies in this direction, Dremlyuk (1980) made the choice of parents based on selection for yield taking account for combining ability in early segregating generations. Mohammed (1980) compared different male testers (R lines, populations, and F_1 hybrids) for combining ability with A lines and found populations to be as satisfactory as the other two testers. Hookstra et al. (1983) in sorghum produced top crosses by crossing male sterile lines with six random mating populations and identified superior populations to initiate inbreeding to develop improved R lines. In the early-generation testing of B lines for combining ability, the recombinants produced in early inbreeding progenies of B line crosses of different genetic sources were superior (ICRISAT and Texas). The top cross-derived hybrids were significantly superior with respect to per se performance to check and gave better results for days to flowering than the average flowering B and R lines. Likewise, Ejeta and Rosenow (1993) in a R × R study produced 10 restorer lines (PI561846–PI561855) derived from crosses between elite US restorer lines and drought-tolerant restorer lines from the Sudan. These R × R derivatives were an excellent source of combining ability, yield stability, and drought tolerance. Madhusudhan and Patil (1996a) crossed a random sample of 97 F_3 sorghum segregants/lines of the cross 3660B × MR750 to a stable male sterile tester. The derived F_1s expressed significant genetic variance, and 21 derived hybrids showed significant yield heterosis, implying transgressive segregation for combining ability in desirable direction in early F_3 generation (Figure 7.4).

In a further attempt, Madhusudhan and Patil (2000) determined the possibility of broadening the variability created for combining ability beyond the first stage. Here, two new second-stage F_2 populations were developed by further crossing two F_3 lines (of original lines 3660B × MR750). Differences were noticed among these two second-stage F_2 populations for the release of variability for combining ability in desirable directions based on the evaluation of derived F_1 hybrids.

Madhusudhan (2002a), in another similar study, obtained derived F_1 hybrids (male sterile line × segregants of a cross) that exhibited desirable transgressive segregation for grain yield. This finding confirmed that good combiners could be traced to segregating progenies from a cross between two elite lines. Vasal et al. (2004) also provided information on population improvement strategies for crop improvement. They suggested that among the various methods of intrapopulation improvement, selection in early self-generation is suitable for improving combining ability. On similar lines of research, for the improvement of parental lines of a hybrid for combining ability through early-generation selection, many studies have also advocated the use of B × R crosses (Rao et al., 1982; Atkins, 1988; Nayakar, 1985; Desai, 1991; Nayakar et al., 1994; Patil and Pandit, 1995; Madhusudhan and Patil, 1996b,c; Madhusudhan and Patil, 2000; Madhusudhan, 2002a) in grain sorghum and Madhusudhan (2002b) in pearl millet. All these findings reveal that in a crop like sorghum, having autogamous nature (5–30%), hybridization followed by selfing and assessment of early segregating generation for variation in combining ability could be beneficial.

7.1.4 Hybrids

Hybrids are the means of maximizing the best complementary combination of desired genes from two deliberately selected parents. But ultimately, it was the superiority of the F_1 hybrid for harvestable products that encouraged plant breeders to use heterosis for the improvement of crop performance. The most important characteristics of interest to the plant breeder where hybrids can be superior to their parents (that is, show heterosis) include increased grain or biological yield, earliness in maturity, enhanced grain size, and fruits, as well as better nutritional quality. Early flowering and superior early vigor are commonly observed with hybrids, in addition to higher grain and fodder yields (Quinby, 1974). The reasons for lower productivity of postrainy season sorghum (0.7 t/ha) are many—including low temperatures, terminal drought, shoot fly damage, and lack of farmer-accepted improved cultivars with adaptation to highly variable production environments. Though strong efforts have been made to develop hybrids with wider adaptability to varied production environments, the results were not encouraging.

Heterosis: Many efforts have been made to introgress farmer-preferred traits such as bold, lustrous, and semicorneous grain types and juicy stalks into the hybrids targeted for postrainy season cultivation by crossing improved Indian landraces as pollinators with the established exotic and elite female parent, CK 60A (*milo* or A_1 cytoplasm). However, the resulting hybrids lacked marked heterosis, had threshing difficulties, and were too tall, which were not amenable characteristics for increasing plant population per unit area. Several studies in the past have indicated only modest levels of heterosis for economic traits in postrainy season sorghum, as most of the parents utilized in postrainy hybrid programs were related by descent. The hybrids, CSH 7R and CSH 8R, developed from the improved parents and released in 1977, were not acceptable to farmers despite their high heterosis, for they lacked grain luster, resistance to shoot fly, and lodging (Rao, 1982). Later, CSH 12R (released in 1986), CSH 13R (released in 1991), CSH 15R (released in 1995), and CSH 19R (released in 2000) also could not progress well compared to the varieties and failed to make any impact on farmers' fields. However, studies by Rana et al. (1997) have indicated appreciable levels of heterosis for grain yield and other agronomic traits. The hybrids developed and released did not attract farmers, as these hybrids lacked matching grain quality, and shoot fly resistance was comparable to the most popular landrace variety, M 35-1. Also, poor seed setting, mainly caused by low temperatures prevailing during anthesis, is observed in hybrids. Greater yield heterosis was observed in derivative × tropical (African) varietal crosses due to the diversity of genes (Rana et al., 1985). Rana and Murty (1978) also reported that increases in number of seeds per panicle branch in short compact headed varieties (tropical) and increases in the panicle branches in the long panicle type (temperate) by introgression of genes from African germplasm result in yield heterosis. Large heterotic response for grain yield and HI were accompanied by stalk rot and shoot fly susceptibility (Rao, 1982). For increasing the grain yield within the limits of the available water supply, the choice of female parent for hybrid production should be made for both LA and photosynthetic rate, and the selection of pollinators should be made for maximum seed number per panicle (Krieg, 1988).

Hybrid parents: The combining ability of compact-headed Indian landraces was found to be relatively poorer than very long panicle types of exotic origin. It was hypothesized that lack of reinforcement between genes responsible for primary axis length and those contributing to girth resulted in lack of marked heterosis for ultimate grain yield (Rao, 1970). Pollinators and female lines from exotic germplasm contributed to poor grain quality. The derivatives *durra-caudatum* (*zera zera*) crosses, as pollinators developed for postrainy season, and cytoplasmic nuclear male-sterility (CMS) lines, developed for *kharif* season from *kafir-caudatum* (*zera zera*) crosses, did not attract the attention of farmers as they lacked grain luster, size and shape, and fodder yield comparable to M 35-1, despite their superiority under late sowings in postrainy season with yield heterosis of 45–64%. Landrace pollinator-based hybrids, where many of the desirable attributes of landraces are inherited favorably in their hybrids, possess moderate levels of shoot fly resistance and desirable grain quality traits. However, they lack lodging resistance and have moderate yielding ability (Reddy et al., 1983). Most of the landraces, including M 35-1, showed segregation for fertility restoration/sterility maintenance ability, indicating the need to select for restoration ability within the landraces. This also explains the partial restoration observed when bulk pollen of M 35-1 was used by many researchers. Jayanthi (1997) showed that shoot fly resistance in both parents (or at least in seed parents) is needed in order to realize higher frequency of shoot fly–resistant hybrids in postrainy season.

7.1.4.1 Alternate CMS Systems and Races

Fertility restoration by landraces was poorer on *durra*-derived A_1 CMS lines than on *caudatum*-derived A_2 CMS lines. When both A_1 and A_2 CMS lines were based on *caudatum*, fertility restoration was higher with A_1 than A_2. This finding has a bearing on developing CMS lines involving *caudatum*-based germplasm lines adapted to postrainy season and testing for fertility restoration in hybrids. The hybrids involving *caudatum*-based female parents and *durra*-based landraces showed high heterosis for grain yield, but grain quality was poor.

7.1.4.2 Male-Sterility Systems for Hybrid Sorghum Breeding

In sorghum, genetic male sterility was reported by Ayyangar and Ponnaiya as early as 1937 in the MS 1716 variety, and the male sterility was observed to be due to empty anther sacs. This was probably the first record of occurrence of male sterility in sorghum. It was a simple recessive Mendelian character in sorghum and designated as *ms.* About the same time, genetic male sterility was also reported by Stephens (1937), and the sterility was due to incompletely developed anthers in black hull kafir. Another case of genetic male sterility reported in the "Day" variety, which was used as a female parent in three-way cross-hybrids in sorghum (Stephens et al., 1952). Kajjari and Chavan (1953) also reported the male sterility conditioned entirely by nuclear genes. Efforts to produce hybrids using three-way crosses were abandoned since cytoplasmic nuclear male sterility systems were soon perfected.

Cytoplasmic genetic male sterility in sorghum: Commercial production of hybrid seed in sorghum (*S. bicolor* (L.) Moench) relies on the CMS system. Male sterility is due to sterile cytoplasm and fertility is restored by a restorer gene, which is usually dominant nuclear gene either in the homozygous or heterozygous condition. The cytoplasm in the restorer line can be either sterile or fertile. In recent years, heterosis breeding has received much attention in sorghum, as with other often cross-pollinated crops, after the discovery of at least seven sterile cytoplasm types (namely, A1, A2, A3, A4, Maldandi, VZM, and G1). Using the respective maintainer sources and various restorer lines, several sorghum hybrids have been released for *rabi* season. Important among them are CSH7R, CSH8R, CSH12R, CSH13R, CSH15R, and CSH19R.

Sources of cytoplasmic genetic male sterility in sorghum: Male sterility conditioned by the interaction between nuclear genes and cytoplasmic factors was reported for the first time in sorghum by Stephens and Holland (1954) in crosses between dwarf yellow sooner milo with Texas black hull kafir. Schertz and Ritchey (1977) identified a sterile with IS 12662C cytoplasm as different from *milo*. This was released as A_2 (Schertz, 1977; Rosenow et al., 1980; Schertz et al., 1981b). IS 1112C or converted Nilwa from India is the strain that furnished the A_3 cytoplasm (Quinby, 1980). It belongs to *Durra-subglabrescens* group and *durra* race. Worstell et al. (1984) identified the female with cytoplasm from Nigerian line IS 7920C belonging to the conspicuum group and guinea race. It provided a different restoration reaction and is designated as A_4.

Additional sources of cytoplasmic sterility have been reported by Hussaini and Rao (1964) in durra (G2, VZM1, and VZM2). At the Raichur indigenous sterile line, M 31-A is said to owe its origin to induced mutation. Nagur (1971) gave *Sorghum dochna* as the source of cytoplasm in male sterile M 35-1 and *Sorghum cernum* as the source of cytoplasm of male sterile M 31-2. The cytoplasm in the male sterile of M 35-1 is A_2 and the cytoplasm in M 31-2 has not yet been identified but is probably A_2 cytoplasm (Quinby, 1982). Alternative sources reported by the Indian workers are basically from traditional local varieties. Main male sterile types included Maldandi (M 35-1A, M 31-2A), Guntur (G1A), and Vizianagaram (VZM2A) lines. In the last few decades, Indian agriculturists and plant breeders under the All India Coordinated Sorghum Improvement Project have developed several male sterile lines based on three Indian male sterile cytoplasms designated as Maldandi, Guntur, and Vizianagaram. These cytoplasms are of Indian origin (race *durra*) and have been identified separately in the regions of Maldandi, Guntur, and Vizianagaram. They have been tentatively grouped as Indian A_4 types, and the Indian A_4 group is different from the American A_4 group (Sane et al., 1996).

7.1.5 General *Rabi* Breeding Concepts

7.1.5.1 Parental Line Criteria

The requirements of seed parents are different from those of restorer parents, especially when hybrids are based on CMS systems. Seed parents have a number of particular per se requirements, besides stable and perfect male sterility, even under high temperatures typical of semiarid tropics. Stable male-sterile lines could be selected by evaluating the lines in environments where the day temperatures exceed 40°C during flowering. The other requirements of seed parents are: desired maturity and plant height, high-yielding and nontillering and good panicle exertion, seed set, and seed size and seedling vigor. Apart from these, they must have several other yield stabilizing traits such as disease, insect pest, and lodging resistance.

Restorer parents should perfectly restore the fertility in the hybrids even under low temperatures, typically experienced during postrainy seasons in India. Efficient and stable restorers could be selected by evaluating their test crosses or hybrids under conditions where night temperatures are below 13°C during the flowering phase. In order to ensure enhanced outcrossing during hybrid seed production, restorer should be taller than seed parents with profuse and prolonged pollen grain production ability. Restorer parents must also possess many of the other traits required for seed parents that are heritable in hybrids (Reddy, 1997).

7.1.5.2 Germplasm Base and Selection for Hybrid Parents

High genetic variability for the specific traits characteristic of seed parents and restorer parents are the prerequisites for selection and development of elite hybrid parents. The *caudatum* race has been extensively exploited to develop high-yielding male-sterile lines as well as restorer lines. The available heterotic seed parent, 296B, bred by the Indian National programs, though based on the *durra* race, it has several seed production problems such as sensitivity to low temperature, leading to poor seed set and chaffy ear heads, as well as susceptibility to most insect pests and diseases. While the use of the *durra* race directly introduces high sensitivity to low temperatures, the direct use of *guineas* in developing hybrid parents, produce hybrids with clasped glumes, an undesirable trait.

The *caudatum*-based restorers are exploited fully for over a decade, suggesting the need to diversify restorers to further enhance yield levels in their hybrids. An evaluation of hybrids developed by crossing five representative lines from each of the landraces (*caudatum, durra, guinea, bicolor,* and *kafir*) as pollinators to six common male sterile lines (A_1) bred at International Crops Research Institute for the Semi-Arid Tropics (ICRISAT) in 1991 postrainy and 1992 rainy seasons, showed greater contribution of *guinea* and *caudatum* restorer lines to grain yield in hybrids across the seasons, primarily due to increased grain mass. The *kafir* and guinea restorer-based hybrids showed heterosis for earliness in both seasons, while those based on the *durra* race exhibited heterosis for earliness only in postrainy season. Hence, as a short-term strategy, CMS-based seed parents and restorers need to be diversified by creating separate gene pools through crossing between *guinea*-based B-lines and *durra*-based B-lines and between *caudatum*-based R-lines and *guinea/durra*-based R-lines for various selected traits. However, banking on the limited variability by crossing among already-developed B-lines leads to a narrow genetic base of the resulting hybrids, although better parents could be quickly developed through this approach. However, for sustained and continuous improvement, fresh variability should be created by introgressing new sources of desired genes from landraces into elite agronomic background.

7.1.5.3 Relationship of Mean Performance and Combining Ability and Heterosis

An abundance of literature shows that heterosis is greater in present-day hybrids than those developed earlier. From a trial to compare the yields of old and new sorghum hybrids in Texas, Miller and Kebede (1981) found that at least 40% of the yield increase had been realized in 30 years. This is largely attributed to improvement of the parents for performance per se. Doggett (1988) claimed that about half of the yield increase could be ascribed to better parents. This is not surprising, because improved hybrid performance comes primarily from additive gene action (Kambal and Webster, 1965; Miller and Kebede, 1981). For the same reason, early-generation (in F_4 or F_5) testing for combining ability to capitalize additive gene action with a certain level of per se eliteness for yield and yield constraints together with desired grain qualities are the keys to breeding hybrid parents. Early-generation general combining ability (*gca*) tests are not intended to definitely identify the best general combiners (this cannot be done without extensive testing), but to increase the probability of retaining them for detection in later tests, as *gca* is primarily a function of additive gene action, which can be fixed through selection and inbreeding. Opinion is divided about when to begin testing for combining ability in seed parent development: before, during, or after seed parent development. Given an uncertainty of maintainer reaction in the CMS system, it is better to test for combining ability after confirming maintainer reaction of the test plants. Parental per se performance and *gca* in sorghum is strongly correlated with hybrid performance (Quinby and Karper, 1946; Murty, 1991; Murty, 1992; Bhavsar and Borikar, 2002). This is also evident from the past experience of Indian National Agricultural Research Systems (NARS) in hybrid cultivar development. For example, after the release of first four hybrids (CSH 1 to CSH 4), development of superior (in per se performance) restorer parents (CSV 4, CSV 5, and PD 3-1-11) and seed parents (2077A, 2219A, 36A, and 296A) have brought significant yield improvement in the subsequently released rainy season hybrids (CSH 5, CSH 6, and CSH 9), as well as postrainy season hybrids (CSH 7 and CSH 8R). The improvement in CSH 9 over CSH 5 or CSH 6 could be attributed mainly to the superiority of its seed parent 296A since CS 3541 is a common restorer parent (Rana et al., 1985). New inbreds tend to be more vigorous than their predecessors. In particular, they are better equipped with defensive traits, such as disease and insect resistance, and tolerance to abiotic stresses, such as heat and drought or mineral deficiencies and toxicities (Doggett, 1988). It seems likely, therefore, that increased yield in grain sorghum hybrids gradually will depend less on heterosis per se and more on complementation of defensive traits from parents that confer yield stability. It is logical that high average yield depends on the stability of performance across years and/or locations under all expected stresses, as well as the ability to maximize yield under favorable environments. Therefore, improving the parental gca and/or per se performance for yield and yield constraints such as biotic and abiotic stresses, together with desired grain qualities are the keys to breeding hybrids parents.

7.1.5.4 Parental Diversity and Heterosis

It has been established that divergent parents give rise to higher frequency of heterotic hybrids, and these heterotic hybrids generally throw a broad spectrum of variability in segregating generations in sorghum.

Shinde et al. (1983) studied heterosis in postrainy season sorghum and reported that crosses between local parents exhibited maximum heterosis (41%) compared to improved × improved varieties (13%) that might be due to greater genetic diversity present in the local strains by virtue of natural selection acting on them. Temperate × tropical crosses have played an important role in the past for the development of superior restorer parents (CSV 4, CSV 5, and PD 3-1-11) and seed parents (2077A, 2219A, 36A and 296A) of rainy season hybrids (CSH 5, CSH 6, and CSH 9), as well as postrainy season hybrids (CSH 7 and CSH 8R). Greater yield heterosis was observed in derivative × tropical (African) varietal crosses due to diversity of genes (Rana et al., 1985). Rana and Murty (1978) have also reported that an increase in number of seeds per panicle branch in short compact-headed varieties (tropical) and increase in the panicle branches in the long panicle type (temperate) by introgression of genes from African germplasm result in yield heterosis. F_1s made on *caudatum*-based seed parents with *durra*-based pollinators resulted in high heterosis under postrainy season condition.

7.1.5.5 Breeding Hybrid Parents for Yield and Defensive Traits

Simultaneous selection for resistance and grain yield in early-generation segregants starting from F_4 and converting those with maintainer reaction into male sterile lines would enable to develop male sterile lines with resistance to pests and diseases in the shortest possible period of 4–5 years (Reddy et al., 2003). Also, selecting for resistance on family basis and selecting individual single plants within the chosen resistant families based on the grain yield has been found to be most effective. Considering the independence of antibiosis (Singh and Rana, 1994) and the differences in the pattern of inheritance of resistance to flower and peduncle damage plus dead heart formation, breeding seed parents for resistance to stem borer should involve these three traits (i.e., foliar and stem damage and percentage of dead hearts). A paired plot technique with comparisons of infested and noninfested plots has been used successfully to identify genotypes with resistance to stem borer (Reddy et al., 2003). Large G × E interaction for the development of trichomes, an important component trait conferring shoot fly resistance in both rainy and postrainy seasons (Jayanthi, 1997), suggests breeding lines should be screened for shoot fly resistance in the season for which the hybrids are targeted. Since component traits conferring shoot fly resistance are each controlled by additive gene action, resistance genes need to be deployed in both hybrid parents in order to obtain resistant hybrids. For increasing the grain yield within the limits of the available water supply, the choice of female parent for hybrid production should be made for both LA and photosynthetic rate, and the selection of pollinators should be made for maximum seed number per panicle (Krieg, 1988).

7.1.6 Exploitation of Heterosis in Sorghum

The phenomenon of heterosis was observed in sorghum as early as 1927 (Conner and Karper, 1927). Commercial exploitation of heterosis has been possible in sorghum owing to (i) the availability of a stable and heritable CMS mechanism (Stephens and Holland, 1954), enabling large-scale, economic hybrid seed production and (ii) sufficiently high magnitude of heterosis across a range of production environments for economic characters justifying the replacement of currently adapted homozygous or pure-line varieties. The hybrids besides being superior for grain yield and other traits of interest are stable across environments in all the types of sorghums; grain sorghum (rainy and postrainy seasons), forage sorghum, and sweet sorghum.

7.1.6.1 Heterosis in Postrainy Season Grain Sorghum

Although several hybrids have been developed and released for postrainy season cultivation, the area covered with hybrids is almost negligible. Lack of appropriate hybrids with acceptable grain quality adapted to different agro-ecological situations of postrainy season characterized by terminal drought, low temperature, and biotic stresses like shoot fly infestation are major constraints for higher productivity. The very low rate of productivity increase in postrainy season calls for a change in production strategy, including breeding appropriate hybrids.

Landrace pollinator-based hybrid approach: The hybrids developed and released by national programs did not attract farmers, as these hybrids lacked matching grain quality and shoot fly resistance comparable to the most popular landrace variety, M 35-1. The landrace pollinator-based hybrids developed from a few female parents then available prior to the 1980s lacked marked heterosis, had threshing difficulties, and were too tall; tallness was not amenable for increasing plant density and hence this approach was not pursued further.

Shoot fly resistance in landrace pollinator-based hybrids: Landrace pollinator-based hybrids involving cytoplasmic male sterile lines and diverse restorers were assessed for their reaction to shoot fly by Jayanthi (1997). Hybrids from resistant parental lines, apart from being superior to the hybrids involving susceptible parental lines, showed uniformity in recovery following shoot fly incidence. Hybrids of resistant females and landrace restorers were superior to hybrids involving susceptible parental lines for seedling vigor in low temperatures, trichome density, uniformity in recovery after shoot fly incidence, acceptable plant height, and bold grain, which are important for postrainy season sorghum. This has clearly shown the requirement of shoot fly resistance in both the parents or at least in seed parents in order to realize higher frequency of shoot fly resistant hybrids in postrainy season.

Temperature sensitivity in landrace pollinator-based hybrids: Seed-setting ability (under bagging) in hybrids at low temperatures is critical to the postrainy season hybrids and requires greater attention to ascertain the differences among the landraces for their ability to restore fertility in hybrids, especially in low temperatures, as normally observed in postrainy season. A highly significant positive correlation ($r = 0.34^{**}$, $P = 0.01$) was observed between average minimum temperature starting from 5 days before to 5 days after 50% flowering and percentage seed set under bagging in hybrids, indicating the strong influence of low temperature on suppressing the self seed set. The seed set on unbagged panicles in hybrids and bagged heads in B- (except *durra*-based) and R-lines involved in the hybrids was normal. The reduced seed set in hybrids under bagging may be attributed to lack of sufficient viable pollen.

Fertility restoration in landraces: Individual plant progenies of M 35-1 that restored A_1, also restored A_2 CMS lines. However, the frequency of restoration and restoration ability of progenies of M 35-1 were dependent on the race from which the CMS lines were derived. Most of the landraces, including M 35-1, showed segregation for fertility restoration/sterility maintenance ability, indicating the need to select for restoration ability within the landraces. This also explains the partial restoration observed when bulk pollen of M 35-1 was used by many studies. Both A_1 and A_2 CMS lines could be used for producing hybrids for postrainy season.

When individual plant progenies of M 35-1 were test-crossed onto A_1 CMS lines (ICSAs 73, 83, 101, and 102), derived either from a *durra* race (M 35-1) directly or from a population with *durra* germplasm introgressed, and A_2 CMS lines (MR 750A_2, and MR 840A_2), derived from *zera-zera* germplasm, the results showed the following:

- There was segregation for fertility restoration/sterility maintenance in M 35-1.
- The frequency of restoration was less on A_1 than on A_2.
- The progenies that restored on A_1, also restored on A_2.
- The progenies showing partial restoration on A_1, fully restored A_2 CMS lines.

An intensive study of fertility restoration (seed setting percentage under bagging) in 18 selected postrainy season landraces was carried out at ICRISAT by evaluating individual plant test crosses made with each of the selected landraces onto *zera zera* based on A_1 and A_2 CMS lines. Unlike the previous study, restoration frequency was higher on the A_1 (65%) than on the A_2 (56%) CMS lines. Most of the landraces, including M 35-1, showed segregation for fertility restoration/sterility maintenance ability, indicating the need to select for restoration ability within the landraces. This also explains the partial restoration observed when bulk pollen of M 35-1 was used by many studies. However, hybrids of landraces IS 23496, Swathi, and IS 18361 showed almost full restoration on A_1 CMS lines, as did hybrids of Swathi, IS 10955, IS 33844, IS 5476, IS 18361, and IS 18372 on A_2 CMS lines. The mean percentage of seed set in hybrids obtained by crossing 10 common restorers onto four isonuclear A_1 and A_2 male sterile lines showed that the average restoration was higher than 80% for all R-lines (confirming that they are restorers of both A_1 and A_2 CMS systems), although there were significant differences among them.

A_1 supported significantly more restoration than A_2 in this study with isonuclear male sterile lines, unlike the previous study with M 35-1 progenies where CMS lines were not isonuclear. However, the fertility restoration level with A_2 male sterile lines was sufficiently high (>85%) to ensure a full seed set. These studies clearly demonstrated that both A_1 and A_2 CMS lines could be used for producing hybrids for postrainy season.

Role of nuclear genome (*durra* and *caudatum* types) on fertility restoration: Fertility restoration levels were lower on *durra*-derived A_1 CMS lines than *caudatum*-derived A_1 CMS lines. This finding has a bearing on developing CMS lines involving *caudatum*-based germplasm lines adapted to postrainy season and testing for fertility restoration in hybrids.

Performance of selected landrace pollinator-based hybrids: The selected landrace pollinator-based hybrids were significantly superior to M 35-1, CSH 12R, or both, and were on par with CSH 13R for grain and fodder yields. They were significantly earlier in maturity and had a larger grain than CSH 13R. The hybrids involving *caudatum*-based female parents and *durra*-based landraces showed high heterosis for grain yield. The *caudatum*-derived female lines were adapted to rainy season. Rao et al. (1986) argued that by eliminating temperature sensitivity (in relation to development) in both male and female parents, greater success could be achieved in breeding hybrids for postrainy season. However, male sterile lines bred from landraces adapted to postrainy season, as indicated earlier, showed temperature-induced restorer inefficiency. Also, the hybrids developed from landraces crossed to female lines derived from M 35-1 did not show high heterosis.

Heterosis in landrace pollinator-based hybrids: Landrace-based hybrids were evaluated in an advanced hybrid trial (AHT) during the 1993 postrainy season at ICRISAT, Patancheru, and in a landrace advanced hybrids and parents trial (LRAHPT) during the 1995 postrainy season at Nandyal (Andhra Pradesh, India) and ICRISAT, Patancheru. Superiority over checking for grain yield in AHT was recorded in 79% of the hybrids (range 3.8–49.2%) and in all the hybrids in LRAHPT (range 23.2–81.3%). For fodder yield, 45% of the hybrids (range 0.4–40%) in AHT and 97% of the hybrids (range 0.3–118%) in LRAHPT were heterotic.

7.1.6.2 Extent of Heterosis for Agronomic, Physiological, and Biochemical Traits

Heterosis for yield and yield components: The range of heterosis and heterobeltiosis for various agronomic traits are given in Table 7.1. Values of heterobeltiosis and heterosis for a trait does not correspond to the same study. Hence, some heterobeltiosis values are higher than the heterosis values.

Heterosis for days to 50% flowering: Hybrids that flowered earlier than their parents were recorded (Atkins, 1979; Kanaka, 1979; Giriraj and Goud, 1981; Patil et al., 1983; Palanisamy et al., 1983; Gururaj Rao et al., 1993; Biradar, 1995; Ganesh et al., 1997; Iyanar et al., 2001) in several studies. Heterosis over the mid-parent was observed by Indi and Goud (1981), Desai et al. (1985), Kide et al. (1985), Shivanna and Patil (1988) and Belawatagi (1997). Positive heterosis over the better parent was documented by Subba Rao et al. (1976a), Pandit (1989), Senthil and Palanasamy (1993), Badhe and Patil (1997), and Tiwari et al. (2003), while negative heterosis for the mid-parent for the trait was reported by Franca et al. (1986), Naik et al. (1994), Rodriguez et al. (1994), Madhusudhan and Patil (1996c), Tourchi and Rezai (1996), Lokapur (1997), Nguyen et al. (1997), Navabpour and Rezaie (1998), Salunke and Deore (1998), Pattanashetti (2000), Pawar (2000), Ravindrababu and Pathak (2001), Kulkarni (2002), Karthik (2004), and Patil (2004). The useful negative heterobeltiosis in the desired direction was observed by Chaudhary et al. (2003).

Heterosis for plant height: A very high degree of heterosis expression for the character was documented by several authors (Vasudev Rao, 1973; Indi, 1978; Lodhi et al., 1978; Kanaka, 1982; Bhagmal and Mishra, 1985; Sahib et al., 1986; Berenji, 1988; Chinna and Phul, 1988; Shinde and Borikar, 1991; Pedrosoperez, 1994; Senthil and Palanasamy, 1994; Biradar, 1995; Biradar et al., 1996; Ganesh et al., 1996; Tourchi and Rezai, 1996; Badhe and Patil, 1997; Navabpour and Rezaie, 1998; Salunke and Deore, 1998; Ravindrababu et al., 2001; Kulkarni, 2002; Patil, 2004).

The maximum extent of relative heterosis was reported by Franca et al. (1986), Jebaraj et al. (1988), and Chaudhary (1992). Both profitable heterosis and heterobeltiosis for the trait

TABLE 7.1 Heterosis and Heterobeltiosis for Yield and Yield Components in *Rabi* Sorghum

Trait	Heterosis	Reference	Hetero-beltiosis	Reference
Days to 50% flowering	10.5–11	Deshpande (2005), Krishna Murthy et al. (2010)	4.4–15.3	Deshpande (2005), Umakanth et al. (2006), Krishna Murthy et al. (2010)
Days to maturity	10.1	Deshpande (2005)	10.9	Deshpande (2005)
Plant height	36.7–43.4	Deshpande (2005), Krishna Murthy et al. (2010), Prabhakar and Raut (2010)	2.4–47.2	Deshpande (2005), Umakanth et al. (2006), Krishna Murthy et al. (2010), Prabhakar and Raut (2010)
Number of leaves	9.6–10.6	Deshpande (2005), Krishna Murthy et al. (2010)	7.7–8.2	Deshpande (2005), Krishna Murthy et al. (2010)
Leaf length	18.3	Deshpande (2005)	11.3	Deshpande (2005)
Leaf breadth	22.5	Deshpande (2005)	33.4	Deshpande (2005)
Root length	30.1	Deshpande (2005)	17.3	Deshpande (2005)
Peduncle length	20.4	Deshpande (2005)	10.7	Deshpande (2005)
Panicle length	12.3–40.6	Deshpande (2005), Krishna Murthy et al. (2010), Prabhakar and Raut (2010)	17.4–31.5	Deshpande (2005), Krishna Murthy et al. (2010), Prabhakar and Raut (2010)
Panicle breadth	22.1–46.1	Deshpande (2005), Krishna Murthy et al. (2010)	10.1–33	Deshpande (2005), Krishna Murthy et al. (2010)
Number of primaries	32.5	Deshpande (2005)	28.2	Deshpande (2005)
Panicle weight	64.7–163.4	Deshpande (2005), Krishna Murthy et al. (2010), Prabhakar and Raut (2010)	60.4–150.5	Deshpande (2005), Krishna Murthy et al. (2010), Prabhakar and Raut (2010)
1000-Grain weight	17.6–48.5	Deshpande (2005), Krishna Murthy et al. (2010), Prabhakar and Raut (2010)	4.9–44.6	Deshpande (2005), Umakanth et al. (2006), Krishna Murthy et al. (2010), Prabhakar and Raut (2010)
Number of grains per panicle	68.3	Deshpande, 2005	62.5	Deshpande (2005)
Grain yield	55.2–85.7	Deshpande (2005), Krishna Murthy et al. (2010), Prabhakar and Raut (2010)	6.6–76.5	Deshpande (2005), Umakanth et al. (2006), Krishna Murthy et al. (2010), Prabhakar and Raut (2010)
Fodder yield	71.8–103.4	Deshpande (2005), Krishna Murthy et al. (2010)	45.4–73.6	Deshpande (2005), Krishna Murthy et al. (2010)
Total chlorophyll content	23.6	Krishna Murthy et al. (2010)	18.7	Krishna Murthy et al. (2010)
RWC (%)	9.3	Krishna Murthy et al. (2010)	7.9	Krishna Murthy et al. (2010)
Free soluble sugar content	45.8	Krishna Murthy et al. (2010)	18.8	Krishna Murthy et al. (2010)
Protein content of seed	7.7	Krishna Murthy et al. (2010)	7.2	Krishna Murthy et al. (2010)

was proposed by Ghorade et al. (1997). Pronounced hybrid vigor with appreciable heterobeltiosis was noticed by Franca et al. (1986), Senthil and Palanasamy (1993), Ganesh et al. (1996), Lokapur (1997), Salunke and Deore (1998), Pawar (2000), and Chaudhary et al. (2003). Subba Rao et al. (1976a), Patel et al. (1987), Desai (1991), Madhusudhan and Patil (1996c), Belawatagi (1997), and Chaudhary et al. (2003) recorded significant economic heterosis over the commercial check. Patel et al. (1990) observed only negative heterosis for the trait, while Indi and Goud (1981), Shivanna (1989), Naik et al. (1994), and Rodriguez et al. (1994) obtained both positive and negative heterosis over standard checks. Reports of heterosis to a limited extent for plant height were also on record (Bhale and Borikar, 1982; Lazanyi and Bajai, 1986). Similarly, Vasudev Rao and Goud (1977) and Giriraj and Goud (1981) observed nonsignificant heterosis in the genetic material evaluated.

Heterosis for number of leaves per plant: Positive heterosis for the trait was reported by Chavda and Drolsom (1970), Mani (1981), Kanaka (1982), Giriraj and Goud (1983), Nayeem and Bapat (1984), Bhagmal and Mishra (1985), Pandit (1989), Desai (1991), and Pattanashetti (2000). The studies of Franca et al. (1986), Shivanna (1989), and Gururaj Rao et al. (1993) indicated the expression of heterobeltiosis for the trait. Patil and Bapat (1991) recorded heterobeltiosis in a medium range of −11.50 to 14.89%, for leaf number per plant, while Kulkarni (2002) reported a wide range of average heterosis (−14.87 to 28.57%) and heterobeltiosis (−22.83 to 23.28%) for the trait.

Heterosis for leaf length: Limited heterosis for the trait was observed by Kanaka (1982). Lazanyi and Bajai (1986) presented information on combining ability for forage components in hybrids derived from maintainer lines of sorghum. Significant heterosis was observed for leaf length in the hybrids. Berenji (1988) reported larger leaves in the hybrid that gave higher yield levels. Mallick and Gupta (1989) derived information on combining ability and identified the cross MAUT1 × IS165 with high heterosis for leaf length. Ravindrababu et al. (2002) found the presence of a low level of heterosis with a range of − 9.34 to 14.93% for average heterosis and −15.64 to 10.02% for heterobeltiosis in a study comprised of 45 hybrids derived from 10 homozygous sorghum lines crosses in a half-diallel fashion. Contrarily, Tiwari et al. (2003) obtained the highest magnitude of heterosis for leaf length in the cross 880 × PI81041 in a diallel cross involving 10 diverse genotypes.

Heterosis for leaf breadth: Based on a 7 × 7 diallel analysis, Kanaka (1982) reported that heterosis exists for leaf breadth. Lazanyi and Bajai (1986) also reported the presence of limited heterosis for leaf width based on a diallel study in five maintainer and five male sterile lines. Similarly, Berenji (1988) observed large leaves in the hybrids giving higher yields. A low range of relative heterosis (−16.98 to 27.57%) and heterobeltiosis (−24.22 to 24.09%) for breadth of leaf was documented by Ravindrababu et al. (2002) based on evaluation of 45 hybrids. The crosses GSSV 148 × SR 670 and ICSV 705 × GSSV 148 recorded maximum heterosis for the trait among the hybrids. In another similar study, Tiwari et al. (2003) identified the cross SPV 383 × CSV 189 possessing the highest magnitude of heterosis for width of leaf based on a combining ability study involving 10 diverse genotypes of sorghum.

Heterosis for root length: Among the various traits, root length is the most important trait for evaluating the genotypes for drought tolerance. Deep-rooted plants yield more under moisture stress conditions (Mambani and Lal, 1983). Comparatively, very few studies have reported significant heterosis for root length (Ashby, 1937; Robbins, 1941; Paddick, 1944; Whaley, 1952; Sarkissan and Srivastawa, 1967; Damodar et al., 1978). Blum et al. (1977) evaluated three hybrids and their parents in hydroponic vessels and emphasized the presence of significant levels of heterosis for seminal root length in all the hybrids. Omra and Hussein (1987) studied 27 hybrids and their parents in barley and observed significant heterosis for root length. In sorghum, Maiti (1996) recorded the highest heterosis for root length in CSH-1 and 22E. Similarly, Azhar et al. (1998) obtained significant mid-parent heterosis for the trait based on evaluation of four hybrids of sorghum. Of the different hybrids studied by Patil (2004) in a 4 × 4 diallel study in *rabi* sorghum, the highest magnitudes of average heterosis and heterobeltiosis were observed in the cross CSV 8R × DSV4. The heterosis over the mid-parent and better parent ranged from −13.77 to 34.18% and −26.19 to 19.26%, respectively.

Heterosis for peduncle length: Vasudev Rao and Goud (1977) noticed partial dominance for the trait with significant standard heterosis in the hybrids. Further, Giriraj and Goud (1983) noticed a wide range of heterosis over the mid-parent with values ranging from 2.10% to 87.64%. They reported maximum heterobeltiosis up to 54.39%. Sahib et al. (1986) and Sridhar (1991) also reported significant average heterosis for peduncle length. Pandit (1989) and Desai (1991), in studies related to the evaluation of derived hybrids for recombinational variability, noticed significant heterosis over the better parent and commercial checks. In another similar study related to heterosis in derived F_1s, Madhusudhan (1993) observed standard heterosis of up to 35.61% for peduncle length.

Heterosis for panicle length: Appreciable positive heterosis for head length was evidenced by several studies (Desai et al., 1980; Indi and Goud, 1981; Mani, 1981; Kanaka, 1982; Giriraj and Goud, 1983; Nagabasaiah, 1985; Prezcabrera and Miller, 1985; Sahib et al., 1986; Jebaraj et al., 1988; Shivanna, 1989; Senthil and Palanasamy, 1994; Ganesh et al., 1997; Nguyen et al., 1997; Navabpour and Rezaie, 1998; Pattanashetti, 2000). On the other hand, negative heterosis for length of panicle was reported by Shinde et al. (1983) and Dinakar (1985). Perezcabrero (1986) concluded that heterosis for panicle length contributed significantly to the expression of heterosis for grain yield. Pronounced hybrid vigor with significant mid-parent heterosis for panicle length was reported by Franca et al. (1986), Nimbalkar et al. (1988), and Biradar et al. (1996). Tiwari et al. (2003) documented the highest magnitude of heterosis for panicle length in the cross 880 × FTB24.

Heterosis for panicle breadth: Only limited literature exists with respect to heterosis for this trait. Nagabasaiah (1985) made genetic analysis of 10 quantitative traits in F2 generation of a seven-parent diallel set in sorghum. He reported significant heterosis for panicle breadth. In a study of heterosis in 60 hybrids of *rabi* sorghum, Salunke and Deore (1998) found pronounced hybrid vigor with high heterobeltiosis for panicle breadth. Murumkar (2002) performed a combining ability analysis of newly established male sterile and restorer lines in *rabi* sorghum. Significant positive heterosis was reported over the mid-parent (33.18%), better parent (32.89%) and standard check (24.62%) in the best cross 53A × Rb32. The range of heterosis for the trait was −13.75 – 32.89%. Similarly, Tiwari et al. (2003) also studied the heterotic response for yield and component traits in grain sorghum in a diallel cross involving 10 diverse genotypes. A significant and highest magnitude of heterosis for width of panicle was noticed in the cross KIJ77 × PI81041.

Heterosis for number of primaries per panicle: Favorable heterotic response in the desired direction was reported for the trait (Govil and Murty, 1979; Indi and Goud, 1981; Giriraj and Goud, 1985; Nandawankar, 1990; Biradar et al., 1996). Nayeem and Bapat

(1984) emphasized that the increased yield levels in CSH5 was due to the significant heterosis for the number of primary branches in the hybrid. Nimbalkar et al. (1988) and Shinde and Borikar (1991) also observed highly significant positive heterosis over the mid-parent for the trait. The maximum positive heterosis with a magnitude of 53.46% over the mid-parent was noted by Shivanna (1989). On the contrary, Dinakar (1985) reported both positive and negative heterosis for the trait. Heterobeltiosis in the positive direction was evidenced by Sahib et al. (1986), Patel et al. (1987), Sankarapandian et al. (1994), Biradar et al. (1996), Badhe and Patil (1997), and Lokapur (1997), while Pandit (1989), Biradar (1995), and Esha (2001) recognized significant average heterosis, heterobeltiosis, and economic heterosis for the trait. Based on an evaluation of 32 F1 hybrids in postrainy season, Umakanth et al. (2003) identified the cross IS23399 × NR11349 as the best, with greatest heterosis over the mid-parent (43.51%) and superior parent (33.30%) for number of primaries per panicle. Similarly, Patil (2004) interpreted that DSV5 × M35-1 was the best cross, sharing the highest relative heterosis (27.10%) for the trait.

Heterosis for panicle weight: Subba Rao et al. (1975) obtained a higher magnitude of useful heterosis and heterobeltiosis for head weight in American × African hybrids (namely, IS511 × IS9183, IS511 × IS9294 and IS3151 × IS9183). Heterosis to the extent of 37% for the trait was observed by Berenji (1988). Highly significant and positive heterosis in desirable directions over the mid-parent for ear weight was reported by Nagabasaiah (1985), Chinna and Paul (1988), Nimbalkar et al. (1988), Shivanna (1989), Nandawankar (1990), Sheriff and Prasad (1990), Shinde and Borikar (1991), Veerabadhiran et al. (1994), Biradar (1995), Pillai et al. (1995), Biradar et al. (1996), Madhusudhan and Patil (1996c), Tourchi and Rezai (1996), Navabpour and Rezaie (1998), and Kulkarni (2002). The highest range of heterosis over the mid-parent (96.34%) and heterobeltiosis (65.90%) was recorded for panicle weight by Gururaj Rao et al. (1993) and Ganesh et al. (1996). Gite et al. (1997) identified two hybrids (namely, MS101A × GMPR4 and 53A × GMPR4) with the highest degree of useful heterosis over commercial checks for panicle weight.

Heterosis for 1000-grain weight (grain size): Many earlier studies reported significant heterosis for test weight in grain sorghum (Vasudev Rao and Goud, 1977; Rana and Murty, 1978; Rao et al., 1978; Govil et al., 1979; Rana and Ahluwalia, 1979; Dabholkar and Baghel, 1980; Singhania, 1980; Srihari and Nagur, 1980; Kanaka, 1982; Giriraj, 1983; Khidse et al., 1983; Dabholkar et al., 1984; Patil and Thombre, 1984; Shinde and Kulkarni, 1984; Desai et al., 1985; Giriraj and Goud, 1985; Patil et al., 1985; Sahib et al., 1986; Kulkarni and Shinde, 1987; Berenji, 1988; Dinakar et al., 1988; Shivanna and Patil, 1988; Cheng et al., 1989; Reddy and Joshi, 1990). Subba Rao et al. (1975) reported positive heterosis for 1000-grain weight in the combining ability studies in sorghum among American × African crosses. The highest heterosis in terms of increase over superior parent was observed in the cross IS1055 (India) × IS 454 (United States) for test weight (Subba Rao et al., 1976a). Perezcabrero (1986) recorded negative values for 300-grain weight in the 48 F_1 hybrids generated by using 16 seed parent lines and 3 inbred pollen parents. Similarly, in another study of 99 F_1 hybrids derived from 11 male sterile lines and 9 restorers, an average heterosis for 1000-grain weight was recorded in the negative direction (Franca et al., 1986). Chinna and Paul (1988) reported positive and significant heterosis for grain weight after evaluating 40 hybrids. Nimbalkar et al. (1988) noticed significant heterosis for grain weight. Shinde and Borikar (1991) studied heterosis involving maldandi cytoplasm in sorghum and found increased heterosis for test weight. Jagadeshwar and Shinde (1992) made combining ability analysis of F_1 hybrids of eight sorghum varieties and reported significant heterosis for the trait. Nayakar et al. (1994) tabulated data for 10 crosses made between two male and five female lines. Studies indicated higher heterosis for grain weight. Pedrosoperez et al. (1994) generated information on heterosis for a test weight. In the hybrid 1831 × UDG110, significant positive heterosis was recorded. Positive heterosis values for the trait was also observed by Rodriguez et al. (1994). Senthil and Palanasamy (1994) studied diverse cytosteriles of sorghum and revealed positive heterosis for grain weight. Sheriff and Prasad (1994) sought significant results for heterosis for a test weight based on combining ability studies in six genotypes and their F1 hybrids. Similarly, Veerabadhiran et al. (1994) noticed average heterosis for a test weight in the cross 2077A × CO26. Dabholkar et al. (1995) crossed seven genotypes of sorghum in diallel fashion and obtained positive heterosis values for the trait. Biradar et al. (1996) found heterosis with respect to seed weight involving popular *rabi* cultivar M35-1 in two crosses. Madhusudan and Patil (1996) obtained transgressive segregants in F_2 and noticed heterosis for seed weight in the 21 derived F_1 hybrids. Rao and Patil (1996) observed positive heterosis levels for a test weight based on a study of inbreeding and back cross-generations. Badhe and Patil (1997) noticed heterosis in the positive direction over the mid-parent and better parent for test weights. Ganesh et al. (1997) studied the per se performance of 42 *S. bicolor* hybrids and reported significant heterosis for 100-grain weight. Increased heterosis levels in 100-seed weight was also reported by Ghorade et al. (1997) based on 11 × 11 diallel set analysis. In a combining ability study of newly developed male sterile and restorer lines of sorghum, high combiner crosses showed significant heterosis for the trait (Gite et al., 1997). Mangush and Andryushchenko (1998) studied heterosis for yield-related traits in grain sorghum and an appreciable amount of heterosis was observed for test weights. Navabpour and Rezaie (1998) conducted an experiment on heterotic pattern evaluation in sorghum. A higher value for average heterosis was observed for the trait. Salunke and Deore (1998) obtained pronounced hybrid vigor for 1000-grain weight in 60 hybrids of *rabi* sorghum. Scapim et al. (1998) observed positive heterosis for test weights in two crosses and advanced inbreeding generations. Tourchi and Rezai (1996) also reported that heterosis led to the improvement in test weight by evaluation of 19 top crosses (19 male sterile lines × mixture of 120 restorer lines). Hovny et al. (2001) made 30-grain sorghum crosses and reported that test weights showed a high amount of heterosis over the better parent. Iyanar et al. (2001) opined that out of the 40 hybrid combinations evaluated, the crosses 2077A × SPV1192, 296A × TNS30, and 2077A × CO26 displayed significant heterosis for test weights. Ravindrababu and Pathak (2001) evaluated a diallel set comprising 10 genetically diverse parents and observed significant heterosis for test weights. Umakanth et al. (2002) obtained an appreciable amount of heterosis for test weights in 32 F_2 hybrids created by crossing eight lines and four testers. Similarly, Chaudhary et al. (2003) conducted a trial of 28 sorghum hybrids along with their parents and identified the best desirable combinations with a high degree of heterotic effect. Most of the hybrids exhibited positive heterobeltiosis for 1000-grain weight. Tiwari et al. (2003) obtained noticeable heterosis for test weights in a 10 × 10 diallel study. The highest magnitude of heterosis was observed for the trait in the cross FTB24 × CSV189.

Heterosis for number of grains per panicle: A wide range of heterosis for the trait was reported by earlier researchers (Srihari and Nagur, 1980; Singh, 1982; Giriraj, 1983; Desai et al., 1985; Perezcabrera and Miller, 1985; Kasenko, 1986; Pandit, 1989; Sheriff and Prasad, 1990; Shinde and Borikar, 1991; Gururaj Rao et al., 1993; Biradar et al., 1996; Swarnalata et al., 1996; Badhe and Patil, 1997; Mangush and Andryushchenko, 1998; Navabpour and Rezaie, 1998; Salunke and Deore, 1998). Significant heterosis over the superior parent was observed by Vasudev Rao (1973), Desai et al. (1980), Singhania (1980), Dinakar (1985), Geeta and Rana (1988), Blum et al. (1990), Patel et al. (1990), Desai (1991), and Chaudhary et al. (2003). The highest average or relative heterosis for the trait was shown by Franca et al. (1986), Perezcabrero (1986), Berenji (1988), Hewenan (2001), and Patil (2004). Heterobeltiosis up to 128% was recorded by Lokapur (1997) in the cross SB101A × DKR9502.

Heterosis for grain yield per plant: A wide range of heterosis in hybrids for grain yield per plant (−22.67 to 147 parents) over the parents has been evidenced by several studies (Indi and Goud, 1981; Nayeem and Bapat, 1984; Dinakar, 1985; Nimbalkar et al., 1988; Pandit, 1989; Nandawankar, 1990; Chaudhary, 1992; Gururaj Rao et al., 1993; Chen, 1994; Biradar, 1995; Ganesh et al., 1996; Kulkarni, 2002; Karthik, 2004). Subba Rao et al. (1976b), in a 9 × 9 diallel cross involving some Indian and exotic elite lines, reported the highest heterosis in terms of increase over the superior parent for grain yield in the cross IS 401 (USA) × IS3572 (Sudan). In another study of 5 × 5 complete diallel analysis, Vasudev Rao and Goud (1977) obtained the maximum heterosis for grain yield in IS 2226 (dwarf introduction) × Karad local (tall). Rana and Murty (1978) attempted a partial diallel cross among seven long panicle (LP) parents and three compact panicle (CP) parents. They recorded yield heterosis in LP × LP crosses. Rao et al. (1978) made a 14 × 14 diallel without reciprocals involving 11 early, dwarf American lines and 3 late, tall African lines. Results revealed the highest heterosis for grain yield in American × African cross IS2797 × I38622 (68.6%), followed by American × American cross IS84 × IS511 (47.7%). Grain yield was one of the most heterotic traits in exotic × Indian crosses, as evidenced by Singhania (1980). Srihari and Nagur (1980) evaluated a total of 21 diallel crosses from seven parents. The hybrid combination, Swarna × CS3541, showed the highest heterobeltiosis value. Kanaka and Goud (1982) performed a 7 × 7 diallel analysis and reported that heterosis for grain yield ranged from −11.36% (LSR1 × FR169) to 74.72% (IS84 × RCR408). Singh (1982) conducted a diallel experiment among 12 entries and reported significant heterosis for the yield parameter. Desai et al. (1983) observed heterobeltiosis for grain yield per plant with a range of −5.59 to 99.98% in the 80 F1s tested. Giriraj (1983) presented information from an analysis of crosses involving four tall, late and four dwarf, early lines. Heterosis for grain yield was exhibited mainly due to the heterotic effect expressed for 100-grain weight. The work of Khotyleva and Neshina (1983) also revealed that the highest heterosis for yield was obtained by crossing the lines L803, L922, and L904. In the crosses of temperate × tropical types, heterosis was reported for the yield character owing to the diversity of genes (Rana et al., 1983). Guzhov and Malyuzhenets (1984) evaluated crosses between 36 varieties and identified eight best hybrids characterised by high heterosis for grain yield. The F1 hybrids from a six-parent diallel, without reciprocals, showed heterosis for grain yield (44% above the midparental value and 21.9% above the value for the best parent) (Desai et al., 1985). High positive heterosis values were recorded for yield by an analysis by Perezcabrera and Miller (1985) of data in 16 seed parent lines, 3 inbred pollen parents, and 48 F1 hybrid combinations. Franca et al. (1986) made the genetic analysis of some agronomic traits in grain sorghum and noticed high positive heterosis for yield per panicle in postrainy season, indicating the adoption of the parents to the particular season. Patil and Thombre (1986) crossed *rabi* variety SPV151 with three varieties grown in *rabi* season (*rabi* × *rabi* crosses) and concluded that the SPV151 × R16 combination produced the highest heterosis for grain yield. Perezcabrero (1986) crossed tropically adopted and temperately adopted elite inbred lines and observed a heterosis for grain yield that depended on heterosis for grains per panicle and panicle length. Heterosis for the trait averaged to 29% was reported by Sahib et al. (1986) based on an analysis of data from a 10 × 10 diallel cross. Mahdy et al. (1987) realized an increase in grain yield with significant heterosis by crossing among the superior families. Berenji (1988) recorded the greatest magnitude of heterosis over mid-parental value in a cross between four CMS lines and eight restorer lines. Chen (1988) produced five homo-cytoplasmic and five hetero-cytoplasmic lines using 6-grain sorghum varieties with four types of cytoplasm. About 50% of these lines showed heterosis for grain weight and grain yield. Geeta and Rana (1988) found that among the characteristics, heterosis over the better parent was the highest for grain yield (27.3%). Jebaraj et al. (1988) obtained 30 hybrid combinations involving six male sterile lines and five indigenous restorer lines of *S. bicolor* and recorded the highest relative heterosis for a single-plant yield in the crosses 2758A × K6 and CK60A × K6. Kishan and Borikar (1988) compared A_2 sorghum cytoplasm that was derived from Ethiopian source IS126626, with *milo* cytoplasm (A_1) and two other isosteriles, by crossing individually with 11 male parents (five exotic and six Indian lines). The grain yield per plant and heterosis for this trait were higher in A_2 hybrids (28.9 g and 39.9%, respectively). Shivanna and Patil (1988) recognized a high-yielding heterotic combination (SPV86 × Afzalpur local with 89.2 grains per plant). Kishan and Borikar (1989) evaluated isosterile (A_1, A_2, and A_4) females and a set of five fertility restorers. They reported the highest heterosis for grain yield and recommended use of A_2 cytoplasm in heterosis breeding. Mehetre and Borikar (1992) produced 15 hybrids from crosses between 5 cytoplasmic male sterile lines (Maldandi cytoplasm) and 3 fertility restorer lines. The cross IS84A × SPV503 exhibited significant positive heterosis. Reddy and Joshi (1993) recorded the highest heterosis (87.5% over better-parented value in F_1) for grain yield per plant in crosses involving six varieties (five Indian and one exotic).

Chen (1994) studied heterosis for yields in 26 hybrids with nuclei and cytoplasm from different parents. The maximum heterosis and heterobeltiosis values recorded for yield were 59.88% and 26.74%, respectively. Similarly, in a line × tester analysis, Manickam and Das (1994) noticed the highest heterosis for seed yield (150%) in a cross TNAUMSA1 × Co21. Ombhako and Miller (1994) carried out trials in single crosses and three-way crosses. Heterobeltiosis ranging from −45% to 210% was reported. Rodriguez et al. (1994) sought information in heterosis in 12 F_1 hybrids. Positive heterosis values were obtained for the grain yield. A high magnitude of heterosis in grain yield was realized in grain yield in crosses between high × low combiners (2217A × K3) (Sankarapandian et al., 1994). Similarly, Veerabadhiran et al. (1994) identified the crosses 2077A × CO26 and 296A × IL101 with significant heterosis for the yield component. Salunke and Pawar (1996) observed an appreciable amount of heterosis for grain yield

(73.2%) by a line × tester analysis. Swarnalata et al. (1996), in order to improve the maintainer lines for postrainy season, obtained heterosis in the range of 1.5%–65.6% for grain yield. Ghorade et al. (1997) exhibited profitable heterosis and heterobeltiosis for the grain yield by evaluating 32 hybrids. Lokapur (1997) reported very high heterosis in MS2986A × DKS9502 cross. Nguyen et al. (1997) reported a high positive heterosis in grain yield for more than half of the hybrids studied in a diallel mating system with six early maturing sorghum parents. Regarding the inheritance studies of seed size and grain yield in sorghum, Mahajan et al. (1998) observed significant heterosis for grain yield in the cross 296A × CS3541. Mangush and Andryush Chenko (1998) determined the highest heterosis for grain yield in grain sorghum hybrids. Salunke and Deore (1998) produced hybrids with high heterotic value for grain yield. The combination 116A × RS67 exhibited higher heterosis out of a total 60 hybrids evaluated. Tourchi and Rezai (1996) crossed 19 randomly selected male-sterile lines to broad genetic base tester (a mixture of 120 restorer lines). They concluded that heterosis led to improvement in grain yield. Axtell et al. (1999) described hybrid development in sorghum in Niger, and they concluded that exploitation of heterosis led to sustained improvement in grain yields. Hovny et al. (2000) described high (57.94%) heterosis over the better parent in the cross JCSA88005 × MR812. Thawari et al. (2000) suggested an exploitation of heterosis for yield in the crosses AKMS3513 × AKMS3113. The hybrids (namely, 104A × BRJ198 and 104A × RC47) showed average heterosis and heterobeltiosis and standard heterosis (Esha, 2001). Hovny et al. (2001) developed 30-grain sorghum crosses and reported high estimate of heterosis over the better parent (71.28%). Iyanar et al. (2001) estimated the highest heterosis value for grain yield per plant in the resultant 40 hybrid combinations between 4 lines and 10 testers, while the highest heterosis over the better parent for the trait to the extent of 102.90% was recorded by Prabhakar (2001a,b). Madhusudan (2002a) observed that 20% of total crosses exhibited significant heterosis over the mid-parent. Ravindrababu et al. (2002) crossed 10 homozygous sorghum lines in half diallel (excluding reciprocals) fashion and observed exploitable heterosis for grain yield in the cross IS18551 × IS2312. Smilovenko (2002) studied the inheritance of traits in sorghum hybrids. Medium early ripening lines A105, A176 and A278 were used as the mother form and medium ripening lines L19, L28 and L52 were used as the father form. The highest increase of heterosis was observed in grain yield per plant in the crosses A105 × L52, A278 × L19 (178.5–248.6%). Chaudhary et al. (2003) conducted a trial for examining hybrid vigor involving diverse cytosteriles and sorghum. Out of 28 sorghum hybrids evaluated 10 crosses showed positive heterobeltiosis for the grain yield per plant, while the highest magnitude of heterosis for the trait was observed in the cross SPV383 × 6243 in a diallel cross involving 10 diverse genotypes of sorghum (Tiwari et al., 2003). Umakanth et al. (2003) found significant heterosis over the mid-parent and better parent in 24 and 14 hybrids, respectively. Patil (2004) conducted 4 × 4 complete diallel analysis (with reciprocals). The highest manifestation of heterosis for grain yield was noticed in the crosses CSV8R × DSV4 and M35-1 × DSV4.

Heterosis for fodder yield: Significant heterosis for fodder yield per plant in sorghum was reported by many studies (Arkin and Monk, 1979; Sharma, 1980; Rathore and Singhania, 1987; Patil and Bapat, 1991; Naik et al., 1994; Raut et al., 1994; Scapim et al., 1998). Shouny et al. (1990) obtained stable positive heterobeltiosis for forage yield. Similarly, Reddy

(1993) reported a consistent heterosis for forage yield in F_1–F_4 generations resulting from a half-diallel cross. Pathak and Sanghi (1992) observed highly significant but varying degrees of heterosis for fodder yield. Jayaprakash and Das (1994) recorded significant positive values for relative heterosis, heterobeltiosis, and standard heterosis. A high magnitude of heterosis in fodder yield was realized in the cross combinations 2077A × K3, 2077A × K7, 2217A × K3, 3660A × SPV 544, 3660A × M25-1, 2077A × KS7078 in the absence of *gca* effects (Sankarapandian et al., 1994). Desai et al. (2000) observed high heterosis over the superior parent and standard checks in the crosses (namely, ICSA77 × IS3312 and ICSA77 × IS3314). Similarly, Laxman (2001) reported high positive useful heterosis for dry fodder yield. Ravindrababu et al. (2002) obtained considerable heterobeltiosis in the cross IS18551 × Malwan. Patil (2004) also reported significant heterosis over the better parent to the extent of 40.08% in the cross DSV5 × CSV8R. A range of −9.76 to 150.46% standard heterosis over the check was reported by Karthik (2004).

7.1.6.3 Combining Ability in Relation to Heterosis in Sorghum

The magnitude of heterosis is determined by the combining ability of the parents. There are two types of combining ability (namely, gca and sca). Sprague and Tatum (1942) defined *gca* as the average performance of lines in a series of hybrid combinations and *sca* as the deviation of certain crosses from the average performance of the lines. Their results depicted that *gca* was primarily due to genes that are additive in their effects and *sca* was due to deviations from the additive model caused by dominance and epistatic interaction. Similarly, Henderson (1952) also opined that *sca* in a real sense was the outcome of intra-allelic interaction (dominance) and interallelic interactions (epistasis), while Griffing (1956) reported that gca, on the other hand, involved additive effects, additive × additive interaction, and higher-order interactions. Kramer (1959) was the first to study general and sca status for yield and other related traits in sorghum.

Several experimental procedures (Tysdal et al., 1942; Allard, 1960) are known for combining ability analysis—namely, open-pollinated test, top cross-test, poly cross-test, and single cross-test, of which single cross-test is the most practical with self-pollinated crops. Diallel cross-mating (Hayman, 1954; Griffing, 1956) is commonly employed to study the combining ability. Line × tester analysis (Kempthorne, 1957) is also a popular method for self-fertilized crops. A number of reports on combining ability in sorghum are reported in the literature (Table 7.2). Most of the studies used diallel mating design with or without reciprocal, while a few have used line × tester mating design. With a few exceptions, all the studies showed significant gca and sca effects for grain yield, indicating that both additive and nonadditive gene actions are important for this trait. The relative proportion of gca to sca variances were found to vary in different studies. Sorghum cultivars were identified as possessing high gca for yield (Table 7.2). The crosses involving good × good and good × poor combiners show more heterosis. The estimate of interaction component gca × environment and sca × environment is significant in some of the studies made. However, the interaction component due to gca × environment variances showed higher values compared to the sca × environment variances. It shows that the additive variance × environment interaction was more than nonadditive variance × environment interaction. Hence, diverse good combiners should be used in hybridization, and the

TABLE 7.2 Sorghum Varieties Identified as Good or Poor Combiners for Yield

Good combiner	Poor combiner	Reference
Grain yield		
M148-138, SPV 570	104B	Deshpande (2005)
SL 19B, SLR 13, SLR 24, SLR 30	SL 12B, SLR 39	Prabhakar and Raut (2010)
104A, BRJ 416	P2A	Biradar et al. (2000)
117A, 104A, RSLG 112, RSLG.206, SPV 1320 and CSV 14R	–	Kadam et al. (2000)
SPV 438, SPV 86, SPV 41	SPV 43, SPV 422, M 35-1	Jagadeshwar and Shinde (1992)
Fodder yield		
SPV 570, M 35-1	104B	Deshpande (2005)

resultant material should be exposed to different environments for identifying the stable high-yielding genotypes (Jagadeshwar and Shinde, 1992).

In addition to yield and yield components, combining ability studies have been carried out for several other traits—namely, plant height, days to flower, low temperature germinability, postanthesis cold tolerance, grain quality traits, grain size, and drought tolerance (Jagadeshwar and Shinde, 1992; Biradar et al., 2000; Kadam et al., 2000; Deshpande, 2005; Prabhakar et al., 2013). Like yield and yield components, gca and sca effects were important in different crosses in the inheritance of these traits.

7.2 BREEDING FOR RESISTANCE TO ABIOTIC STRESSES

7.2.1 Drought Resistance

Drought limits the agricultural production by preventing the crop plants from expressing their full genetic potential (Jiban Mitra, 2001). Drought is the most significant cause of crop yield loss, especially in areas where sorghum is grown. Water serves many vital roles in the plant, including acting as a solvent, a transport medium, and an evaporative coolant (Boyer, 1982). Consequently, water limitation causes a decrease in whole plant growth and photosynthesis, wilting, and stomatal closure, and is associated with changes in carbon and nitrogen metabolism (Sanchez et al., 2002). Although sorghum possesses excellent drought resistance compared to most other crops, improving its drought resistance would increase and stabilize grain and food production in low-rainfall, harsh environmental regions of the world (Rosenow et al., 1997).

7.2.1.1 Drought-Tolerance Mechanisms

Crop productivity in dryland agriculture is regulated by the extent of water capture by the crop, the efficiency with which the crop exchanges water for CO_2 via transpiration to produce biomass, and the fraction of total biomass that ends up in the grain (Passioura and Angus, 2010). Here, DS is discussed in the agronomic sense (namely, a reduction in grain yield attributable to plant water deficit). Drought tolerance in sorghum is specific to the GS; that is, sorghum genotypes with good tolerance during one of the developmental stages might be susceptible to drought during the other GSs, making them complex to work with (Reddy et al., 2009). Plants perceive water loss by triggering a cellular signal transduction pathway where physical stress is converted into a biochemical response. One of the major signals, the plant hormone abscisic acid (ABA), is known to induce various early and later responsive genes involved in a signaling cascade for the regulation of downstream biochemical protective mechanisms (Shinozaki and Yamaguchi-Schinozaki, 1997) in addition to promoting stomatal closure, which dramatically reduces foliar transpiration. The accumulation of soluble sugars (sucrose, glucose, and fructose) is strongly correlated with the acquisition of drought tolerance in plants (Hoekstra and Buitink, 2001). It is a drought-resistance mechanism that delays premature senescence under soil moisture stress during grain filling and maturity, and it is also associated with stalk rot resistance. The mechanisms of drought tolerance in sorghum are escape, avoidance, and tolerance, as described next (Reddy et al., 2009):

1. *Drought escape:* Early maturity is a well-known drought-escape mechanism through which the crop completes its life cycle before the onset of severe moisture deficits and is often associated with reduced yield potential. Short-duration sorghums have lower evapotranspiration rates due to their having smaller LA and smaller root density than long-duration ones (Blum, 1979a). To some extent, yield loss can be overcome by increasing the plant density.
2. *Drought avoidance:* Drought avoidance is a mechanism for avoiding lower water status or to maintain a relatively higher level of hydration in tissues during drought by maintaining cell turgor and cell volume either through aggressive water uptake by an extensive root system (Manschadi et al., 2006; Hammer et al., 2009), or through reduction of water loss from transpiration and other nonstomatal pathways (Ludlow and Muchow, 1990), such as through the plant cuticle. Most cultivated dryland sorghum genotypes show epicuticular wax values close to the maximum (Jordan et al., 1983). Plants avoid low leaf water potential (LWP) by one or more mechanisms, such as a change in rooting pattern, an increased root growth for maximizing water uptake, or an adjustment in LA for optimization of the use of absorbed water for the production of dry matter (Seetharama et al., 1982a,b). Sufficient genetic variability was recorded for the root attributes, transpiration efficiency and transpiration regulation (Hammer et al., 1997; Kholova et al., 2010a), and canopy development (Borrell et al., 2000a; Kim et al., 2010; van Oosterom et al., 2011).
   A. *Root attributes:* Water uptake during key stages of the crop life cycle is crucial, with small water-saving differences generating large yield differences (Zaman-Allah et al., 2011b) and significant variation exists for the total amount of water extracted from the soil column (Vadez et al., 2011a; Talwar and Vadez, 2011). For maximizing extraction of moisture from the soil, the requirements are (i) deep penetration of roots, (ii) adequate root density through the soil profile, and (iii) adequate longitudinal conductance in the main roots (Fischer et al., 1982). Even though terminal DS is common, crops are not necessarily limited by a deficiency of stored soil moisture, but by an inability of the crop either to fully extract water stored deep

in the profile or extract it fast enough for yield formation (Jordan et al., 1983). However, the growth of roots into deeper soil layers is a function of both genotype and environment; the interaction between the two often makes it difficult to distinguish genotypic differences in root growth. Root length density (L_v, in cm/cm^3) usually decreases with depth. In environments where the crop is grown on stored soil moisture, high L_v in the surface layers is not required. In this case, efforts should be directed toward increasing L_v at depth. On the other hand, if the crop is targeted for an environment where rainfall occurs in short spells during the growing season, then high L_v in the surface layers (<0.5 m depth) is advantageous. Besides moisture, other factors such as profile nutrient distribution, temperature, relative humidity, and soil strength of topsoil also influenced rooting depth (Guohua et al., 2010). Limitations in plant hydraulic conductance, likely linked to aquaporin activity (Parent et al., 2009; Kholova et al., 2010b; Gholipoor et al., 2010), may reduce transpiration at times of high atmospheric vapor pressure deficit, and thus influence TE. By selecting for smaller metaxylem vessel diameters in the seminal roots, Richards and Passioura (1981) developed wheat genotypes that could use water more slowly in early GSs. Simulation studies in sorghum indicated that yields of deeper-rooted genotypes were at least 20% more than control genotypes in 1 out of every 3–5 years across locations (Jordan et al., 1983). Screening and selection for rooting depth on a large scale are expensive and laborious processes. A selective herbicide method (Khalfaoui and Havard, 1993), root pulling resistance method (Ekanayaka et al., 1985), and apparent sap velocity method (Ketring, 1986) have been suggested for various crops. Recently, the genetic associations between variations in nodal root angle and water capture and yield in sorghum (Singh et al., 2010; Mace et al., 2012) have been reported. Genetic variability in root characteristics of sorghum grown in solution culture is not expressed to the same degree in the field during drought (Jordan and Miller, 1980). Hence, suitability of the model system to the field situation must be demonstrated before large-scale evaluation of germplasm for root attributes is undertaken.

B. *Shoot attributes:* Grain yield is dependent upon the amount of water available to sustain growth during the late flowering and grain-filling stages (Zaman-Allah et al., 2011a). In terminal stress environments, this is greatly influenced by the dynamics of canopy development, which is regulated via maturity (Ravi kumar et al., 2009), tillering (Kim et al., 2010), leaf size and appearance (Chenu et al., 2008; van Oosterom et al., 2011), leaf conductance (Kholova et al., 2010b) and partitioning among plant organs (van Oosterom et al., 2008). Genomic regions controlling these attributes have been identified in sorghum and pearl millet (Feltus et al., 2006a,b; Kholova et al., 2011). Canopy structure is determined by the leaf size, leaf shape, leaf surface characteristics and reflectance properties, leaf angle, and geometrical arrangement of leaves. Crop cultivars with more erect and narrow leaves and lower light extinction coefficient values—and hence higher critical leaf area index (CLAI; the leaf area index that intercepts 95% of the incoming solar radiation)—generally have higher CGRs. If the LAI of a crop canopy is high, a higher leaf angle can result in increased canopy photosynthesis because the intercepted radiation is distributed across a greater sunlit LA (Tollenaar and Lee, 2006).

Stomata, by adjusting their apertures, play a major role in regulating water loss so as to match the evapotranspirational demand to the water-supplying capacity of the roots, but this operates when the fraction of transpirable water falls to 0.3 or less (Sinclair and Ludlow, 1986). Genetic variability (Markhart, 1985) and heritability for stomatal characteristics are high, indicating the feasibility of genetically manipulating this trait (Buttery et al., 1993). Also, it may be possible to select for optimum stomatal aperture size by selecting for higher TE. TE is known to vary considerably among diverse sorghum lines (Vadez, 2011a,b; Talwar and Vadez, 2011), and due to complexity in measuring the variation, it remains unutilized in breeding programs. Other morphological features, such as a thick cuticle or wax deposits on the leaf surface, can reduce water losses from the leaf surface and thus minimize residual transpiration rate (Jefferson et al., 1989).

a. *Leaf WP*. Gaosegelwe and Kirkhan (1990) have suggested that LWP might be used as an easy and fast way to screen sorghum genotypes for drought avoidance. The physiological adaptations effective in improving tolerance to moisture stress were found to vary with plant GS in sorghum (Ackerson et al., 1980). Before flowering, plants avoid dehydration largely by maintaining higher LWP, while after flowering, plants avoid dehydration by maintaining higher turgor at a given level of moisture stress. This could be partly responsible for the classification of drought tolerance before and after flowering in sorghum (Rosenow et al., 1983). The most evident control of LWP is at the root system. Small root resistances and a large root-length density contribute to the maintenance of a higher LWP (Blum, 2011). Genotypic differences in sorghum root growth have been noted under moisture stress. Blum (1979b) has shown that early-maturing sorghum genotypes not only escape drought, but also avoid it because of reduced transpiration as a result of increased root length accompanied by reduced LA (high root length to LAR). Because of the high sensitivity of LA expansion to changes in turgor, the LA expansion can be used as the criterion for screening the genotypes for drought tolerance.

Leaf firing was found to be a simple phenotypic trait that allows large populations to be screened (Andrews et al., 1983). Leaf rolling is an established symptom of wilting in cereals (Kadioglu and Terzi, 2007), and delayed leaf rolling under DS is being used as one component of a selection index for drought tolerance (avoidance) in sorghum (Matthews et al., 1990). Greater leaf rolling was indicative of reduced LWP in different sorghum genotypes. Under relatively mild stress, delayed leaf rolling may be associated with sustained plant growth and production. However, under severe drought and heat stress conditions, greater leaf rolling may be associated with better chances for recovery when moisture stress is relieved. Stricevic and Caki (1997) showed a predawn LWP of -0.5 MPa as the threshold value for scheduling irrigation, because physiological processes were significantly decreased below this value, suggesting that those genotypes that maintain predawn LWP above this level can be considered as drought-tolerant. However, breeding programs are slow to adopt this trait.

b. *Osmotic adjustment (OA)* can be defined as the active accumulation of solutes within the plant tissue (either in roots or shoots) in response to a lowering of soil WPs, which helps maintain turgor of both shoots and roots as plants experience water stress (Morgan, 1984). This could lead to a lowering of OP, which provides the driving force for extracting water from low WPs. OA can play a major role in determining the drought resistance of a given genotype by (i) maintaining turgor over fluctuating soil WPs, (ii) maintaining stomatal conductance and thus photosynthesis, (iii) maintaining growth, (iv) increasing dehydration tolerance, and (v) increasing the extraction of soil water (Ludlow, 1987).

A wide variety of organic solutes accumulate in plant tissues during water and salt stress and contribute to OA (Gorham et al., 1985). The relative contribution of organic and inorganic solutes to OA varies among crop species. For example, in sorghum, organic solutes play a major role in OA, whereas in sunflowers, inorganic ions contribute a major share of OA (Jones, 1980). Reduction in solute potential (SP) can also occur through changes in turgid weight/dry weight ratio, reducing the osmotic volume without accumulating additional solutes (Ludlow, 1980a).

Also, OA is associated with stimulation of root growth in cereals by providing additional carbon (Morgan and Condon, 1986). It helps in maintaining stomatal conductance and photosynthesis over a longer period. However, in some cases, OA is not associated with improved root growth. The impact of this additional carbon in determining yield depends on the GS at which the crop experiences the water deficit. If water deficit occurs during the heading or grain-filling stage, the additional carbon available due to OA may play a crucial role in preventing spikelet sterility, and in increasing grain set and seed filling, thus improving HI (Pierce and Raschke, 1980).

OA is positively correlated with yield under drought conditions in sorghum (Morgan, 1984). Based on this, Tangpremsri et al. (1995) concluded that the adverse effect of water stress could be reduced by selecting sorghum genotypes with high OA. Some reports indicate no relationship between OA and growth or yield under field conditions (Blum et al., 1989a,b). Also, in some cases, improved OA resulted in smaller cell size (Ackerson, 1981), and thus small organs and small plants. The smaller size of plant organs, as either source or sink, can result in lower potential yield. A negative correlation between OA and yield also has been reported (Grumet et al., 1987). The increase in solutes that occurs with a reduction in WPL eventually reaches a limit, and this varies among crop species (Turner and Jones, 1980). Flower et al. (1990) concluded that while drought-tolerant sorghum varieties had better OA (and consequently less leaf rolling) under stress than that of susceptible varieties, these responses did not influence growth under very dry and hot conditions. The ecological habitat of the genetic materials, the GS, and growth conditions can influence the degree of OA (Girma and Krieg, 1992). Girma and Krieg (1992) determined the mechanisms contributing to diurnal changes in OP using a single sorghum hybrid grown under variable water supplies. They found that net solute accumulation accounted for 42% of the diurnal change in OP before flowering and 45% of the change during grain filling in water-stressed sorghum leaves.

Studies on OA have been accelerated by the use of the pressure chamber method and analysis of pressure volume graphs to measure water, osmotic, and turgor potentials. Thermo-couple psychrometry have also aided in the measurement of water and OPs. Variation in OA among sorghum genotypes was found to range from zero to 1.7 Mpa (Blum and Sullivan, 1986). Landraces from dry habitats compared to those from humid regions have a greater capacity for OA (Blum and Sullivan, 1986). OA has a direct positive effect on yield under moisture stress (Ludlow et al., 1990) and is largely ascribed to increase in root size, root length density, and soil moisture extraction (Tangpremsri et al., 1991a,b). Postrainy season cultivars such as M 35–1 and CSH 8 have a greater capacity to decrease their OP under stress than the rainy season cultivar CSH 6 (Seetharama et al., 1982a,b). Two independent major genes (OA I and OA 2), with some minor effects, have been reported to control the inheritance of OA in sorghum (Basnayake et al., 1995). Little if any progress in breeding for drought tolerance using either OA or any of the other physiological traits has been documented in sorghum, partly because of poor understanding of these traits conferring drought tolerance (Bohnert et al., 1995) and lack of procedures to impose reproducible stresses and rapid methods to measure these traits (Santamaria et al., 1990). It appears that individual physiological traits identified to date are not sufficiently related to overall drought tolerance under field conditions to merit selection based on them (Rosenow et al., 1997c).

3. *Drought/dehydration tolerance Drought tolerance* is a mechanism by which plants maintain metabolism even at low WP. Dehydration results in irreversible disruption of cellular organization and metabolism, and most crop plants belong to the dehydration-intolerant category. Severe dessication represents a small proportion of the total instances of drought. However, enhancement of dehydration tolerance, which results in continued leaf growth and decreased senescence during mild or moderate drought, could have a positive effect on agricultural production. This trait is considered as the most difficult to improve through conventional plant breeding. Seedlings are subjected to severe levels of moisture and heat stress and then membrane damage can be assessed by leakage of electrolytes (Leopold et al., 1981). Since then, extensive research has led to identification of component traits, sources of genes, and field management practices to approach this complex problem. *Antioxidant capacity* is the ability of plants to detoxify reactive oxygen species that cause cellular injury, such as lipid peroxidation or protein and nucleic acid modification (McKersie and Leshem, 1994). Water stress had been hypothesized to induce membrane-lipid peroxidation by means of activated oxygen species (Zhang and Kirkham, 1994). Zhang and Kirkham (1996) compared the degree of lipid peroxidation between sunflower and sorghum, using malondialdehyde (MDA) content as an indicator of the prevalence of free radical reactions in tissues. MDA content was found to be much higher in sunflower than sorghum under both watered and unwatered conditions. Furthermore, an increase in MDA content in sorghum occurred later in the drought stage compared to an earlier onset of increase in sunflower, suggesting a higher level of lipid peroxidation and more severe oxidative stress in sunflower than in sorghum during drought.

Apart from these physiological adaptations, certain biochemical compounds and micronutrients, such as increased levels of glycine betaine and proline levels (Wood et al., 1996) and grain K and Fe contents (Abu et al., 2002), were associated with drought tolerance in sorghum. However, the usefulness of dehydration tolerance can be realized only if it is placed in a genetic background that has other mechanisms related to the maintenance of production under moisture-deficit environments. Dehydration tolerance coupled with stomatal adjustment contribute to greater drought resistance of sorghum in semiarid areas (Ludlow, 1980b). High dehydration tolerance is associated with slow rates of growth and development, and most plants with good dehydration tolerance lack dehydration-avoidance strategies (Ludlow, 1980b). Nevertheless, it is considered that moderate dehydration tolerance is desirable in crops of semiarid environments (Steponkus et al., 1982).

7.2.1.2 *Integrated Traits Assisting Crop Performance*

A. *Morphophysiological traits:* Identifying the morphological or physiological traits imparting tolerance to DS has been given high priority in many crop plants (Craufurd, 1993). Sorghum is said to be drought-tolerant, yet at certain critical stages, moisture stress causes considerable reduction in growth and yield (Griffin, 1966). In arid and semiarid environments, soil moisture in the seed bed can be suboptimum, which causes reduced germination and emergence and results in low yields (Saxena, 1987). Rapid root development (especially primary roots) and growth would facilitate successful establishment of seedlings when soil moisture is suboptimal after sowing (Assay and Johnson, 1983). These roots must be capable of exerting considerable growth pressure, possibly through alterations in root diameter or OA in roots. Also, dehydration tolerance of the leaf tissue and the meristems may play an important role in protecting the seedling and thus maintaining the plant stand. Rapid early growth and canopy development will reduce surface soil evaporation, thus increasing moisture available for transpiration. However, if the crop is raised entirely on stored soil moisture, then early growth vigor needs to be balanced with the rate of moisture use to ensure that enough moisture is left for the grain-filling period. Simulation modeling could assist in estimating the degree of growth vigor required for a given production environment. LA expansion is more sensitive to drought than photosynthesis and transpiration (Muchow, 1985b). Under rainfed conditions where intermittent drought patterns are common, the ability to retain LA during the period of moisture deficit plays an important role in determining the productivity of the genotype. The ability to retain LA is an integrative function of many traits, such as deep root system, other root traits related to moisture extraction, stomatal control, OA, dehydration tolerance, and leaf movements. The glossy character was correlated with the reduction or elimination of wax deposits on the leaf surfaces, while hairiness might occur in either glossy or nonglossy genotypes. Measurements showed that cuticular transpiration of glossy leaves was often more than double that of nonglossy leaves. Comparisons among sorghums showed that nonglossy lines had higher stomatal densities than glossy lines (Traore et al., 1989).

B. *Mobilization of preanthesis stored reserves:* Because translocation is more tolerant than photosynthesis and respiration to moisture deficit, the ability to store and mobilize large quantities of carbohydrates for OA, or for grain filling under terminal drought, should improve the ability of a cultivar to perform under drought conditions (Blum et al., 1983a,b). Also, ABA plays a major role in inducing mobilization after anthesis.

Under conditions of terminal moisture deficit, net photosynthesis decreases, and so the proportion of translocation of stored soluble carbohydrates for grain filling becomes larger. To quantify the contribution of stored preanthesis reserves for grain yield under terminal drought environments, green LA is removed by spraying chemicals such as magnesium or sodium chlorate (4% active ingredient) 14 days after flowering, thus forcing plants to rely entirely on stem reserves for grain filling (Blum et al., 1983a,b). Although buildup of higher levels of nonstructural carbohydrate reserves during the preflowering stage is advantageous, it adversely affects yield potential under optimal conditions and also leads to sink limitation. The accumulation also predisposes the stem to fungal infections or lodging, causing weakening of the stem. This has been shown in sorghum, in which genotypes that mobilize large quantities of carbohydrates stored in the stem during seed filling were more sensitive to charcoal rot (*Macrophomina phaseolina*) under conditions of terminal DS (Rosenow et al., 1983).

7.2.1.3 *Agronomic Factors Influencing Plant Responses to Drought*

A. *Soil type:* Grain yield decreased in shallow soil as compared to medium black soils (51.6%), due to reduction in LAI, TDM, grain number per panicle, and 1000-grain weight. The adaptability of varieties (namely, Sel.3 and RSLG 262) to specific soil depths was noted in sorghum and showed less reduction in grain yield in shallow soil, while the M 35-1, RSLG-262, and RSP-3 varieties performed better under medium black soil. The RSLG-262 genotype was suitable for both soil types (Jirali et al., 2007). Silicon application Si_{200} (200 mL/L of potassium silicate per kilogram of soil) improved the drought tolerance of sorghum through the facilitation of root growth and enhancement of water uptake ability (Hattori et al., 2005; Ahmed et al., 2011), thereby increasing LAI, chlorophyll content, leaf dry weight, shoot dry weight, root dry weight, net assimilation rate, relative growth rate, and WUE by 28–40%.

B. *Irrigation:* Two lifesaving irrigations, one at ear head initiation and the other at the flowering stage, gave 88% and 65% more grain and fodder yields of rabi sorghum without irrigation, respectively (Thorve et al., 2009).

C. *Fertilizer:* Potassium affected leaf WP and maintenance of turgor potential in water stress conditions. Under severe water stress, OP ranged from −0.3 MPa without applied K to −1.7 MPa with the highest K rate. Potassium fertilizer strongly affected the leaf rolling index and RWC maintenance in water stress conditions, and the highest K rate increased the leaf rolling index by 39% (without K) to 47% (Arjomand et al., 2000).

7.2.1.4 *Biochemicals Contributing to Drought Tolerance*

Stress-induced changes in peptidylpropyl cis-trans isomerase (PPIase) activity of different tissues indicated a significant increase in leaf- and root-PPIase activity in the drought-tolerant cultivar ICSV 272, and the PPIase activity of the two tissues decreased appreciably in the susceptible cultivar SPRU 94008B. The PPIase activity in different tissues was due to the presence of both cyclophilins (Cyps) and FK506-binding proteins (FKBPs). The effect of water stress on Cyp- and FKBP-associated PPIase activity was differential and tissue-dependent. The differential effect of DS on PPIase activity in the two cultivars was independent of WP, suggesting different regulatory pathways in the drought-tolerant and susceptible cultivars of sorghum (Sharma and Singh, 2003). Immonodetection studies revealed the presence of a 20-kD, cyclophilin-like protein (designated as SorgCyp20) in the leaves and seeds of sorghum (Sharma and Singh, 2003). In most cultivars (seeds), K and Fe concentrations were more in the tolerant cultivar compared to the susceptible type. The concentration of Fe decreased with maturity in the tolerant group, and it increased with maturity in the susceptible group (Assar et al., 2002). In both the root endodermis and leaf epidermis, silicification was higher in the drought-tolerant cultivar (Gadambalia) than in the drought-sensitive cultivar (Tabat) (Lux et al., 2002). The involvement of auxin in various aspects of growth and development is well known. Recent studies suggest that many members of these gene families involved in auxin-signaling cascades are also involved in stress/defense responses (Ghanashyam and Jain, 2009; Jain and Khurana, 2009). Three genes, SbIAA1, SbGH3-13, and SbLBD32, were highly induced in sorghum under all four treatments: indole-3-acetic acid (IAA), brassinosteroid (BR), salt, and drought (Wang et al., 2010).

7.2.1.5 Genetic Improvement Strategies

A. *Characterization of the drought environments:* Characterization of the target production region is the first and crucial step in undertaking a genetic improvement program aimed at improving yield and yield stability in drought-prone environments (Robertson, 1988). Production environments can be grouped into subsets of environments using canonical variate analysis (cluster analysis) (Malhotra and Singh, 1991), by considering such factors as soil water balance, temperature regimes, the potential evapotranspiration of the growing environment (using long-term climatic data), or G × E interactions (Seif et al., 1979). Thus, the mean and variance of moisture deficits likely to occur during the growing season and length of the growing season can be calculated. This assists in estimating the intensity and duration of stress that the crop is likely to experience, and at which crop GS this will occur, thus guiding the development of a relevant screening methodology for evaluating germplasm lines. This strategy also will help in identifying the environments that have similar drought patterns, to guide multilocational testing of genotypes developed for specific drought patterns. Recent development of geographic information system (GIS) techniques allows computerized mapping, and thus ready visualization, of such iso-drought environments. New statistical methods for spatial design and analysis are available (Chan and Eccleston, 2003). Smith et al. (2001) reported heritability increases from 6% to 46% from the adoption of the spatial method of analysis over a randomized block analysis.

B. *Screening methodologies:* In most cases, screening for drought resistance involves evaluating the genotypes under field conditions with and without irrigation. Line source (LS) irrigation is used to evaluate the genotypes under a gradient of water deficit (Hanks et al., 1976). Moisture response curves are generated by regressing the yield of individual entries against water applied or environmental yield, to calculate the stability of a genotype across a range of moisture availabilities using the stability analysis procedures of Finlay and Wilkinson (1963) and Eberhart and Russell (1966). However, this approach does not consider the confounding effects of phenology on yield or the effects of yield potential on the slope of regression, and hence on the intercept and drought resistance. Fischer and Maurer (1978) proposed a drought susceptibility index (DSI), based on relative yield, to account for the confounding effects of yield potential of genotypes. To account for time to flowering, associated drought escape, and yield potential, Bidinger et al. (1987a,b) proposed a drought resistance index (DRI), which is based on the residual variation in grain yield adjusted for experimental error. The index is positively correlated with yield under drought, independent of yield potential and time to flowering.

Field screening at more than one location with different water regimes can be used to ensure stress at different stages of growth. Rainout shelters and irrigation gradient systems can be used to supplement field screening. Two main strategies are employed by breeders for improving drought resistance in sorghum: (i) direct selection for yield in drought environments and (ii) indirect selection for traits known to be associated with drought.

7.2.1.5.1 INDIRECT SELECTION

a. **Using 25% polyethylene glycol (PEG) method:** This method is used at the seedling stage. It can reduce the WP equivalent to natural drought conditions so that water absorption by roots is affected (Singh and Chaudhary, 1998).
b. **Using 0.3% solution of potassium iodide:** This method also is used at the seedling stage. It can simulate leaf senescence and abscission equivalent to natural drought conditions (Singh and Chaudhary, 1998).
c. **In vitro screening method:** The immediate response by the different organs of a plant to water stress is a decrease in turgor. This may be partially or fully adjusted by accumulation of solutes. The change in size and shape of pollen grains when subjected to in vitro osmotic stress using PEG was considered as a measure of OA. PEG (discriminative osmotic stress) at 40% discriminated the pollen grains of rabi genotypes (i.e., retained their size) from the kharif genotypes (i.e., showed shrinkage). There is a close correspondence between intrinsic OA in pollen grains and high leaf OA. In addition, the study indicates that OA is a drought-adaptive trait and could have evolved in the rabi genotypes by virtue of their regular exposure to moisture stress (Patil and Ravikumar, 2011).
d. **High-temperature challenge method:** Identification of the nonsenescent (stay-green) trait requires the right intensity of DS at the right developmental stage to visually evaluate lines in the field. By means of a 30-min high-temperature challenge to leaf tissue during flowering of WW sorghum and a 30-min room temperature recovery, stay-green lines can be readily identified. Tissue with higher intercellular sucrose concentrations exhibited higher chlorophyll fluorescence yield following the temperature challenge. Stay-green lines maintained higher dawn leaf sucrose levels than the senescent lines (Burke et al., 2010).

7.2.1.5.2 DIRECT SELECTION

The plants are grown directly in drought-prone areas during dry season. Production of total biomass, yield, and its components can be used as criteria for selection.

a. **Germination and seedling emergence:** A hot dry seedbed environment with a soil surface temperature often greater than 55°C during crop establishment due to the absence of subsequent rain after initial planting is a common occurrence in most regions of semiarid tropics (SAT; Gupta, 1986; Hoogmoed and Klaij, 1990; Peacock et al., 1993). Seedling death can occur at one of three defined stages (germination, emergence, and postemergence) during crop establishment. Initial selection for coleoptile length of test lines at 5 days after planting in germination boxes, followed by selection in raised brick tanks using charcoal and heavy kaoline (to simulate higher or less than normal temperatures, respectively) in 12-cm planting depth based on plant counts (as a reflection of mesocotyl length), is highly effective in identifying lines with long mesocotyl length that is necessary for emergence in deep sowing. Useful genetic variability was noted for seedling emergence (10–50%) at 5 DAS among 166 sorghum genotypes grown in alfisols with limited soil moisture during hot, dry summer season and under differential irrigation (5–30 mm) using a LS sprinkler system (Seetharama et al., 1982a,b). Diminishing soil water availability after germination due to dry spells after initial rains during sowing greatly affects seedling growth and survival. Selection of breeding lines in the sandy soil–filled brick tanks spread uniformly with charcoal powder at 125 g/m^2 (which induces high soil surface temperatures) based on the seedling emergence counts on the 6th day after planting was effective in identification of lines with high seedling emergence under high surface soil temperatures (Reddy, 1985a,b). Significant cultivar differences for seedling emergence at low soil moisture conditions (Soman, 1990) and subsequent growing ability have been reported in sorghum (O'Neill and Diaby, 1987).

b. **Postemergence/early seedling stage:** Sustained seedling growth following emergence depends on the capacity of the seedling to elongate, produce leaves, and become autotrophic. Postemergence seedling death due to moisture stress under field conditions is primarily caused by the prevalent high soil surface temperatures, at least in the first 10 days following sowing, and only after that does water deficit start to take effect (Peacock et al., 1993). It was established that selection of breeding lines for recovery from severe seedling drought, induced after germination for 24 days, followed by termination of drought in the 29th day after planting in polyvinyl chloride (PVC) vases based on recovery scales, number of plants recovered/vase, and number of green leaves/vase, was effective to screen for recovery from seedling drought (Reddy, 1985a,b).

c. **Midseason/pre-flowering stage:** The preflowering response is expressed when plants are stressed during panicle differentiation prior to flowering. Symptoms of midseason or preflowering DS susceptibility include leaf rolling, uncharacteristic leaf erectness, leaf bleaching, leaf tip and margin burn (leaf firing), delayed flowering, poor panicle exertion, panicle blasting and floret abortion, and reduced panicle size (Rosenow et al., 1997a–c). Tolerance of preflowering DS is indicated by the alternative condition in each instance. When screening for midseason stress, curtailing irrigation 3 weeks after sowing for over 45 days in a rain-free season was found to provide the required level of DS for effective screening for preflowering DS (Reddy, 1985a,b). Excellent sources of tolerance of preflowering drought have been identified (Rosenow et al., 1997a–c). Good correlations between scores for leaf firing and the ability to recover and agronomic traits were reported (Seetharama et al., 1982a,b).

d. **Terminal/postflowering stage:** Postflowering stress is due to inadequate soil moisture during the grain-filling stage, especially during the latter portion of grain fill. Symptoms of postflowering DS susceptibility include premature plant (leaf and stem) death or plant senescence, stalk collapse and lodging, and charcoal rot (caused by *M. phaseolina*), along with a significant reduction in grain size, particularly at the base of the panicle

(Garud et al., 2002). The cultivars that remain green under postflowering drought are referred to as having the *stay-green* trait. Therefore, selection of breeding lines for the stay-green trait was useful for screening for terminal drought tolerance. The stay-green trait is expressed only when the materials at the postflowering stage are exposed to severe moisture stress. Comparison of yields on shallow vertisols or on partially saturated deep vertisols with yields with an irrigated control has been advocated to screen genotypes for terminal drought tolerance in receding moisture conditions. An LS sprinkler irrigation technique developed at Utah State University was followed at ICRISAT for screening sorghum genotypes for terminal drought tolerance. Each side of the LS formed one replication. The field was uniformly irrigated until the crop reached the boot stage, and the LS was used at 50, 61, and 77 DAS to create a gradient of soil moisture (stress). The amount of water received across the plot was measured via catch cans placed at crop height. LS was also used to study the effect of soil moisture stress on charcoal rot incidence (Seetharama et al., 1987). The rows of plants farthest from the LS showed disease earlier than those nearest. This was apparent for each of the three parameters of disease spread: percentage of soft stalks, number of nodes crossed, and length of fungal spread (Seetharama et al., 1987).

C. **Screening nurseries:** Sandy soil or shallow soil sites are best suited for preflowering evaluation, while heavier and deeper soils are best for evaluating postflowering stress (Rosenow et al., 1997). At Purdue University, specific nurseries were used in dry environments, such as Mexico, to screen for preflowering and postflowering drought response (Rosenow et al., 1997). In Sudan, nurseries in the past (Ejeta, 1987) have been designed to specifically evaluate for either preflowering or postflowering stress. In the public breeding program in Queensland, Australia, breeding progeny are routinely rated in regular field breeding nurseries for premature leaf and plant senescence at or near physiological maturity. The stay-green trait from IS 12555 (SC35) has been successfully used in Australia to develop postflowering DS resistance and lodging resistance in parental lines and commercial hybrids (Henzell et al., 1992; Henzell and Hare, 1996).

Conventional breeding for stay-green has been primarily based on two sources for this trait, B 35 and KS19 (Mahalakshmi and Bidinger, 2002). KS19 is a selection from a cross of short Kaura, an improved landrace cultivar from northern Nigeria, with combine Kafir 60 (Henzell et al., 1984). B35 (PI534133) was selected from a converted (dwarf, height, early-flowering) version of IS12555, an Ethiopian landrace (Rosenow et al., 1983, 1997). KS19 has been commercially used primarily in the breeding program of the Queensland Department of Primary Industries (QDPI), while B35 is widely used in both public- and private-sector breeding programs in the United States (Mahalakshmi and Bidinger, 2002). The partially converted (B35) and fully converted (SC35C-14E) versions of IS12555 (Rosenow et al., 1983) have provided the major and best sources of the trait used in the QDPI program (Henzell et al., 1997). QL41 was the key line developed, and it was derived from the cross QL33 × B35 (Henzell et al., 1992). It has a high level of expression of stay-green and crosses of it with QL38 and QL39 (sorghum midge-resistant lines) formed the basis of the female stay-green and midge-resistant gene pool in the QDPI programs, although less progress has been made in developing such germplasms in the male population (Henzell et al., 1997).

FIGURE 7.5 Field under screening for preflowering drought stress.

While most commercial sorghum hybrids in the United States possess good tolerance to preflowering DS, only a few have good postflowering resistance (Nguyen et al., 1997). In spite of the availability of the simply inherited stay-green trait associated with terminal drought tolerance, progress in enhancing postflowering drought resistance has been slow (Rosenow et al., 1997). This is because the expression of this trait is strongly influenced by environmental factors and limited number of sources of stay-green are currently used in sorghum breeding programs (Mahalakshmi and Bidinger, 2002).

The stay-green trait expresses best in environments in which the crop is dependent on stored soil moisture, but where this is sufficient to meet only part of the transpiration demand. Sufficient expression of trait for selection is thus dependent upon the occurrence of a prolonged period of DS of sufficient severity during the grain-filling period to accelerate normal leaf senescence, but not severe enough to cause premature death of the plants (Mahalakshmi and Bidinger, 2002). Because of this precise requirement for the trait expression, field environments do not offer ideal conditions for selection; therefore, identification of quantitative trait loci (QTLs) conferring the stay-green trait and the molecular markers tightly linked to these QTLs are powerful tools to enhance drought resistance (Crasta et al., 1999; Xu et al., 2000; Mahalakshmi and Bidinger, 2002). Recent identification of several (e.g., IS22380, QL 27, QL 10, and E36 × R 16 8/1) tropically adapted lines with stay-green expression equivalent to those of the best temperate lines B35 and KS19 (Mahalakshmi and Bidinger, 2002) is further expected to hasten the process of mapping QTLs and their subsequent introgression into agronomically elite lines. Rapid progress has been achieved in mapping QTLs affecting the stay-green trait and their potential use in improving the drought tolerance (Figures 7.5, 7.6, and 7.7).

D. **Screening for drought tolerance using lysimeter facility:** Lysimeters are PVC tubes of 25-cm diameter and 2.0-m length, filled with soil in 1:1 alfisol:vertisol. The tubes were set in 2.0-m-deep trenches so that the top of the cylinder was at the ground level,

FIGURE 7.6 Field under screening for postflowering drought stress.

FIGURE 7.7 Stay-green resistant (left) and susceptible (right) genotypes.

which avoided sun exposure on the lysimeters. The trenches are set outdoors under natural conditions. The PVC end plate is placed on top of four screws 3 cm from the bottom of the cylinders to prevent the soil from seeping through. The end plate did not fit the cylinder tightly, allowing water drainage. The soil was sieved to a particle size of less than 1 cm. The cylinders were filled with soil in three increments of 40 kg of dry soil. After the addition of each 40-kg increment, the soil level in several cylinders was checked to ensure that they were similar, and all tubes were watered. About 8 L of water were added to each 40-kg increment to reach water-holding capacity. Finally, all cylinders were topped up with dry soil to ensure that they were filled to the same level. This top-up varied between 500 g and approximately 1 kg (i.e., less than 1% variation across cylinders). Weighing of the cylinders indicated that all saturated and freely drained cylinders weighed between 164 and 165 kg. The lysimeters were filled with soil that had been fertilized with diammonium phosphate (DAP, 18–46–0 NPK) and muriated potash (0–0–60 NPK), both at a rate of 200 mg/kg soil. The soil also contained sieved and sterilized farm manure at a rate of 1:25 to prevent micronutrient deficiency. The tops of the cylinders were equipped with a metal collar and rings that allowed the cylinder to be lifted. Weighing of the cylinders was done by lifting the cylinders with a block-chained pulley, and an S-type load cell (Mettler-Toledo, Geneva, Switzerland) was inserted between the rings of the cylinder and the pulley. The scale (200-kg capacity) allowed repeated measurements and gave an accuracy of 20 g on each weighing. The lysimeters were separated from one another by a distance of approximately 5 cm. Therefore, the crop of sorghum was planted at a density of approximately 11 plants/m^2; the plant population is very similar to the field planting (row-to-row distance of 60 cm and plant-to-plant spacing of 15–20 cm). This allowed an accurate assessment of the water extraction pattern of a crop cultivated in conditions similar to the field. The tubes were arranged in two sets of two adjacent trenches 2 m deep and 1.75 m wide, each set being separated by a 1.5-m path and each trench within each set being separated by a 20-cm concrete wall. Three trenches were used for the terminal DS treatment and three trenches for the WW treatment. The purpose of the trial was to assess the pattern of water use, water extraction ability, and transpiration efficiency in plants grown under both fully irrigated conditions (WW) and under terminal drought conditions (DS). Generally, four to five seeds were sown in each cylinder. Plant stands were thinned to two seedlings per cylinder at 14 DAS and then to one plant per cylinder at 21 DAS. All plants were kept in fully irrigated conditions until 28–30 DAS. After the regular irrigation at 28 DAS, the cylinders were covered with a 2-cm layer of low-density polyethylene beads to prevent soil evaporation. Preliminary testing indicated that the beads prevented more than 90% of the soil evaporation, so differences in cylinder mass were primarily due to plant transpiration. Weighing of the cylinders was first done at 30 DAS, and then subsequently every week. The first weighing at 30 DAS gave the saturated weight of each cylinder. The cylinders in this experiment were distributed. Transpiration was calculated for approximately 1-week intervals between 30 DAS and maturity. This allowed the water use before and after anthesis to be calculated. Daily transpiration values were calculated for each plant by dividing the transpiration of each time interval between weighing by the number of days in each interval. Then, preanthesis transpiration was the sum of the daily transpiration values until anthesis, plus water used in the first 28 DAS, which was estimated to be 1.5 L for all genotypes. This was based on dry weight estimates for plant biomass of 15 g at 28 DAS, and on the assumption of a TE of 10 g/kg water transpired at this early stage of crop development. The water use after anthesis was the sum of the daily transpiration values from anthesis until maturity. At harvest, leaf, stem, and panicle weights were taken after drying for 3 days in a forced-air oven set at 70°C. Dried panicles were then threshed to determine grain yield. HI was calculated as the ratio of grain yield divided by the total aboveground biomass (the aggregated dry weight of stems, leaves, and panicles). TE was calculated as the ratio of the total aboveground biomass to the sum of transpiration values between 30 DAS and maturity (Figure 7.8).

FIGURE 7.8 Lysimeter facility for screening for DS at Indian Institute of Millets Research, Hyderabad.

E. Physiological trait-based selection

Cuticular resistance was positively related to the amount of leaf epicuticular wax. Infrared aerial photography can be used for identifying drought avoidance in large-field screening nurseries. Drought tolerance was revealed in the ability of leaf cell membranes to function after stress. A method for assessing the relative injury to cell membranes by desiccation is available. Free proline accumulation at a given leaf WP was related to seedling recovery rate after desiccation (Blum, 1979a). Gaosegelwe and Kirkhan (1990) suggested that WP, rather than stomatal resistance, might be used as an easy and fast way to screen for drought-sensitive genotypes, while Voigt et al. (1983) suggested that sorghum lines that can maintain higher rates of CO_2 uptake under stress environments should be considered as valuable germplasms for breeding programs. Stricevic and Caki (1997) showed a predawn leaf WP of −0.5 MPa as the threshold value for scheduling irrigation, as physiological processes were significantly decreased below this value. Tangpremsri et al. (1995) concluded that the adverse effect of water stress can be reduced by adopting sorghum genotypes with high OA. However, selection for high OA needs to ensure that OA is not solely due to small head size. Studies on OA have been accelerated by the use of a pressure chamber and analysis of pressure volume curves to measure water, osmotic, and turgor potentials. Thermocouple psychrometry have also aided in the measurement of water and OPs (Parsons, 1982). Also, accumulated inert biochemical end products in proportion to the stress level can be used. However, the metabolites should be totally inert, and the assay should be simple and rapid.

7.2.1.6 Trait Associations

The response of sorghum genotypes to postanthesis drought and aluminium toxicity was tested, and the results indicated that multiple stress tolerance can increase sorghum productivity (Okiyo et al., 2010).

The grain yield was positively correlated with the RWC of leaves, LAI, biomass at harvest, grain number per panicle, test weight, earhead exertion, stay-green at physiological maturity, per-day production of grain, and fodder yield; and it was negatively correlated with leaf temperature, CSI, and stomatal frequency under DS (Nirmal and Patil, 2008), and also with long roots, coleoptile length, higher root/shoot ratio, LA, and leaf dry matter (Ali et al., 2009a). There was a positive association between RWCs and cell membrane stability, but both of these traits were negatively correlated with residual transpiration and excised leaf weight loss. Head width, head weight, grain yield per plant, and fresh and dry shoot weight were the most important characters explaining 77.65% of the total variation (Ali et al., 2011). Stay-green and grain yield were positively associated in a range of studies conducted in both Australia (e.g., Borrell et al., 2000a) and India (e.g., Al-Naggar et al., 2007), highlighting the value of retaining green LA under terminal drought. Grain number is generally the main determinant of differences in grain yield while grain size is a secondary yield determinant and is often negatively associated with grain number, as observed in sorghum grown under terminal DS in southern India (Borrell et al., 1999). Borrell et al. (1999) also found that reducing the rate of leaf senescence from 3% to 1% loss of LA per day resulted in doubling grain size from about 15 to 30 mg. Thus, stay-green can potentially increase grain yield by improving grain number and grain-filling ability. Stay-green hybrids have been shown to produce significantly greater total biomass after anthesis, retain greater stem carbohydrate reserves, maintain greater grain growth rates, and have significantly greater yields under terminal DS than related but senescent hybrids (Borrell et al., 2000b).

The hybrids obtained through selective fertilization produced a significantly higher mean grain yield compared to hybrids obtained through nonselective fertilization (PEG at 36% was applied to stigma and stylar tissue 1 h before pollination for pollen selection; Ravikumar et al., 2003). Pollen selection for osmotic stress tolerance in sorghum results in superior progeny in moisture stress environments, and Ravikumar et al. (2003) demonstrates the transmission of the selected trait from pollen generation to progeny.

A total of 20 landraces from India, Mali, and Sudan were evaluated in hydroponic conditions in a growth chamber where a water stress treatment was imposed by adding PEG-8000 to the nutrient solution, giving a solute WP of −0.5 MPa, compared with a control solution at 003 MPa. Drought resistance, in terms of relatively less growth inhibition under stress, was higher in races from dry regions than in races from humid regions. Of all the physiological variables measured [carbon exchange rate (CER), transpiration, transpiration ratio (CER/transpiration), leaf diffusive resistance, leaf WP, and OA], only OA under stress was generally correlated with average rainfall at each race's origin, indicating greater OA in landraces from drier regions. Races with a greater capacity for OA were characterized by smaller plants with high rates of transpiration and low rates of leaf senescence under stress. The CER per unit LA increased as live LA decreased under stress due to leaf senescence. Thus, drought-resistant races under stress tended to have lower CER per unit of live LA (but not per plant) than susceptible races (Blum and Sullivan, 1986).

The SPAD chlorophyll meter reading (SCMR) was significantly correlated with total dry matter and grain yield, but not associated with the number of days to flowering. The highly significant relationship between DSI and percent change in SCMR under water deficit conditions indicated that higher chlorophyll concentration is vital for adaptation to water deficit conditions during the postflowering period (Talwar et al., 2009a,b).

The genotypes with LAI around 3.0, higher dry matter accumulation at maturity with higher partitioning efficiency, and higher photosynthetic rate at flowering gave a higher grain yield in the receding soil moisture conditions (Channappagoudar et al., 2007).

7.2.1.7 Genetics of Drought Tolerance

The characteristics (namely, stay-green, root volume, LAI, plant height, and HI) in sorghum showed high values for phenotypic and genotypic coefficients of variation (PCV and GCV), and showed higher estimates of heritability and expected genetic gain, indicating the presence of additive gene effect (Vinodhana et al., 2009). The GCV was highest for excised leaf weight loss followed by grain yield, suggesting considerable scope for selection of these traits (Ali et al., 2009a). Dry root weight exhibited the highest GCV among seedling traits, while excise leaf weight loss among the flag leaf–related characters (Ali et al., 2009b).

The sca was more important than gca variance for most of the traits under DS and nonstress, except for LA and HI under nonstress. Dominance was appreciably greater and more affected by environment than additive variance for all cases except for LA and RWC under drought, where additive was larger than dominance. The degree of dominance a was overdominance ($a > 1.0$) in all cases, except for RWC (no dominance, i.e., $a = 0$) under drought. Narrow sense heritability estimates were generally higher under nonstress than under water stress and ranged from 12.5% (1000-grain weight) to 50.5% (LA) under nonstress and from 6.4% (RWC) to 35.4% (LA) under water stress (Al-Naggar et al., 2007).

Blum et al. (1989a,b) studied physiological assessment of genotypic variation for drought resistance in sorghum using 26 sorghum genotypes for the characteristics leaf WP, HI, stomatal frequency, and grain yield and found that biomass and grain yield increased, while HI decreased with longer growth duration. Khizzah et al. (1995) studied the number of genes involved in the expression of high WP and reported that the inheritance was under the control of multiple gene with dominant and recessive effects.

Preponderance of nonadditive gene action under moisture stress conditions was observed for HI (Palanisamy and Subramanian, 1984), grain yield per plant, panicle length, panicle weight, plant height (China and Phul, 1988; Patil, 1990; Gururaja Rao et al., 1994; Salunke, 1995; Poor and Rezai, 1996; Salunke et al., 1996; Desai and Shukla, 1997; Salunke et al., 2001; Umakanth et al., 2002; Rajguru et al., 2004; Murumkar et al., 2005; Jhansi Rani et al., 2007), and 100-grain weight (Gururaja rao et al., 1994; China and Phul, 1988). However, predominance of additive gene action was reported for HI (Swarnalatha and Rana, 1988), grain yield (Dhoble and Kale, 1988; Borole, 2002), stomatal number (Nandawankar, 1990), panicle length, plant height, and CGR (Poor and Rezai, 1996). Importance of both additive and nonadditive gene action in the inheritance of plant height, number of leaves per plant, panicle length, and yield per plant was also reported (Amsalu and Bapat, 1990; Mishra et al., 1992; Patel et al., 1993; Dhole, 2004). Wenzel (1990) revealed the positive correlations between the characterizations related to growth rate (top dry matter and LA) between those related to drought resistance in the study of the inheritance of drought resistance characters in grain sorghum seedling. Top dry matter and LA were significantly negatively correlated with relative moisture loss. The preponderance of additive genetic variation and the potential for improving drought resistance were verified by the high estimates of heritability. Sankarapandian et al. (1993) reported high heritability for the characteristics green LA, stomatal count, number of closed stomata, and dry matter production for all the stages in an association of physiological characters with drought resistance. Sankarpandian and Bangarusamy (1996) revealed the significant interactions between genotype and stress treatment for stomatal count, root length, number of closed stomata, dry matter production, and grain yield per plant.

Drought tolerant in gadambalia, a drought-tolerant sorghum cultivar, is associated with higher water extraction efficiency, fewer nodel roots per plant, fewer metaxylem vessels per nodel root, smaller LA, and well-developed sclerenchyma (Saliah et al., 1999).

Bichkar (2005) reported predominant additive gene action for adaxial stomatal density, whereas nonadditive gene action was found to be important in the inheritance of relative leaf water content, CSI, aboveground dry matter biomass, 100-grain weight, stomatal density and grain yield per plant. The correspondence between per se performance and gca effects in desired direction was not consistent. Appreciable amount of heterobeltiosis and standard heterosis among hybrid was observed for all the traits.

Padhye (2006) reported the positive association of grain yield with transpiration rate and transpiration rate with photosynthesis rate, because photosynthesis rate is the major cause and grain yield is the effect. Since higher transpiration rate itself decides low WUE, negative correlation was observed between grain yield and WUE, though nonsignificant.

Thul (2007) reported that a preponderance of dominant gene effects, along with dominance × dominance nonallelic gene interaction effects, were important in the inheritance of relative leaf water content, CSI, stomatal density on the upper surface, harvest index, total aboveground dry matter, and grain yield per plant. Additive and dominant gene effects were found to be important, along with additive × additive gene interactions in governing stomatal density (lower surface). Complementary gene action played an important role in the inheritance of all the traits imparting to drought resistance. Exploitation of heterosis or development of composite variety would be effective in the improvement of these traits.

In the study of genetics of some physiological parameters related to drought-tolerant in *rabi* sorghum, Kandekar (2008) reported a predominance of additive and dominant gene effects for the traits RWC, CSI, adaxial and abaxial stomatal density, plant height, total aboveground dry matter, HI, and grain yield per plant. The dominance × dominance gene interaction effects were also found to be predominant in the inheritance of traits relative to leaf water content, stomatal count, plant height, total aboveground dry matter, and grain yield per plant, with duplicate epistasis indicating that hetirosis breeding would be effective for improvement of these traits.

Khot (2008) reported a preponderance of additive and dominant gene action in governing the following characteristics: days to 50% flowering, CSI, number of leaves, panicle length, harvest index, dry matter per plant, and grain yield per plant, whereas the magnitude of dominant gene action was higher for most of these traits. The presence of both additive × additive and dominance × dominance gene interactions were predominant for panicle length, HI, dry matter per plant, and CSI. Duplicate epistasis was observed for all the characters in most of the crosses.

7.2.1.8 Breeding Approaches for Drought Tolerance

The development and use of crop cultivars adapted to water-stressed conditions is a long-term solution for improving and stabilizing crop productivity. Sorghum will likely become much more important in arid and semiarid regions of the world as the demand for limited fresh water and global warming trends increase. Drought resistance in sorghum is a complex trait affected by several interacting plant and environmental factors. The GS at which moisture stress occurs is very important in determining the response or reaction of sorghum to water stress. Two distinct drought responses (namely, preflowering and postflowering drought responses) have been described in sorghum and are probably controlled by different genetic mechanisms (Rosenow, 1987). Although somewhat difficult to combine, some sorghum hybrids containing both preflowering and postflowering drought resistance have been developed (Rosenow et al., 1996). Strategies for genetic enhancement of crop plants for drought tolerance have been widely discussed (Reddy et al., 1980). It has been postulated that the genes for yield and tolerance of stresses are different, at least at some of the loci, and therefore drought tolerance can be improved without sacrificing substantial yield.

Four basic approaches to the breeding for drought tolerance and resistance have been proposed. The first is to breed for high yields under optimal conditions; that is, to breed for yield potential and then to assume that this will provide a yield advantage under suboptimal conditions. The second is to breed for maximum yield by empirical selection in the field in the target drought-prone environment. The success of this approach depends entirely on how variable the target environment is. It works well in the Indian postrainy season environment, which is very predictable, but not in the rainy season environment, which is highly unpredictable. The third approach is to incorporate the selected physiological and morphological mechanisms conferring drought tolerance into traditional breeding programs. For example, Blum (1983) has recommended selection in the F_5 and F_6 generations for yield and yield components under optimal production conditions and simultaneous selection of duplicate samples under moisture-stress conditions.

The fourth breeding approach involves identifying a key trait that confers drought tolerance at specific GSs and its introgression into the high-yielding background. This method involved selecting (through pedigree selection) breeding materials for specific traits, such as (i) longer mesocotyl length for emergence under crust, and grain yield under drought-prone and yield-potential areas for early seedling stage drought; (ii) and for grain yield under drought-prone and yield-potential areas alternatively for midseason drought; and (iii) for stay-green and nonlodging and grain yield under drought-prone and yield potential areas alternatively for terminal drought. Crosses were made between high-yielding adapted lines and lines were selected for high yields under drought, one or more drought-related traits, or both. Selections from F_2 onward were made by evaluating the segregating material in alternate generations under specified drought (early, midseason, and terminal stage) and in yield potential environments. The F_5/F_6 pure lines are evaluated for drought yields, potential yields, and specific drought-related traits. Testing for yield under mild stress was adequate as the rankings of genotypes for potential and drought yields were similar, since the drought-tolerant lines selected under mild stress had high yield potential in nonstress environments. These practical investigations are in very good agreement with those of theoretical considerations (Rosielle and Hamblin, 1981), indicating a general increase in mean yield in both stress and stress-free environments if selection is practiced for mean productivity (average yield in stress and stress-free environments). Fischer and Maurer (1978) proposed an empirical DSI to screen the genotypes for drought tolerance under field conditions:

$$\mathrm{DSI} = Y[1 - \overline{Y}/\overline{YP}]/D$$

where D is the drought-stress intensity, Y is the yield in stress conditions, and YP is the yield in irrigated conditions:

$$D = 1 - [\overline{Y}/\overline{YP}]$$

where $\overline{Y}$ is the mean yield of all genotypes under stress conditions and $\overline{YP}$ is the mean yield of all genotypes in irrigated conditions.

The DSI estimates indicate a reduction in yield of a genotype under DS conditions relative to mean reduction in yield of all genotypes under DS. Also, the sum of ratios of the yield of a particular genotype in stress (rainfed) and the ratio of yield in relieved stress (irrigated) provides information on mean relative performance (MRP). The higher the MRP, the better the performance under stress:

$$\text{MRP} = [Y/\overline{Y} + YP/\overline{YP}].$$

This index, however, has a weakness: it identifies both drought escaping and drought-tolerant genotypes as tolerant.

Heterosis breeding: Rao and Khanna (1999) reported the superiority of sorghum hybrids over their parents for LA and dry matter production under both preflowering and postflowering DS. The increased performance of hybrids than their parents is due to greater growth rates and greater total biomass production and higher HI (Gibson and Schertz, 1977), with or without an apparent increase in leaf photosynthetic rates (Sinha and Khanna, 1975) and an average heterosis of 54% (Haussmann et al., 2000). Bhale et al. (1982) found that some sorghum hybrids showed heterosis for proline accumulation (known to confer drought tolerance) under moisture stress. There is evidence suggesting that the wider adaptability of hybrids is due to their relative tolerance to a wide range of abiotic stresses, including soil moisture stress and related factors. As growing conditions become stressed, the yield difference between hybrids and varieties becomes larger (by about 30%), favoring hybrids (House et al., 1997). Further, for increasing the grain yield within the limits of the available water supply, the choice of female parent for hybrid production should be made for both LA and photosynthetic rate, and the selection of pollinators should be made for maximum seed number per panicle (Krieg, 1988). The improvement of per se performance and combining ability of parents for agronomic traits and grain yield under DS should be given strategic importance, considering that parental per se performance and gca in sorghum is strongly correlated with hybrid performance (Bhavsar and Borikar, 2002) Importance of the improvement in parental lines to increase the hybrid performance has been shown by Doggett (1988). He claimed that about half of the yield increase could be ascribed to better parents. Though most cereal breeders acknowledge the benefits of heterosis in providing superior performance of hybrids when grown under stress conditions, they have been reluctant to adopt sorghum hybrids. The belief is that hybrids are adapted to and, therefore, profitable only under high-yielding favorable environments, where modern production practices are employed and production inputs are available (Rosenow, 1987).

Breeding for drought escape: In regions where end-of-season DS is common, such as those on the Indian peninsula in rainy and postrainy season environments, the most effective way to reduce losses due to drought is through developing early maturing genotypes to enable them escape end of season drought (Blum, 1979a; Turner, 1979). Under the terminal drought typically experienced by postrainy season sorghum in India, early maturing improved sorghum cultivars such as CSH 1 (100 days and 4 t/ha), CSH 6 (95 days and 3.2 t/ha), and NK 300 (88 days and 4 t/ha) produced better grain yields than long-duration cultivars such as M 35-1 (105 days 1.9 t/ha) and SPV 86 (108 days and 3 t/ha) (Seetharama et al., 1982a,b). Selection for improved productivity under conditions of moisture stress resulted in a genetic shift to early flowering. Most of these studies have also confirmed the positive association between the long growth duration and yield potential, under normal or stress-free conditions. It is, therefore, evident that while exploiting drought escape as a solution, some of the potential yield must be sacrificed in return for improved stability under moisture stress (Blum, 1988). The reduced yield potential in early genotypes may be compensated to some extent by increasing plant density. Under terminal water stress during the postrainy season, short-duration sorghum genotypes produce equal grain but less dry matter than late cultivars. Moisture stress is known to cause differing degrees of reduction in grain yield depending on the stage of the crop and frequency, duration, and severity of moisture stress. While sandy-soil and shallow-soil sites are best suited for preflowering field evaluation of stress response, heavier and deeper soils are best for evaluating postflowering DS (Rosenow et al., 1997c). Several effective and reliable screening techniques have been developed during the late 1970s and early 1980s at ICRISAT and facilities in Australia, and drought-tolerant sources at different GSs (as mentioned earlier) from the germplasm and breeding lines have been identified. This led to several attempts to breed sorghum for drought tolerance based on either the plant responses at these GSs or physiological response traits conferring drought tolerance.

Breeding for the stay-green trait or terminal drought tolerance: Considerable progress has been made in genetic enhancement of sorghum for drought tolerance at postflowering stage, unlike at other stages. The stay-green trait is now considered an important postflowering drought-tolerance trait (Rosenow et al., 1997c). Genotypes possessing the stay-green trait maintain a greater green LA during grain filling and extend photosynthesis in upper-canopy leaves after physiological grain maturity under postflowering drought compared to their senescent counterparts (Rosenow et al., 1977). Recent studies have shown that stay-green is not only the result of small sink demand, but also higher leaf nitrogen status (Borrell and Hammer, 2000). Genotypes with the stay-green trait take up more nitrogen from the soil, which increases SLN. This delays the onset and reduces the rate of senescence (Ngugi et al., 2010). While low sink demand enables plants to maintain photosynthetic capacity and ultimately results in higher grain yield and lodging resistance (Borrell et al., 2000a), higher leaf nitrogen status retards the decline in protein content of the aging leaves (Humphreys, 1994) and increases transpiration efficiency (Borrell et al., 2000b). The stay-green sorghums accumulate more soluble sugars in stems than do senescent counterparts, both during and after grain filling (McBee et al., 1983). Stay-green genotypes are less susceptible to lodging and more resistant to charcoal rot, and retain higher levels of stem carbohydrates than nonstay-green genotypes (Mahalakshmi and Bidinger, 2002). Also the rate rather than the onset of leaf senescence was the most important component of stay-green (Harris et al., 2007).

Inheritance of stay-green: While Walulu et al. (1994) have concluded that the stay-green trait in B 35 is influenced by a major gene that exhibits varied levels of dominant gene action depending on the environment in which evaluations are made, its inheritance in Q 141, which is derived from B 35, appeared to be multigenic (Henzell et al., 1992). Greater green leaf–area duration during grain filling appears to be a product of different combinations of three distinct factors: green LA at flowering, time of onset of senescence, and subsequent rate of senescence (van Oosterom et al., 1996; Borrell et al., 2000a). All three factors appear to be inherited independently (van Oosterom et al., 1996), and thus sources expressing these components can be combined easily in breeding programs (Borrell et al., 2000a). This is supported by the identification of multiple QTLs affecting the stay-green trait.

The drought tolerance effect seen with sorghum can be attributed to certain physical and physiological characteristics. These include the following (Du Plessis, 2008):

- Efficient absorption of water through the well-developed and branched root system and the limited transpiration through the small LAs.
- The ability of the leaves to fold up efficiently during hot and dry conditions.
- Thin waxy layer that covers the epidermis of the leaf and protects the plant from desiccation.
- Stomata close quickly to limit the loss of water.
- The ability of sorghum to enter a dormant stage when conditions become unfavorable and to resume growth when the situation changes.

7.2.1.8.1 LEAF SENESCENCE

Leaf senescence, the terminal phase of leaf development leading to death, is the result of aging, water deficit, nutrient deficiency, shading, insects, disease, extreme temperatures, hormones, carbohydrate deficiency, or physical damage (Tollenaar and Daynard, 1978; Guo et al., 2004). During leaf senescence, chlorophyll, proteins, lipids, and nucleic acids are degraded and nutrients are removed and recycled for growth elsewhere. The chloroplast is one of the first organs targeted for breakdown, while the nucleus and mitochondria are degraded last (Quirino et al., 2000). Besides the deterioration of existing proteins and the downregulation of gene expression, recent studies show that the onset of leaf senescence involves de novo synthesis of proteins and the expression of a complex array of genes whose products are involved in senescence-related changes (Gepstein, 2004). The main functional categories of senescence-related expressed sequence tags (ESTs) are macromolecule degradation, nutrient recycling, defense and cell rescue, transcriptional regulation, and signal transduction (Guo et al., 2004).

Leaf senescence is regulated during development and modulated by hormones, metabolism, and stress. Many pathways are involved in senescence. The hormonal control of senescence involves ethylene, jasmonic acid (JA), ABA, cytokinin, BRs, and gibberellic acid (GA) (Woo et al., 2001). Cytokinin can delay leaf senescence, and a decrease in the level of cytokinin can lead to premature senescence (Masferrer et al., 2002). Treatment of senescing leaves with cytokinin reactivates protein synthesis and increases chlorophyll and photosynthetic activity (Thomas and Howarth, 2000). Extracellular invertase is an essential component of cytokinin-mediated delay in leaf senescence (Lara et al., 2004). BRs enhance leaf senescence (Yin et al., 2002) while gibberellins, a group of tetracyclic diterpenes, affect stem elongation, reproduction, and delay leaf and fruit senescence. The effects of sugars on leaf senescence are controversial, and C:N ratios and ABA are more likely to affect senescence than absolute sugar levels (Gibson, 2005). During leaf senescence, lipids are degraded and converted to phloem-mobile sucrose for transport out of the leaf. Antisense suppression of SAG101, a gene that encodes an acyl hydrolase that releases oleic acids from triolein, retards leaf senescence and its overexpression promotes leaf senescence (Yoshida, 2003). The loss of enzymes involved in degrading lipids delays leaf senescence (Fan et al., 1997).

7.2.1.8.2 STAY-GREEN

Stay-green is used to describe genotypes that have delayed leaf senescence as compared to a reference genotype (Thomas and Howarth, 2000). The benefit of delayed senescence is that stay-green plants can assimilate more carbon and nitrogen than senescent plants because they retain more photosynthetically active leaves for a longer time. A stay-green phenotype can occur in three ways (Thomas and Smart, 1993). Type A stay-green phenotypes have a delayed onset and a normal rate of senescence following its onset. Type B stay-green phenotypes initiate leaf senescence normally, but the rate of senescence is comparatively slower. Type C stay-green phenotypes retain chlorophyll despite the normal onset and progression through senescence. Type C stay-green phenotypes have mutations in the chlorophyll degradation pathway and are thus considered cosmetic stay-greens. Many crop plants express stay-green phenotypes.

7.2.1.8.3 SORGHUM STAY-GREEN

Stay-green is an important postflowering drought-resistance trait (Rosenow et al., 1997). Genotypes possessing the stay-green trait maintain a greater green LA under postflowering drought than their senescent counterparts (Rosenow et al., 1977). Contrary to earlier reports that the leaves can remain green due to a lack of assimilate demand because the plants have small panicles under postflowering drought (Henzell and Gillerion, 1973; Rosenow et al., 1983). Recent studies have shown that they stay-green not only because of small sink demand, but also due to higher leaf nitrogen status (Borrell and Douglas, 1997; Borrell, et al., 1999; Borrell and Hammer, 2000) and transpiration efficiency (Borrell et al., 2000b), resulting in maintenance of photosynthetic capacity and ultimately, higher grain yield and lodging resistance (Borrell et al., 2000a). Greater green leaf-area duration during grain filling appears to be a product of different combinations of three distinct factors: green LA at flowering, time of onset of senescence, and subsequent rate of senescence (van Oosterom et al., 1996; Borrell et al., 2000a). Further, all three factors appear to be inherited independently (van Oosterom et al., 1996), and thus sources expressing these components can be combined easily in breeding programs (Borrell et al., 2000a). Stay-green hybrids have been shown to produce significantly greater total biomass after anthesis, retain greater stem carbohydrate reserves, maintain greater grain growth rates, and have significantly greater yields under terminal DS than related but senescent hybrids (Henzell et al., 1997; Borrell et al., 1999; Borrell et al., 2000b). Excellent sources of resistance to both preflowering and postflowering drought have been identified. High levels of both types of resistance are generally not found in the same genotype. However, some genotypes possess good levels of resistance to both types (Rosenow et al., 1997). Stay-green lines in sorghum show increased resistance to disease and insects, reduced lodging, and greater biomass production (Rosenow and Clark, 1981; Rosenow, 1983). Furthermore, stay-green genotypes do not show reduced yield under fully irrigated conditions; thus, stay-green genotypes can be grown on both irrigated and nonirrigated land (Borrell et al., 2000a).

Several physiological traits such as heat tolerance, desiccation tolerance, OAs, rooting depth, and epicular wax (Downes, 1972; Sullivan, 1972; Sullivan and Ross, 1979; Turner, 1979; Jordan and Monk, 1980; Kramer, 1980; Jordan and Sullivan, 1982; Peacock and Sivakumar, 1987; Levitt, 1992; Krieg, 1993; Ludlow, 1993) are known to improve drought resistance; based on this fact, screening techniques for drought resistance have been reported (Christiansen and Lewis, 1982; Garrity et al., 1982; Seetharama et al., 1982a,b; Blum, 1983, 1987; Jordan et al., 1983; Ejeta, 1987). However, little if any progress using specific physiological traits has been documented, partly because interaction of various physiological mechanisms involved in drought tolerance is still poorly understood (Bohnert et al., 1995). It appears that individual physiological traits identified to date are not sufficiently related to overall drought response or field performance to merit selection based on them (Rosenow et al., 1997).

Specifically, stay-green has been associated with reduced lodging (Mughogho and Pande, 1984), lower susceptibility to charcoal rot (Mughogho and Pande, 1984), higher levels of stem carbohydrates both during and after grain filling (Mc Bee, 1984), and improved grain filling and grain yield under stress (Rosenow and Clark, 1981). During postanthesis drought, genotypes possessing the stay-green trait maintain more phenotypically active leaves than genotypes not possessing this trait (Rosenow et al., 1983; McBee, 1984). It also improves the quality and quantity of the stover for cattle feed (McBee et al., 1983). Thomas and Smart (1993) reported that two different types of functional stay-green can be distinguished involving either a delayed onset of leaf senescence or slower rate of senescence. Stay-green has been described as a reduced progressive senescence in sorghum (McBee, 1984), resulting in greater functional LA during grain filling and in an extension of the photosynthetic capability in the upper-canopy leaves after physiological grain maturity. Sorghum thus reduces the need for translocation of stored assimilates from the stem during grain filling and extends the period of active assimilation past maturity. As a result, nonsenescent sorghum accumulates more soluble sugars in the stem than does senescent sorghum, both during and after grain filling (Duncan et al., 1981; McBee et al., 1983). The higher concentration of stem sugars improves the digestible energy content of the stover, or, if translated into growth of axial branches (Vietor et al., 1989), increases the market price of the stover as fodder in some areas. Stay-green is thus a particularly valuable trait in dual-purpose (grain plus fodder) sorghum in semiarid environments (van Oosterom et al., 1996). Because of these benefits, selection for enhanced stay-green has been an important component of breeding for improved drought tolerance and improved grain yield in breeding programs (Henzell et al., 1992; Rosenow et al., 1983) for many years. Phenotypic variations for tolerance to water stress during preflowering stages have been observed in sorghum (Rosenow and Clark, 1981). Evaluation of 72 nonsenescent (stay-green) genotypes of sorghum at ICRSAT identified several tropically adapted lines (e.g., IS 22380, QL 27, QL 10, E 36-1 • R 16 8/1) with stay-green expression equivalent to those of the best temperate lines, B35 and KS 19.

Genetic studies of stay-green have generally indicated a complex pattern of inheritance (Rosenow, 1983). Stay-green is a nuclear gene-controlled quantitative trait (Walulu et al., 1994). Conventional breeding for stay-green has been based primarily on two sources of this trait: B35 and KS 19 (Mahalakshmi and Bidinger, 2002). In the B35 genetic background, the stay-green trait appears to be dominant in F1 hybrids, whereas in others (e.g., R9188), it appears to be recessive (Rosenow, 1984; Rosenow et al., 1988). Further, Walulu et al. (1994) reported that the stay-green traits in B35 are influenced by a major gene that exhibits varied levels of dominant gene action depending upon the environment in which evaluations are made. Tenkouano et al. (1993) have proposed that inheritance of stay-green for B35 genetic background may be less complex than previously thought. B35 is widely used in both public- and private-sector breeding programs. The inheritance of the onset of senescence was additive, but a slow senescence rate was dominant over a fast rate (van Oosterom et al., 1996). But nonsenescence in sorghum, measured as green LA retention, was reported to be regulated by both dominant and recessive epistatic interactions. Under drought conditions, stay-green is controlled by dominant action of major genes (Walulu et al., 1994). Different sorghum research groups across the world have identified QTLs for stay-green. Sanchez et al. (2002), in their studies on QTL analysis for drought resistance in sorghum by using recombinant inbred lines (RILs) and near-isogenic lines developed from B35 × Tx7000, identified four major genomic regions associated with the stay-green trait. An integrated interactive sorghum linkage map with approximate positions of stay-green QTLs based on all available published information has been developed (Hash et al., 2003).

7.2.1.8.4 CHLOROPHYLL CONTENT

The relationship between the stay-green rating and total leaf chlorophyll content was studied by Xu et al. (2000) in 98 F7 RILs of cross B35 and Tx7000 evaluated in replicated field trials under limited- (postflowering stress) and full-irrigation (nonstress) conditions. After scoring the stay-green trait of stressed plants, the total leaf chlorophyll content was measured with a chlorophyll meter (SPAD values) and a spectrophotometer method. The SPAD value had a significant linear relationship with total leaf chlorophyll ($R^2 = 0.91$) and with visual stay-green rating (with $R^2 = 0.82$). The results indicate that visual stay-green ratings were a reliable indicator of leaf senescence and should be useful to sorghum breeders in evaluating progeny when breeding for drought tolerance. Genotypic differences in delayed onset and reduced rate of leaf senescence were explained by differences in SLN and N uptake during grain filling. Leaf nitrogen concentration at anthesis was correlated with onset and rate of leaf senescence under terminal water deficit. SLN is positively correlated with photosynthetic capacity in maize (Sinclair and Horie, 1989) and sorghum (Muchow and Sinclair, 1994). The effects of the components of the SLN, LNC and SLW on grain yield were examined in B35 and KS 19 stay-green genetic backgrounds (Borrell and Hammer, 2000). The results suggested that leaf nitrogen status was more critical than leaf thickness for yield determination under postanthesis drought, since greater LNC was associated with increased retention of green LA for both sources of stay-green, yet SLW was associated with stay-green only for B35 sources. The strong association between LNC at anthesis and grain yield under drought suggests that measuring LNC at flowering could be used to screen for drought in sorghum-breeding programs. Chapman and Barreto (1997) have shown that a SPAD chlorophyll meter can be used to estimate LNC in maize, and preliminary studies in sorghum have found good correlation with SLN. In addition, the chlorophyll meter could be used to rate stay-green in breeding lines during the latter half of the grain-filling period instead of the visual approach that is currently used. The total chlorophyll content at 75 DAS was positively and significantly correlated with proline content, LA ratio at 75 DAS, CGR at 60–75 DAS, and grain yield per plant under drought conditions, and hence, it needs to be considered an important parameter for increasing grain yield under drought conditions (Ghorade et al., 2014).

If stay-green is a consequence of the balance between N supply and demand, then a hypothesis was proposed that genomic regions associated with leaf N status (e.g., SLN, LNC, and leaf chlorophyll) will map to one or more of the stay-green QTLs that have already been identified. To test this hypothesis, leaf chlorophyll at anthesis is currently being estimated with a SPAD meter in a set of RILs varying in the B35 source of stay-green (Borrell and Hammer, 2000). In Subudhi et al. (2000), examination of the stay-green QTL profiles of the best and poorest stay-green lines indicated that three stay-green QTLs, *Stg1*, *Stg2*, and *Stg3*, appear to be important for the expression of this trait when the percent phenotypic variation and the consistency in different backgrounds and different environments are considered. A significant epistatic interaction involving *Stg2* and a region on linkage group C was also identified for stay-green and chlorophyll content. It was concluded that Stg2 is the most important QTL controlling stay-green, which explains the maximum amount of phenotypic variation. Awala and Wilson (2005) studied the expression and segregation of stay-green in pearl millet compared to the chlorophyll content of a putative stay-green and normal senescent line. Dominance or overdominance was noticed for the expression of relative chlorophyll content.

Under DS, a reduction in the leaf gas exchange was reported by Kholova et al. (2009), and this might have led to lower biomass accumulation and grain yield (Talwar et al., 2010). A severe DS during the postflowering stage led to loss of chlorophyll, cell electrolyte leakage, flag leaf yellowing, and grain prematuration (Beltrano and Ronco, 2008; Talwar et al., 2009a,b). At the Directorate of Sorghum Research, in Andhra Pradash, India, sorghum genotypes like PEC 17, EP 87, EP 57, SLB 9, SLR 25, and RSLG 262 have been identified as new sources with improved postflowering drought tolerance based on the yield components and green LA retention at physiological maturity (Talwar et al., 2010). A strong relationship has been found between DSI and percent change in SCMR under water deficit conditions, indicating that higher chlorophyll concentration is vital for adaptation to water deficit conditions during the postflowering growth period. Significant and positive relationships of SCMR with total dry mater ($R^2 = 0.67$) and grain yield ($R^2 = 0.42$) suggested that selection using SCMR will have 42–67% probability of selecting genotypes with higher total dry mater and grain yield. Borrell et al. (2000) found that sorghum genotypes with delayed or nonsenescent traits continue to fill their grain normally, even under limited water or moisture stress conditions during the postflowering growth period. Delayed senescence and yield components are positively associated in water-limited environments, as reported in a range of studies conducted in India (Borrell et al., 2001; Reddy et al., 2007a,b) and abroad (Borrell et al., 2000a). Delaying of leaf senescence enhances both RUE and transpiration use efficiency, resulting in higher yield (Richard, 2006).

7.2.1.9 Difficulties Associated with Yield-Based Selection

Drought resistance is complex if measured in terms of crop yield because of the many possible physiological, morphological, and biochemical factors related to drought resistance (Ceccarelli et al., 1991). Genetic gains made in one season may be lost in subsequent seasons due to the variability in the time, duration, and intensity of moisture deficit across the years. Thus, screening and selection based entirely on yield or yield-derived indices may be of limited value for genetic improvement in drought environments (although this criterion is generally used), as the genotypic variance component is low compared with environmental and G × E variance under these conditions (Johnson and Geadelmann, 1989). Yield-based indices may result in the development of genetic materials that have adapted to the selection site but play a limited role in developing genetic stocks or varieties suitable for other iso-drought locations. However, the probability of success in transferring a multigene adaptation, such as drought resistance, by breeding could be improved by selecting for important components of the resistance mechanisms involved rather than selecting at a functional phenotypic level (Jordan and Miller, 1980).

Donald (1968) put forth a concept wherein an ideotype is defined by a set of morphophysiological attributes that are thought to improve the genetic yield potential for a given crop species across a number of production environments. This could be termed the "static ideotype" approach (Lawn and Imrie, 1991). However, this approach denies the breeder the opportunity to exploit specific adaptations (Lawn and Imrie, 1991). Given the range and pattern of moisture stress that can occur in drought-prone environments, a wide spectra of biological models or desired physiological ideotypes need to be developed to suit the specific requirements of various target production environments. This could be termed *dynamic crop plant ideotypes*, wherein it is implicitly recognized that an optimal combination of morphological, phenological, and physiological traits may differ from one production environment to another (Lawn and Imrie, 1991) and would be continuously subjected to validation during the course of a breeding program. Thus, breeders are expected to tailor crop plants for adaptation to specific environments where moisture deficits are well characterized. Such an analytical approach, however, requires considerable knowledge of the physiological and genetic bases for drought resistance. The subsequent sections of this chapter focus on implementing this approach. Compensatory effects and interactions with the environment make it very difficult to assess the value of particular physiological attributes in improving drought resistance (Marshall, 1991). Also, little information is available on how different physiological attributes (mechanisms) interact in determining a given level of drought resistance (Jones, 1980). The effectiveness of a particular trait will depend on the nature of the DS occurring in a particular area and growing season (Ceccarelli et al., 1991). Each physiological pathway to maximization of yield will be effective only when the genotype, environment, and the consequent physiological pathway are correctly matched.

A common approach for assessing the value of traits is by means of isogenic lines or populations (genotypes with similar genetic backgrounds but contrasting in the expression of a particular trait) (Grumet et al., 1987). This approach takes a lot of time, wherein the traits may be expressed differently in different genetic backgrounds. Also, pleiotropic effects could obscure the expression of the physiological trait in different genetic backgrounds.

Different combinations of a given set of traits may result in a similar level of drought resistance (Ceccarelli et al., 1991). One way to overcome these problems is through divergent selection for different traits related to drought resistance (Acevedo and Ceccarelli, 1989). This has a number of advantages over the isogenic line approach: (i) it offers the possibility of assessing the role of individual traits as well as a combination of traits in randomly assorted genetic backgrounds; (ii) it generates information on realized heritability of traits under contrasting environments; and (iii) it allows a comparison of the selection efficiency between a yield-based versus trait-based approach.

Simulation modeling is another possible approach to assess the value of a trait (Muchow and Carberry, 1993). Simulations are performed with the trait present or absent to varying degrees, with all other factors held constant (Hunt, 1993). Also, simulation models play an important role in suggesting hypotheses on which traits are worth pursuing and validating (Ceccarelli et al., 1991). However, the usefulness of simulation modeling depends on the availability of sufficiently robust models for the particular crop and sufficient understanding of the trait and its mode of operation. The simulation crop model in the Agricultural Production Systems sIMulator (APSIM; Hammer et al., 2010) is one of the few in the world that has been developed with the specific intent to model the physiology and genetics of complex adaptive traits in field crops.

7.2.1.10 Conceptual Framework for a Physiological Genetic Approach

A clear conceptual framework is necessary for integrating the analytical approach of crop physiology into the pragmatic approach of breeding a crop for moisture-limiting environments. We believe that the conceptual models that could be used for the rainy season and postrainy season growing environments are different. In the first case, the crop growth model proposed by Duncan et al. (1978) would seem more appropriate; in the second case, the Passioura (1977) model could be used. Although both models can explain and predict yield accurately, the choice of the model for a given environment is determined by the ease with which components of the model can be measured. The crop growth model of (Duncan et al., 1978) describes yield as

$$Y = C \times D_r \times p$$

where Y is the yield of pod or seeds (kg/ha), C is the mean CGR (kg/ha/day), D_r is the duration of reproductive growth (days), and p is the mean fraction of CGR partitioned to yield (Y). This can be derived by dividing the mean rate of yield accumulation (Y/D_r) by mean CGR (C).

Passioura (1977) considers that, under moisture-limited environments, grain yield is determined by the following relationship:

$$Y = T \times \mathrm{TE} \times \mathrm{HI}$$

where Y is the yield (kg/ha), T is the amount of water transpired (mm/ha), and TE is the transpiration efficiency (kg/ha/mm).

Each subcomponent in both models is an integrated function of a number of morphological and physiological attributes (Hardwick, 1988). However, a more direct physiological approach is relevant to genetic improvement considerations. The main ways to achieve this are the "black box" and "physiological ideotype" approaches, described next. The black box approach leads to the identification of potential traits and genetic stocks that would form the building blocks for developing a physiological ideotype that acts as a conceptual framework for the genetic improvement program.

1. **Black box approach:** The black box approach proceeds from established phenotypic differences (i.e., from differences in drought resistance) to the underlying morphological, physiological, and biochemical mechanisms. Genotypes are evaluated in the target environment to establish genetic differences in drought resistance based on yield or yield-derived selection indices. Multivariate statistics can be employed to identify potentially useful combinations of traits that correlate with yield or yield-derived indices (White, 1988). Once a source of drought resistance is identified in the cultivated species or its wild relatives, the next step is to understand the mechanisms underlying drought resistance. Simple and effective means of screening segregating populations for specific physiological or morphological attributes must be developed for a successful breeding program. Different adapted landraces or cultivars have evolved a variety of mechanisms that contribute to yielding ability under a given pattern of moisture availability. Genotypes attain a level of drought resistance through their own combinations of various physiological attributes. For instance, one genotype may attain a given level of drought resistance through its deep root system, whereas another may attain the same level of drought resistance through its higher TE. However, some genotypes may have both attributes. Although different genotypes may show the same level of drought resistance, they can attain this through different physiological mechanisms or attributes (Ceccarelli et al., 1991). These can be identified using the black box approach within the conceptual framework, and this information will assist in identifying potential parents having complementary physiological attributes and help to direct selection in the segregating materials toward these specific physiological attributes.
2. **Physiological ideotype and pyramiding approach:** A physiological ideotype for drought resistance could be defined in terms of specific physiological traits expected to contribute functionally to optimize yield production and stability under moisture-deficit environments. The various steps involved in this kind of approach are as follows:
   1. Define the various physiological traits that have functional significance for productivity in a given crop in the target environment.
   2. Establish the genetic variability and locate sources of high levels of efficiency for each physiological trait through establishing physiological nurseries for screening specific traits similar to disease-screening nurseries.
   3. Establish the genetic basis for each physiological trait under consideration by studying its inheritance and estimating heritability to determine the feasibility of using that particular trait in a breeding program.
   4. Develop genetic markers (RAPD, RFLP markers), if easily identifiable morphological, physiological, or other markers are not readily available for each physiological trait, as this can increase the efficiency of selection from segregating materials.
   5. Identify genotypes for each physiological trait that have good combining ability.
   6. Incorporate relevant traits into agronomically acceptable backgrounds.

However, the physiological-genetic approach should be seen as complementary to the empherical approach, in which the selection is based on yield or yield-derived indices. The difficulties associated with the trait-based approach have been comprehensively highlighted by Lawn and Imrie (1991). However, identification and characterization of genetic stocks have key metabolic efficiency traits and, by introducing these traits into suitably adapted genetic backgrounds, could have a major impact on crop production, and thus crop adaptation to drought-prone environments.

7.2.1.11 Molecular Basis for Stay-Green

The direct selection of drought-tolerance components using conventional approaches has been slow and difficult (Ejeta et al., 2000). Controlled by many genes and dependent on the timing and severity of moisture stress, drought tolerance is one of the more difficult traits to study and manage genetically. Hence, the use of molecular markers and QTL analysis based on carefully managed replicated tests has the potential to alleviate the problems associated with inconsistent and unpredictable onset of moisture stress or the confounding effect of other stresses, such as heat (Ejeta et al., 2000).

The QTL studies (Tuinsta et al., 1996, 1997a, 1998; Crasta et al., 1999; Xu et al., 2000; Ejeta et al., 2000; Kebede et al., 2001) identified several genomic regions of sorghum associated with preflowering and postflowering drought tolerance. The molecular genetic analysis of QTLs influencing stay-green, an important postflowering drought-resistance trait (Tao et al., 2000; Xu et al., 2000; Hausmann et al., 2002), resulted in the identification of up to four QTLs. Subudhi et al. (2000) confirmed all four QTLs (Stg1, -2, -3, -4) that were identified earlier by Xu et al. (2000) by evaluating RIL populations derived from B35 × Tx 700 in two locations for 2 years. Similarly, comparisons of stay-green QTL locations with earlier reports indicated that three of the four stay-green QTLs showed consistency across different genetic backgrounds and environments. They concluded that the Stg2 QTL region is expected to increase our understanding of stay-green, leading to either marker-assisted introgression of this QTL into elite agronomic background or map-based cloning to genetically engineer this locus into improved cultivars.

BTx642 (formerly called *B35*) is a stay-green genotype with a delay in the onset of senescence under water-limited conditions. BTx642 is derived from a cross of BTx406 and IS12555 (Rosenow et al., 1983), a *durra* sorghum from Ethiopia. BTx406 was used as a source of recessive Ma1 and *dw2;* therefore, BTx642 is early-flowering and short in stature, similar to most grain sorghums grown in Texas (Lin et al., 1995). BTx642 is used in many breeding programs in the United States and Australia. Four major stay-green QTL located on three linkage groups were identified using a RI population developed from BTx642 (a stay-green line)/RTx7000 (a senescent line) (Xu et al., 2000). Two QTLs, Stg1 and Stg2, located on LG-03 explain 20% and 30% of the phenotypic variability, respectively (Sanchez et al., 2002). Stay-green QTL Stg3 and Stg4 are, respectively, on LG-02 and LG-05 and account for approximately 16% and 10% of the phenotypic variance (Sanchez et al., 2002). Thus, the ranking of stay-green QTL based on their relative contributions to the stay-green phenotype expressed in this population is Stg2 > Stg1 > Stg3 > Stg4 (Xu et al., 2000).

Seven populations have been utilized to map QTL for stay-green in sorghum. These mapping populations utilized BTx642, QL41, SC56, and E36-1 as sources of

TABLE 7.3 Stay-Green QTL Mapping Studies in Sorghum

Author	Parents	Contain Stg1–4?
Crasta et al. (1999)	BTx642/RTx430	Stg2, Stg4
Tuinistra et al. (1997)	BTx642/Tx7078	Stg1, Stg3
Tao et al. (2000)	QL41/QL39	Stg3
Kebede et al. (2001)	SC56/RTx7000	Stg2, Stg4
Subudhi et al. (2000)	BTx642/RTx7000	Stg1, Stg2, Stg3, Stg4
Xu et al. (2000)	BTx642/RTx7000	Stg1, Stg2, Stg3, Stg4
Hausmann et al. (2002)	E36-1/IS9830	Stg1
Hausmann et al. (2002)	E36-1/N13	Stg2, Stg3

stay-green and RTx430, Tx7078, QL39, RTx7000, IS9830, and N13 as senescent parents (Table 7.3).

In the mapping population of BTx642/RTx430, three major stay-green QTLs (SGA, SGD, and SGG) contributed 42% of the phenotypic variability and four minor QTLs (SGB, SGI.1, SGI.2, and SGJ) contributed an additional 25% of phenotypic variability (Crasta et al., 1999). Only two QTLs are shared between the BTx642/RTx7000 and BTx642/RTx430 QTL studies. SGA and SGJ from the BTx642/RTx430 QTL analysis correspond to Stg2 and Stg4 from the BTx642/RTx7000 QTL analysis. In the BTx642/Tx7078 F5 population, six QTLs for stay-green were identified that accounted for 53% of the variability in stay-green (Tunistra et al., 1997).

Two of these QTLs on LgB and LgG of the BTx642/Tx7078 map correspond to Stg1 and Stg3 of the BTx642/RTx7000 QTL analysis. The RI lines from BTx642/RTx7000 were phenotyped for stay-green in an additional population beyond the Xu et al. (2000) study, and Stg1, Stg2, Stg3, and Stg4 were identified again, with the most important QTL being Stg2 (Subudhi et al., 2000).

These four QTLs explained 53.5% of the phenotypic variation, and the QTLs for chlorophyll content coincided with Stg1, Stg2, and Stg3 (Subudhi et al., 2000). Stg2 and Stg3 interacted together most favorably, explaining 49.8% of the phenotypic variation, and although Stg1 alone accounts significantly for phenotypic variation, it does not act favorably in combination with other stay-green QTLs (Subudhi et al., 2000). QTL mapping using RILs from the cross QL39/QL41, where QL41 is a stay-green line derived from the cross QL33/BTx642 and QL39 is senescent, identified three QTL associated with stay-green (Tao et al., 2000). Despite the small difference in the stay-green phenotype between QL39 and QL41, a stay-green locus was identified that corresponds to Stg3 identified in the Xu et al. (2000) mapping population (Tao et al., 2000).

QTL analysis of F7 RILs from SC56/RTx7000, where SC56 is a *Caudatum nigricans* from Sudan with the stay-green phenotype and RTx7000 is a senescent parent, identified nine QTLs for stay-green (Kebede et al., 2001). Two QTLs from their mapping study, StgA and StgJ, correspond to Stg2 and Stg4 of the Xu et al. (2000) study.

Two populations were made using E36-1, a Guinea-Caudatum hybrid race and stay-green line from Ethiopia, crossed with IS9830, a tall Sudanese Feterita, or N13, a *durra* sorghum from India (Hausmann et al., 2002). Three QTLs for stay-green on Lg A, E, and G were common to both E36-1/IS9830 and E36-1/N13 (Hausmann et al., 2002). The E36-1/IS9830 population have a QTL for stay-green that corresponds to Stg1 of the Xu et al. (2000) mapping population, while the E36-1/N13 QTL analysis found QTLs corresponding to Stg2 and Stg3 of the Xu et al. (2000) mapping population. The dominance of the stay-green trait depends on the origin of stay-green and the genotypes involved in crosses (Walulu et al., 1994). Stay-green in BTx642 shows partial dominance when crossed with RTx7000 (Walulu et al., 1994). When BTx642, RTx7000, and their resultant progeny (F1 or F2) were phenotyped using leaf and plant death scores at maturity, the progeny had better postflowering drought resistance than the parent RTx7000, but less postflowering drought resistance than the other parent BTx642. On a scale of 1–10 (where a score of 1 is 0–10% leaf death and a score of 10 is 90–100% leaf death), BTx642 had a score of approximately 4.5 and RTx7000 had a score of 10 in a rainout shelter (Walulu et al., 1994). The F1 and selected F2 progeny had a score of 7 (Walulu et al., 1994). Similarly, when BTx642 was crossed with the senescent lines BTx623 and BTx378, partial dominance of stay-green was observed, and the stay-green phenotype was more evident in the F1 of BTx642/BTx623 and BTx642/BTx378 than BTx642/RTx7000 (Tenkouano et al., 1993). SC599-11E, a stay-green line derived from IS17459, a Nigricans-Feterita, was crossed with BTx623, and the F1 generation displayed a reduced stay-green phenotype (Tenkouano et al., 1993). Conversely, when crossed with BTx378, the F1 generation had minimal stay-green phenotype. Interestingly, when BTx642 and SC599-11E (both of which possess stay-green) were crossed, stay-green from BTx642 was dominant to SC599-11E (Tenkouano et al., 1993).

The genetic basis of stay-green has been analyzed using six different populations and three sources of stay-green (Tunistra et al., 1997; Crasta et al., 1999; Tao et al., 2000; Xu et al., 2000; Kebede et al., 2001; Hausmann et al., 2002). These studies identified 39 QTLs for stay-green; however, 16 of the QTLs were observed in only one environment or 1 year, suggesting that these QTLs are highly dependent on a specific environment (Tunistra et al., 1997; Crasta et al., 1999; Xu et al., 2000; Tao et al., 2000; Kebede et al., 2001; Hausmann et al., 2002). The remaining 23 QTLs that modulated expression of stay-green were located in 12 regions of the sorghum genome. The colocalization of stay-green QTL from different studies to the same genomic region could indicate that allelic variation of the same gene (or genes) was mapped in different studies. This would not be surprising because three of the QTL studies on stay-green utilized BTx642 or QL41, a genotype derived from QL41, as a source of favorable alleles for stay-green (Crasta et al., 1999; Tao et al., 2000; Xu et al., 2000). In addition, favorable alleles for stay-green for a region conferring stay-green were derived from more than one genotype. This suggests that sorghum germplasm is likely to have more than one favorable allele for many of the stay-green QTLs. A total of 6 of the 12 regions spanning major stay-green QTLs had favorable alleles for this trait derived from BTx642 or QL41, a line derived from the stay-green genotype BTx642. Haplotype analysis of parental DNA of BTx642 showed that the favorable alleles for stay-green in X-Stg1, X-Stg2, X-Stg4, and T-StgI were derived from IS12555; in X-Stg3, favorable alleles could have come from either IS12555 or BTx406; and in C95 StgD and C-StgI2, favorable alleles for stay-green came from BTx406. The identification of favorable alleles for stay-green in BTx406 (a nonstay-green line) was not unexpected because alleles contributing to a specific trait are often identified in populations even though the parental lines do not exhibit a strong phenotype for the trait (Young, 1996). The two specific QTLs of this type, C-StgD and C-StgI2, were expressed in three out of four environments and mapped with relatively high LOD scores (5.0, 2.9) (Crasta et al., 1999). These QTLs were not mapped in the BTx642/RTx7000 (Xu et al., 2000) population, but this can be explained by a lack of allelic diversity in the regions spanning C-StgD and C-StgI2 in this cross.

Stay-green QTL studies based on BTx642/RTx7000 and BTx642/RTx430 identified two of the same stay-green QTLs and six QTLs that were specific to one of the two populations (Xu et al., 2000; Crasta et al., 1999). The explanation for this observation was determined by analyzing the origin and diversity of DNA in the two parents of each population. This analysis showed that C-StgD and C-StgI2 were probably not mapped in BTx642/RTx7000 due to low allelic diversity among the parental lines in these regions of the genome. Second, haplotype analysis showed that favorable alleles for C-StgI1 and C-StgB were derived from RTx430, explaining why these loci were not mapped in BTx642/RTx7000. On the other hand, X-Stg1 and XStg3 were not mapped in BTx642/RTx430, even though BTx642 was the source of the favorable alleles mapped in these loci. X-Stg1 is from 140 to 165 cM on LG-03, and the BTx642/RTx430 map has a 35-cM gap just after 147 cM on LG-03, where cM values are relative to the TAMU-ARS map (Crasta et al., 1999). Thus, the lack of markers likely prevents X-Stg1 derived from BTx642 from being identified in the BTx642/RTx430 population. In the X-Stg3 area from 103 to 145 cM on LG-02, three markers exist in the BTx642/RTx430 map that is in the X-Stg3 region (namely, csu6.3, umc5, and txs2042). It is possible that RTx430 and BTx642 contain a similar allele for stay-green in both of these loci, or that epistatic interactions in the BTx642/RTx430 population prevent the expression of phenotypes modulated by these alleles. A total of 5 QTLs for stay-green were mapped in a population derived from QL41/QL39 (Tao et al., 2000). In this case, QL41 is a stay-green parent derived from BTx642.

While C-StgB corresponds to X-Stg3 and T-StgG overlaps with C-StgI2, no stay-green QTL corresponding to X-Stg1, X-Stg2, or X-Stg4 were mapped in this population. Haplotype analysis of the QL41 pedigree (BTx642/QL33) showed that QL41 obtained DNA from QL33 (a nonstay-green line) in the regions corresponding to X-Stg2 and XStg4. This explains why these QTLs were not mapped in QL41/QL39. The situation for X-Stg1 is less clear. A portion of the DNA in QL41 spanning this locus was derived from BTx642. However, this region of the QL41/QL39 genetic map published by Tao et al. (2000) was not well represented, which possibly explains why this locus was not mapped in their study.

The stay-green trait contributes significantly to the yield of sorghum hybrids in environments subject to terminal drought (Borrell et al., 2000a). Therefore, sorghum-breeding programs have used BTx642 as a source of stay-green in Texas, Kansas, and Australia. Until recently, selection of the stay-green trait in breeding materials was done by screening progeny for the trait, as well as other agronomic indicators. This type of selection is quite different from a QTL mapping study, where all lines are analyzed for expression of the trait regardless of yield or other characteristics of progeny. Therefore, it was of great interest to analyze stay-green sorghum breeding material derived from BTx642 to see which (if any) regions containing favorable alleles for stay-green were present in the breeding lines. The analysis of 13 breeding lines from Texas, Kansas, and Australia showed that 10 of the 13 lines analyzed contained DNA that spanned all or a portion of X-Stg3. In contrast, none of the breeding lines contained DNA from BTx642 containing favorable alleles for X-Stg2. This result was very surprising because among the four stay-green QTLs mapped in BTx642/RTx7000, X-Stg2 explained the largest proportion of the stay-green phenotype. A possible explanation for lack of the X-Stg2 region from BTx642 in the breeding lines was the presence of an allele for lemon yellow seeds in the same region of the genome. Discussion with the sorghum breeders involved in this study revealed that they selected for red seeds, not realizing that this selected against the favorable allele for X-Stg2. It was also apparent that favorable alleles for stay-green in X-Stg4 were not highly represented in the breeding lines, and in no case were more than two of the four favorable alleles for stay-green present in any one line. This suggests that marker-assisted breeding could improve the selection and pyramiding of favorable alleles for stay-green in future sorghum-breeding lines and hybrids.

7.2.2 Cold Tolerance

The ability of genotypes to survive or perform better under low temperatures than other genotypes is called *cold tolerance*. This feature involves increased chlorophyll accumulation, reduced sensitivity to photosynthesis, improved germination, pollen fertility, and seed set.

Lia et al. (1998) reported plant cold tolerance in rice is associated with anther size, number of pollen grains, and diameter of fertile pollen grains at booting stage. However, Sanghera et al. (2001) reported that cold tolerance in rice is associated with high spikelet fertility (>90%) and good panicle exertion under temperate conditions. Ample genetic variation for cold tolerance in rice is available from germplasm collected from high-altitude and low-temperature areas (Shobha Rani et al., 2002).

Cold snaps cause a reaction in the plant that prevents sugar getting into the pollen. Without sugar, there is no starch buildup that provides energy for pollen germination. Without pollen, pollination cannot occur, so no grain is produced. The Commonwealth Scientific and Industrial Research Organisation (CSIRO) found that all the ingredients for starch are present, but they are not getting into the pollen grain when they are needed. A cell layer surrounding the pollen, called the *tapetum*, is responsible for feeding the pollen with sugar. The tapetum is only active for 1–2 days, so if a cold snap occurs at this time, then there is no further chance for pollen growth. But the sugar can't move freely into the tapetum and pass through it to the pollen. Instead, the sugar has to be broken down and then transported in bits to the pollen. Invertase is the catalyst that helps break down the sugar to transport it to the tapetum before it is transported to the pollen (Oliver et al., 2005). Quantities of invertase are decreased in conventional rice when it is exposed to cold temperatures, but they remain at normal levels in the cold-tolerant variety when it experiences cold. By comparing a cold-tolerant strain of rice with conventional rice, CSIRO found that the gene responsible for invertase looks exactly the same in the cold-tolerant variety as it does in conventional rice. So the invertase gene itself does not make the rice plant cold-tolerant, but instead, a mechanism that regulates the invertase gene is different. Early research indicates that the invertase gene is regulated by the hormone ABA. Oliver et al. (2007) has experimented with injecting plants with ABA—the resulting rice plants were sterile, just as they would be if they experienced a cold snap. Also, ABA levels increase when conventional rice is exposed to cold, but they remain the same in the cold-tolerant variety. Recent studies have indicated that the difference between cold-sensitive and -tolerant rice is due to a different ability to control ABA levels. It has also been shown that this mechanism may require interactions with other plant harmones like auxins.

Cold-temperature stress has been reported to have adverse effects on many other crops as well, including the major world cereals such as rice (Nishiyama, 1995; Basnayake et al., 2003; Shimono et al., 2004, 2007; Ali et al., 2006) and maize (Rymen et al., 2007). When encountered early in the season, the stress can result in reduced germination and emergence, poor seedling growth and reduced vigor in sorghum (Tiryaki and Andrews, 2001; Knoll et al., 2008). Likewise, cold stress has been shown to reduce photosynthetic activity in maize, primarily due to impaired chloroplast function (Allen and Ort, 2001; Gomez et al., 2004), ultimately leading to reduced seedling growth and increased cell death. Perhaps due to its impact on growth rate, cold stress often delays phenological development (flowering and maturity) (Quinby et al., 1973; Zinn et al., 2010) and causes spikelet sterility, flower abortion, and reduction in the number of pollen grains intercepted by the stigma in several crop species, resulting in poor seed set and ultimately low grain yield (Khan et al., 1986; Farrell et al., 2001; Lee, 2001; Gunawardena et al., 2003a,b; Oliver et al., 2005; Thakur et al., 2010). Moreover, tremendous yield losses due to cold-temperature stress on major grain crops have also been reported in subtropical and temperate regions (Thakur et al., 2010). In sorghum, low night temperatures during flowering increase the incidence of ergot disease (Stack, 2000).

Current speculation about global climate change is that most agricultural regions will experience more extreme environmental fluctuations (Solomon et al., 2007). Bringing in tolerance to cold temperatures in *rabi* can bring in more area under cultivation of high-yielding hybrids and also lead to sustainable production in the current cultivated areas.

7.2.2.1 Early-Season Cold Tolerance

Owing to its origin in tropical and subtropical regions of Africa, sorghum is well adapted to warm climates and known for its drought tolerance (Doggett, 1988). Low-temperature-induced inhibition of germination and emergence in the field is a common problem encountered in crops that originate from warm environments and that were bred to grow in cooler temperate conditions (such as the northern United States) specifically during early-season sowing (Lu et al., 2007), like sorghum and rice. Even in maize, which is supposed to be less sensitive to cold temperature, early-season cold stress has been identified as one of the environmental challenges (Sezegen and Carena, 2009). The search for new sources of cold tolerance for seedling emergence and their incorporation into sorghum parental lines has received considerable attention (Yu and Tuinstra, 2001; Cisse and Ejeta, 2003; Franks et al., 2006). A Chinese landrace known as *kaoliang* exhibited higher seedling emergence and improved seedling vigor under cool conditions (Franks et al., 2006). Using Shanqui Red as a parental source for early-season cold tolerance, several QTLs for cold germinability and another for seedling vigor with low to moderate contribution to genetic variation were identified (Knoll and Ejeta, 2008; Knoll et al., 2008). Using PI610727, a Chinese kaoliang identified as a source for cold tolerance (Franks et al., 2006), a total of 14 QTLs associated with germination, emergence, and seedling vigor under cold conditions were detected by Burow et al. (2011). These QTLs were located in five sorghum chromosomes (namely, 1, 2, 4, 7, and 9). Though early-season cold tolerance is not a serious constraint in semiarid tropical areas like India, emergence and establishment of seedlings are affected due to cold stress. However, systematic studies on the effect of early-season cold stress on grain yield are lacking.

7.2.2.2 Midseason Cold Tolerance (Cold Tolerance at Anthesis)

Although sorghum is more tolerant of cool temperatures than other warm-loving cereals, it is still sensitive to temperatures lower than 15°C. These temperatures occur during flowering to seed set in postrainy sorghum-growing areas of India. Earlier studies have shown that cold-tolerant cultivars had optimum temperatures for growth that were 6–10°C less than those of cold-susceptible ones, and they maintained respiration that was 20–25% higher.

Postrainy season sorghum is cultivated in conserved moisture in black soils, and sowing is generally done in favorable moisture conditions, which occur after the first week of September. The sowing dates vary from location to location and year to year, and usually extend until the end of October. The minimum temperature declines from about 20°C at sowing to 12°C at flowering and increases to 18°C during the grain-filling period.

The postrainy season adapted landraces are photoperiod-sensitive, thermoinsensitive, and tolerant of moisture stress, and they produce high biomass and possess bold and lustrous grains, but the productivity of these landraces is low. The rainy season–adapted genotypes (mostly Caudatum or Kafir) are very productive and cold-insensitive. However, they lack the grain quality of postrainy-adapted landraces. Since the productivity of postrainy landraces is low, the high-yielding genotypes, especially hybrids, should be targeted. Many of the improved cultivars, particularly hybrids like CSH 13R, CSH-15R, and CSH-19R, are highly productive but relatively photoperiod-insensitive and temperature-sensitive. Because of their temperature sensitivity, growth is reduced, development is delayed, and seed setting is affected.

Cold temperature at flowering seems to be more detrimental to the yield of sorghum. All yield components, including panicle weight, number of seeds per panicle, grain yield, and seed size were severely affected by midseason cold stress (Prasad et al., 2008).

7.2.2.3 Effect of Midseason Cold Stress on Pollen Production and Viability

Interaction of viable pollen production or pollen shedding with temperature was reported by Stephens and Holland as early as 1954. More specifically, when the minimum temperature goes below 10°C for several days during flowering, hybrids that are otherwise male fertile show male sterility (Reddy et al., 2003). Anthers failed to dehisce and release pollen during days when the minimum temperature was below 13°C (Laxman and Rao, 1995). Osuna et al. (2003) noticed that low temperatures reduced the amount of pollen produced, and possibly also modified stigma receptivity in *rabi* sorghum. The severity or extent of such effects depended on the degree of cold tolerance of the cultivars. Cold-tolerant lines produced more pollen, a higher percentage of fertile pollen, a lower percentage of sterile pollen, and a higher amount of seed set under self- and open-pollination than in cold-susceptible genotypes. It was concluded that in susceptible types, low temperatures reduced the number of pollen

FIGURE 7.9 Panicle with poor seed set.

mother cells and their ability to produce pollen, and that difference in tolerance between genotypes suggested that the character was polygenic (Gonalez et al., 1986). Interestingly, stress tolerance in vegetative and reproductive tissues is not always correlated (Salem et al., 2007). In vitro pollen growth assays have been considered as a possible screen to identify germplasm, with the potential for improved reproductive stress tolerance (Kakani et al., 2002, 2005; Salem et al., 2007; Singh et al., 2008). In cases where pollen tube growth is the weak link in temperature stress tolerance, this simple strategy may be highly rewarding. Furthermore, conventional breeding strategies can take advantage of the strong selection pressures on pollen fitness to select for improved stress tolerance (Hedhly et al., 2008).

7.2.2.4 Effect of Cold Stress on Anthesis and Seed Sets

Low anthesis and markedly reduced anther dehiscence occurs due to cold stress, resulting in low amounts of pollen. Although both the anther and stigma have fully extended, the low temperature may have affected the receptivity of the stigma, germination, and growth of pollen tube or fertilization resulting in reduced seed sets and a lower number of seeds per panicle (Downes and Marshall, 1971). Studies conducted on the sister crop (maize) have shown that cold stress imposed just before or at the start of flowering reduced the number of tassel branches and spikelets as well as seed size (Bechoux et al., 2000). The stress was also reported to cause structural and functional abnormalities in reproductive organs that lead to either failure of fertilization or premature abortion of florets, induced flower abscission, pollen sterility, pollen tube distortion, ovule abortion, and thus reduced seed set, ultimately leading to lower yields (Thakur et al., 2010). Earlier

FIGURE 7.10 Panicle with good seed set.

reports have shown that low night temperatures during flowering causes a significant reduction in spikelet and flower fertility in rice (Pereira da Cruz et al., 2006).

Seed-setting ability in hybrids at low temperatures is critical to the success of postrainy season hybrids. This requires greater attention to ascertain the differences among the hybrid parents for their ability to set seeds, especially in low temperatures. Rao et al. (1986) indicated that, if temperature sensitivity is eliminated both in male and female parents, greater success can be achieved in developing hybrids for postrainy season. Variability among the genotypes for seed-setting behavior and reaction to lower minimum temperatures in sorghum was also observed by Reddy et al. (2003). Though quite a few studies were initiated on cold tolerance in winter sorghum, systematic studies on a diverse range of breeding material and their utilization are lacking (Figures 7.9 and 7.10).

7.2.3 Breeding Photoperiod-Sensitive Sorghums

Climate change is a threat to crop productivity in the most vulnerable regions of the world, especially the tropics and semiarid regions, where higher temperatures and increases in rainfall variability could have substantially negative impacts (Parry et al., 2004). Sorghum is mainly produced by small holder farmers under rainfed conditions that have been predicted to be adversely affected by climate change (Folliard et al., 2004; IPCC, 2007). This will have negative impacts on food security and the livelihoods of people in arid regions and semiarid tropics. The characteristics and basis of an ideal sorghum genotype in such ecologies have been stated by several researchers (AGRHYMET, 1992; Vaksmann et al., 1996; Andrew et al., 2000; Sultan et al., 2005; Traore et al., 2007). One trait that has so far been shown to facilitate adaptation of sorghum to variable lengths of the growing season in the arid and semiarid tropics is photoperiod sensitivity (Craufurd et al., 1999; Craufurd and Qi, 2001; Kouressy et al., 2008). Knowledge of photoperiodism is essential for a crop breeder to breed photoperiod-sensitive or -insensitive genotypes. A low degree of photoperiod sensitivity is desirable to obtain the potential yield by adjusting the optimal photoperiod with natural day length (Shinde et al., 2013). Sorghum is classified as a short-day species by Garner and Allard as early as 1923. This trait, therefore, has been given attention in the development of improved and higher-yielding sorghum cultivars with various degrees of adaptation to different climatic patterns in the arid and semiarid tropics (Clerget et al., 2004; Kouressy et al., 2008). A genotype combining photoperiod sensitivity and stay-green traits was revealed as the most stable (Abdulai et al., 2012).

Sorghum landrace varieties show a very large variation for the duration of their vegetative phase, ranging from 50 to 300 days depending on the sowing date (Miller et al., 1968). The sorghum flowering system can sense the differences in day length. Around 200 years ago, sorghum varieties sensitive to photoperiods were introduced from Africa into an area exhibiting distinctive climatological conditions, such as a bimodal rainfall pattern. There, sorghum readily adapted to local farming practices (Gomez and Chanterau, 1997). Breeding photosensitive sorghum for specific tropical environments offers opportunities to increase productivity, enhance grain quality, and maximize agricultural input utilization. This trait, therefore, has been given attention in the development of improved and higher-yielding sorghum cultivars with various degrees of adaptation to different climatic patterns in the arid and semiarid tropics (Clerget et al., 2004; Kouressy et al., 2008). Most of the postrainy sorghum cultivars grown belong to the *durra* race or other intermediate races involving *durra* that are photosensitive.

Photoperiod response in cereals has been shown to be determined by three main components Major (1980):

1. The basic vegetative phase (BVP), defined as the shortest possible time for floral initiation when the plants are not responsive to changes in photoperiod
2. The photoperiod sensitivity (PS), which expresses the varietal linear response to flowering time as plants respond to day length changes
3. Minimum optimal photoperiod (MoP), defined as the photoperiod threshold beyond which the vegetative period is influenced by increases in day length

Photoperiod sensitivity appears to be a key feature that matches flowering time to length of the rainy season and plays an important role in securing the level and quality of harvests (Vaksmann et al., 1996; Deu et al., 2000). Therefore, better understanding of the genetic bases will facilitate the transfer of photocopied sensitivity from local to high-yielding varieties developed for African farmers.

Quinby (1974) described the genetic response of sorghum to daylength for flower initiation. A series of six maturity genes has been recognized to alter flowering time in sorghum: Ma1, Ma2, Ma3, Ma4, Ma5 and Ma6 (Quinby, 1967). The first four maturity genes cause long days to inhibit flowering but allow early flowering with short days. Of these four genes, mutations of *Ma1* cause the greatest reduction in sensitivity to long days. In 1981, Meckenstock (1991) began his work on Maicillos, the tall, photoperiod-sensitive, low-yielding white sorghum ecotypes of Central America. The studies concluded that Maicillos is highly sensitive to photoperiods due to the presence of dominant alleles at the Ma1 and probably the Ma2 loci. Mutations of Ma2, Ma3 and Ma4 generally have a more modest effect on sensitivity to long days (Quinby, 1967). However, even sorghum with the recessive alleles, flower later with long days than short days (Pao and Morgan, 1986). The genes Ma5 and Ma6 represent a special case because only when they are both present in the dominant form will they strongly inhibit floral initiation regardless of day length (Childs et al., 1997). Ellis et al. (1997) studied the effect of photoperiod, temperature, and asynchrony between thermoperiod and photoperiod on development and panicle initiation in sorghum, and it appeared to be an initial photoperiod-insensitive but temperature-sensitive phase (the juvenile phase), followed by a photoperiod-sensitive but temperature-insensitive phase, and then a final phase that is insensitive to both photoperiod and temperature.

A classic study of the effect of tropical photoperiods on the growth of sorghum was conducted by Miller et al. (1968). They concluded that tropical sorghum has lower critical photoperiods than most US sorghum when planted from January to July in Puerto Rico, but the same varieties nonetheless flower in about the same time when planted under day-length conditions of 12.2–11.3 h (mid-September through mid-November).

Lin et al. (1995) identified one QTL located on the linkage group (LG) B of Peirera et al. (1994) and assigned it to the Ma1 gene. Childs et al. (1997) mapped gene Ma3 on LG C of Pereira et al. (1994). QTLs were detected on linkage groups (LG) C, F, and H, respectively (Trouche et al., 1998). On LG H, position was also congruent with a QTL detected for PSS with a major effect (18% of the phenotypic variance). Paterson et al. (1995) established that several RFLP markers of LG B, located near Ma1, were duplicated on LG H. The genotypes involved in these studies were day-neutral and quantitatively short-day sorghums able to flower in the United States.

However, Trouche et al. (1998) was able to detect one QTL on linkage group H for PS, which explained 18% of total phenotypic variation in more photoperiod-sensitive sorghum. On the same linkage group, they located one QTL for direct measurement of photoperiod response. However, they could not succeed in detecting QTLs for MoP and BVP. Recently, Deu et al. (2000) detected and mapped QTLs for BVP and PS components of photoperiod response. Identification of markers linked to QTLs may help in transferring photoperiod sensitivity to elite cultivars for better adaptation to tropical environments (Deu et al., 2000).

7.3 BREEDING FOR RESISTANCE TO DISEASES

7.3.1 Charcoal Rot

Root and stalk diseases of sorghum are caused by several soil-borne fungi appearing in both *kharif* and *rabi* seasons. Among them, *M. phaseolina* (Tassi.) Goid., causing charcoal rot, is predominant in *rabi* season and is widely distributed, causing major losses in grain yield. It was first reported in India by Uppal (1931) to be the cause of seedling blight and hollow stems of sorghum and was identified on maize at the same time in the United States and can now be found on other plant species as well. *M. phaseolina* has a wide host range that includes sorghum, beans, potatoes, legumes, tomato, cotton, tobacco, and other crops (Tarr, 1962). In India, the disease has been reported as epidemic in winter-sown sorghum, causing heavy crop losses (Nagarajan et al., 1970; Bhagwat, 1975; Parameswarappa et al., 1976). It has been estimated to cause up to 48.60% loss in seed weight at Dharwad by Anahosur and Rao (1977) and 15.18–54.59% in different genotypes by Anahosur and Patil (1983). Desai (1998) also recorded charcoal rot incidence up to 52% and lodging up to 52.30% at Bijapur. Almost all the varieties and hybrids cultivated in different parts of India are susceptible to the disease. Loss in grain yield is mostly due to the reduction in size of the grain, and there is a significant reduction in grain yield that ranges from 15–55% (Anahosur and Patil, 1983). The origin of the name of this disease is due to the charcoal appearance of infected areas, where vascular bundles become covered with numerous tiny black microsclerotia of the pathogen (Mughogho and Pande, 1984). The disease is related to soil moisture deficit during the grain-filling stage and is associated with high temperatures and senescence. A definite relation exists between sink source and charcoal rot severity. Plants are prone to susceptibility between the postflowering and grain-filling stages, when food reserves are translocated from stem to ears and the food supply to roots is reduced. Under experimental conditions, 100% lodging and 23–64% grain yield losses in CSH 6 was noticed at three locations in India and another location in Sudan. Jahagirdar et al. (2000) reported the existence of physiological races in *M. phaseolina* and found that Solapur isolate is more virulent than the Bijapur and Dharwad isolates. However, grouping of isolates was not related to the regions of collection (Virupaksha prabhu et al., 2012). Lodging, being the most prominent symptom as plants mature, occurs as a result of damage and weakening of the stalk after disintegration of the pith and cortex by the pathogen, resulting in the lignified fibrovascular bundles becoming suspended as separate strands in the hollow stalk, leading to complete yield loss (Mughogho and Pande, 1984). Resistance can be obtained by selecting for the stay-green postrainy flowering trait (Rosenow, 1994). Considerable progress has been achieved in mapping the stay-green trait, and it has been discussed as a trait related to drought resistance.

Charcoal rot is a stress-related disease caused by *M. phaseolina*, which infects about 500 plant species worldwide (Sinclair, 1982). In sorghum, the pathogen causes three phases of symptoms; namely, seedling blight, root rot and stalk rot. If the infection occurs before the emergence of secondary roots, the plants usually die. However, less severely infected seedlings survive and establish secondary roots, and under favorable conditions, such as sufficient soil moisture, availability to grow as mature plants (Partridge et al., 1984). The disease is systemic in nature and pathogen moves upward through xylem. Under stress conditions (such as moisture, temperature, and photosynthesis) that often coincide with the onset of flowering, the entire host-defense-system is weakened and pathogenic activity of *M. phaseolina* increases manyfold, leading to rapid and extensive rotting of the root and stalk.

Reviews by Tarr (1962), Dhingra and Sinclair (1977, 1978) provide comprehensive information on the biology of *M. phaseolina* and the epidemiology and control of the diseases that it causes in many plant species. *M. phaseolina* is known to survive as sclerotia in corn and sorghum stalk residues for 18 and 16 months, respectively (Cook et al., 1973). *M. phaseolina* is capable of living saprophytically on dead organic tissues, particularly on many of its natural hosts producing sclerotial bodies (Sen and Bandopadhyaya, 1988) (Figure 7.11).

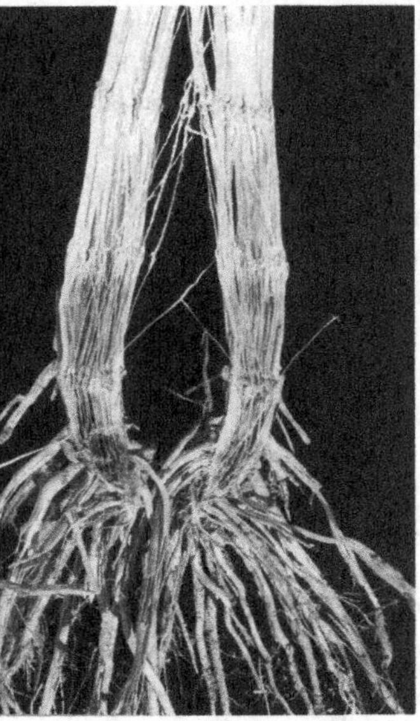

FIGURE 7.11 Stalk showing symptoms due to charcoal rot damage.

7.3.2 Physiological Stresses Influencing Stalk Rots

The stalk rots are associated with lack of nonstructural carbohydrates in the senescing plant tissues. This type of tissue in maize has shown to produce less DIMBOA, which accounts for weak pathogens invading complex inheritance patterns linked to environmental and physiological interactions in plants, predisposing them to stalk and root rot (Dodd, 1977).

- **i.** *Water deficits:* Water stress occurring after flowering increases the charcoal rot intensity up to 90% (Edmunds, 1964). The effect of water deficit on photosynthesis leads to maturity-related stalk rots.
- **ii.** *Leaf destruction:* Reduction in actively synthesizing leaf tissue reduces the amount of carbohydrates available to the plant for cell maintenance and storage in grains. Artificial removal of distal halves of leaves increases the rate of cell death in sorghum (Papelis and Katsanos, 1966) and predisposes the plants to root and stalk rot diseases.
- **iii.** *Light reduction:* Light reduction due to prolonged period of cloud cover, crowded plants causes photosynthetic stress, predisposing plants to root rot.
- **iv.** *Mineral deficiencies:* Supplementation of potassium in potassium deficient soils has decreased stalk rot incidence and lodging in sorghum (Murphy, 1975). Potassium is known to involve senescence and influence carbohydrate production.

7.3.3 Defense Mechanisms Influencing Stalk and Root Rot

In addition to photosynthesis and translocation balance, several other defense mechanisms operating in plant systems are discussed next:

i. Phenolics are well-known antifungal, antibacterial, and antiviral compounds. Several kinds of phenolic compounds occur naturally in plants. Host enzymes like polyphenol oxidase and peroxidase oxidize phenolics to quinines, which are more fungitoxic than phenolics. Anahosur et al. (1985) and Anahosur and Naik (1985) reported that high levels of sugar and phenols in sorghum root and stalk have a close relationship with resistance mechanisms.

ii. Phytoalexins are mostly isoflavanoids, terpenoids, and polyacetylene compounds and are synthesized de novo on infections by pathogens.

iii. Lignins may act as physical barriers to the pathogen. Phenyl alanine and cinnamic acid are the important precursors. Lignification occurs at the site of fungal penetration, and this is resistant to cellulose and macerating enzymes of the pathogen.

iv. Callose is a substance found in sieve tubes; it may prevent the leakage of sieve tube sap or water in cell walls. Penetration of incompatible pathogens into the host tissues results in the production of papillae, which mostly contain callose.

v. Sugars are the precursors for the synthesis of phenolics, phytoalexins, lignin, and callose. Horsfall and Diamond (1957) assigned a major role for sugars in disease resistance. Sugar concentration may affect the toxin-producing ability of the pathogen.

vi. Some of the amino acids are essential for the synthesis of phenolics, phytoalexins, and lignin. Phenyl alanine, tyrosine, and tryptophan are important precursors of defense chemicals.

Screening methodology: The toothpick method (Hsi, 1961) and stem tape inoculation method (Mayee and Garud, 1978) can be used to screen for charcoal rot resistance. However, the most practical and highly dependable method is the sick plot method, using the susceptible check variety SPV 86 or a hybrid like CSH 6. Uniform distribution of inoculum density over the entire sick plot has been a problem with the sick plot technique. Hence, simultaneously with the sick plot technique, the toothpick method is followed for screening.

The stay-green variety E 36-1 has been found to be stable for charcoal rot resistance. Several local genotypes collected in Karnataka (including Bidar local, Chitapurlocal, Kandukur local, Kannoli local, Muddebihal local, and Hountagi local), were also reported to be resistant to charcoal rot (Hiremath and Palakshappa, 1991; Jahagirdar et al., 2002). Only stressed fertile plants developed charcoal stalk and root rot. Root systems of stressed male-sterile plants had a high percentage of roots with latent infections, but they did not develop symptoms (Odvody and Dunkle, 1979).

A few researchers have attempted biological control of the disease. Although the bioagents (namely, *T. viride* and other fungal species) have been found effective in laboratory and pot conditions, their success rate was limited in field conditions with susceptible genotypes. However, in moderately resistant genotypes, seed treatment with Trichoderma was found effective at bringing down incidence level further (Jahagirdar et al., 2001).

The absolute water content (AWC) and water saturation deficit (WSD) of plants varied substantially among the genotypes differing in resistance to charcoal rot pathogen. Cultivars CS-3S41 and PVK-3, which were moderately tolerant of charcoal rot, exhibited higher AWC and lower depletion rate than the susceptible cultivars, CSH-1 and R • 16. After inoculation, AWC drastically declined in susceptible cultivars. WSD increased substantially in inoculated plants of CSH-1 and R-16, and the effects were more pronounced after flowering when the disease expresses (Pedgaonkar and Mayee, 1990).

Role of biotechnology: Little is known about host-pathogen interactions for stalk rot. Developing cultivars screened for other characters than resistance to pathogen per se made much progress. Three main characteristics have been mentioned in the research literature with reference to stalk rot resistance: namely, lodging resistance, nonsenescence, and greenbug resistance (Mughogho and Pande, 1984; Rosenow, 1984; Giorda et al., 1995). The first two may be considered as phenotypic expression for postflowering drought tolerance. Although much progress has been made on charcoal rot resistance by using the nonsenescence trait (Rosenow, 1984; Pande and Karunakar, 1992; Karunakar et al., 1993), additional criteria for resistance to charcoal rot is needed. The relationship of nonstructural carbohydrate (NSC) partitioning and charcoal rot resistance in sorghum was investigated by Tenkouano et al. (1992). They suggested that high-yielding cultivars resistant to *M. phaseolina* could be developed since the developing grain could not be identified as the cause or the beneficiary of stem NSC exhaustion. Most unaffected hybrids that are tolerant of charcoal rot yield nearly as well as the unaffected susceptible ones. Some of the recently used molecular techniques for controlling *M. phaseolina* include antibody-mediated control and the use of secondary metabolites like cyclic hydroxamic acids (Osbourn, 2001).

With the advent of new molecular techniques, the use of molecular markers as a tool in comprehensive genetic analysis of crop plants has become important. Attempts to classify the isolates of this fungus based on colony morphology using specified media and chlorate sensitivity tests have indicated that isolates with similar morphology need not be genetically identical; hence, the use of molecular tools has been suggested to study genetic variation among isolates (Su et al., 2001).

A total of 35 isolates of *M. phaseolina*, a causal organism of charcoal rot in sorghum, were obtained from different agroclimatic regions of major *rabi* sorghum–growing areas (Karnataka, Maharashtra, and Andhra Pradesh). Variation among different isolates was studied using polymerase chain reaction–random amplified polymorphic DNA (PCR-RAPD) markers. From a set of 20 random primers tested, a total of 142 amplicon levels were available for analysis. The OPA-03 and OPA-08 primers showed 100 percent polymorphism. It was possible to discriminate all the isolates with any of the primers employed. Unweighted pair group method with arithmetic mean (UPGMA) clustering of data indicated that the isolates showed varied levels of genetic similarity with a similarity coefficient index of 0.58–0.97, indicating significant levels of variation. The study suggested that the grouping of isolates was not related to the regions of collection in any way. The maximum genetic similarity of 97% was found between Solapur and Mulegaon. The least genetic similarity was observed between Ahmednagar and Mahboobnagar isolates. Rajkumar (2004) observed that 10 isolates of *M. phaseolina* representing the Dharwad and Bijapur regions proved pathogenic to sorghum. Molecular profiling using RAPD markers indicated the genetic differences among the isolates.

Four and five QTLs were reported at two locations for charcoal rot resistance using the mapping population derived from the cross IS 22380 (susceptible) × E 36-1 (resistant; Reddy et al., 2008). Later, using the same population, nine stable QTLs for morphological traits along with three QTLs for biochemical traits (phenol and lignin content) influencing charcoal rot disease incidence were detected using additional gene-based simple sequence repeat (SSR) markers (Patil et al., 2012).

Resistant sources: There have been concerted efforts in the evaluation of sorghum germplasm for the disease. Local varieties like M35-1, 5-4-1, and A1 were highly tolerant (Anahosur et al., 1974; Avadhani and Ramesh, 1979; Gowda et al., 1981). DSV-4 (9-13) is a high-yielding sorghum with built-in resistance and released in Karnataka (as *rabi* variety; Singh et al., 1990). Jahagirdar et al. (2002) found that the local genotypes like Honnutagilocal and Muddehallijola, along with 9-13(DSV-4) and E-36, recorded reduced levels of charcoal rot incidence. Jahagirdar et al. (2006) identified several charcoal rot–resistant lines, including BRJ 356, CRP 3, and CRP 48. The entries RS-29, E36-1, 9–13, DSV-5, 296B, and RSLG 262 were found promising against charcoal rot resistance (Jahagirdar et al., 2006).

Genetics of resistance: Stalk rot was indicated as qualitative in nature by Rosenow and Frederiksen (1982). Subsequent researchers have reported on the polygenic nature of resistance. Both additive and nonadditive gene action controlled the nonlodging of plants infected by charcoal rot (Gururaja Rao et al., 1993). They suggested that the resistance is governed by recessive genes and controlled by five pairs of genes exhibiting a trigenic ratio. Patil et al. (1985) reported that the high levels of sugar and phenols in the stalk of sorghum genotypes may be attributed to the resistance mechanism. Resistance appears to be a polygenic threshold characteristic governed by duplicate epistasis with low heritability (38%; Rana et al., 1982). Rao et al. (1980) reported that resistance to charcoal rot is independently inherited and is not associated with resistance to any other diseases in sorghum. In a study regarding inheritance, resistance to charcoal rot was reported to be controlled by recessive genes by Rosenow and Frederiksen (1982). This was further confirmed by the analysis of F_2 population of an intervarietal cross 1202A × CSV5 by Shinde (1981). They noticed that dominance of susceptibility was controlled by three major genes: two showing duplicate epistasis and one displaying basic complementary interaction with one of the duplicate genes fitting a F_2 ratio of 45 susceptible to 19 resistant plants. Rana et al. (1982a) documented that resistance for charcoal rot was a quantitative characteristic and that the nature of inheritance was additive. In another similar study, Rana et al. (1982b), based on a six-generation mean analysis from three resistant (R) × susceptible (S) crosses, reported that resistance to *M. phaseolina* (Tassi.) Goid. was polygenic, governed by duplicate epistasis, while evidence from R × S F_1s also indicated that resistance was partially dominant. Similarly, Indira and Rana (1983), based on a diallel study, found that resistance showed partial dominance for per cent stalk infections and susceptibility index, while susceptibility was partially dominant for the number of nodes crossed. Contrarily, Venkatarao et al. (1983), based on crosses between resistant and susceptible varieties, indicated that resistance to charcoal rot was recessive and controlled by three complementary genes.

In another study involving six generations of a single cross, Rao and Shinde (1985) observed that although susceptibility was dominant in F1 generation, it also displayed a polygenic nature. Based on generation mean analysis, nonallelic interactions like additive × dominance and dominance × dominance controlled the trait. Likewise, Garud and Borikar (1985) also worked out the genetics of charcoal rot resistance in sorghum. For variance of resistance, dominance components predominated over additive components and overdominance was also indicated. On the other hand, Rao and Shinde (1985) registered that inheritance of resistance is polygenic and both additive × additive and dominance × dominance nonallelic interactions played an important role, as revealed by the results of six-generation mean analysis in the cross 1202A × CSV5. In contrast, Bramel Cox et al. (1988), based on a 10-parent diallel cross, revealed the significance of *gca* effects, indicating the prominence of additive gene action for the resistance mechanism. In two separate studies, Rosenow et al. (1977) and Omer et al. (1985) were of the opinion that stay-green trait in general was a good indication of resistance to charcoal rot, while Anahosur et al. (1987) argued that lodging percent could be used to evaluate disease resistance. Further, Bramel Cox and Claflin (1989) observed that in order to evaluate for disease resistance, the level of postflowering stress was very important, and it needs to be severe to ensure full development of disease.

In another study, Rao et al. (1989) by comparing mean values of resistant and susceptible genotypes found that taller plant stature was associated with charcoal rot resistance. Tenkouano (1990) evaluated four sorghum inbred lines, diallel F1s, F2, and back cross progenies for resistance to charcoal rot. Inheritance of resistance was shown to be largely dependent on nonadditive genetic effects. Further, Tenkouano et al. (1993) determined that both charcoal rot resistance and nonsenescence were controlled by dominant and recessive epistatic interactions between two gene loci and a third locus with modifying effects. They proposed that nonsenescence should not be used as a sole criteria for resistance to charcoal rot in sorghum. However, Garud et al. (2002) proposed that an increase in stay-green is an important factor in charcoal rot resistance. Khanure et al. (1997) evaluated 31 sorghum genotypes and observed sufficient genetic variability for resistance to the charcoal rot. Likewise, Desai (1998) evaluated 20 different sorghum genotypes and recorded wide variations with respect to charcoal rot incidence (5.95–48.20%), lodging percent (7.25–33.45%), mean number of nodes crossed (0.42–2.19), and mean length of spread (6.75–25.22 cm). In another study, Patil et al. (1998) reported a new *rabi* variety, GRS1, which is resistant to charcoal rot as an alternative to M35-1 (susceptible variety) with 13.4% higher grain yield levels than M35-1. Similarly, Narkhede et al. (1999) reported that a new variety, SPV1359, was more tolerant of charcoal rot than M35-1.

7.4 BREEDING FOR RESISTANCE TO INSECT PESTS

Increased genetic vulnerability to different pests, especially shoot fly (*Atherigona soccata* Rond.), corn plant hopper/shoot bug (*Peregrinus maidis* Ashm.), the sugarcane aphid (*Melanaphis sacchari* Zehnt.), and spider mite (*Oligonychus indicus* Hirst.), are major constraints for productivity in the *rabi* season, particularly in the absence of proper pest management strategies.

FIGURE 7.12 Aphid colonies on the lower side of the leaf.

7.4.1 Sugarcane Aphid

The sugarcane aphid, *M. sacchari* (Homoptera: Aphididae), is considered an economically important pest to sorghum in China (Wang, 1961), Taiwan (Chang, 1981a,b; Pi and Hsieh, 1982a), Japan (Setokuchi, 1973), India (Young, 1970), South Africa (van Rensburg, 1973a), and Botswana, while it is common to cultivated sorghum in Zimbabwe (Flattery, 1982). Sorghum responses to *M. sacchari* injury include purple leaf discoloration of seedlings, followed by chlorosis, necrosis, stunting, delay in flowering, and poor grain fill, including quality and quantity yield losses. The sugarcane aphid feeds on the abaxial surface of older sorghum leaves. Leaves below the infected ones are often covered with sooty molds, which grow on the honeydew produced by the aphid (Narayana, 1975).

Sorghum is typically infested soon after plant emergence, but significant infestations usually occur during late growth stages and dry periods (van Rensburg, 1973a). There is a significant increase in the population of *M. sacchari* on sorghum from the boot stage to the soft dough stage (40–70 days after planting) in the spring, and heading to harvesting (60–100 days after planting) in autumn (Fang, 1990). Waghmare et al. (1995) observed population increase and peaks during January, when the postrainy sorghum crop was between the flowering and milk stages, and declined thereafter till maturity.

The aphid incidence was the least (51.72%) in early-sown crop (third week of September) and the highest intensity (68.98%) was witnessed in late-sown crop (third week of October) (Balikai, 2001); that is, sugarcane aphid reaches a peak in the third or fourth week of January in the Karnataka region of India (Balikai and Lingappa, 2002a) and in the second week of December in western Maharashtra in India (Jadhav 1993). The onset of the winter season serves as a trigger for population buildup (Balikai, 2004). Upon aphid population buildup, natural enemies such as Coccinellids (Balikai and Lingappa, 2002a), Chrysopids, and Syrphids (Balikai, 2001) reach their peak (Figure 7.12).

Plant stress due to drought may intensify damage to sorghum by the sugarcane aphid. Plant growth stage and temperature had significant effects on the population buildup of the sugarcane aphid, and dispersal occurs within 6–10 days at temperature regimes of 15.1°C and 31.0°C (Balikai, 2001), 16.0°C and 29.0°C (Mote and Kadam, 1984), 19.5°C and 34.7°C (Narayana et al., 1982), and 22.5°C and 32.5°C, and at 84% relative humidity (RH; AICSIP, 1979–2003) and 18.0–31.0°C (van Rensburg, 1973a,b), but the population died at 35.0°C (Behura and Bohidar, 1983). In addition to the temperature, cloudy weather with increasing humidity can result in aphid colonies completely covering the abaxial surface of all the leaves of sorghum plants (Mote, 1983). The highest rate of population buildup was at 94% and 43% RH and at 11.4°C and 30.0°C temperature in the morning and afternoon, respectively (Waghmare et al., 1995). Aphid density was greater in irrigated than in unirrigated conditions, and its occurrence on sorghum at the milk stage did not affect the grain yields severely, but the fodder quality deteriorated (Balikai, 2001).

The sugarcane aphid affects grain and fodder yields and fodder quality. In sorghum, the losses varied between 12–26% and 10–31% with an overall loss of 16% and 15% for grain yield and fodder yield, respectively (Balikai, 2001). Narayana et al. (1982) and Mote (1983) observed a gradual decrease in the crude protein, crude fiber, and total sugar content of fodder with delayed sowings with increased aphid population. On the contrary, there was an increase in the polyphenol content, thus making the fodder less palatable to the livestock (Balikai, 2001; Balikai and Lingappa, 2003). High content of phosphorus, polyphenol, and potash in sorghum leaves make them resistant, while that of nitrogen and total sugar make them susceptible for aphid (Mote and Shahane, 1994; Balikai and Lingappa 2002b). Sugarcane aphid–infested sorghum grain was significantly associated with the poor preparation of beverages (Pi and Hsieh, 1982b), and reduction in diastatic activity, malt, and abrasive hardness index (van den Berg et al., 2003), as well as causing grain yield reduction and poor quality of forage sorghums (Setokuchi, 1979), similar to decrease in crude protein and increase in neutral detergent fiber (NDF) by the yellow sugarcane aphid, *Sipha flava* (Fukumoto and Mau, 1989).

Sources and mechanisms of resistance: Several sources, levels, and mechanisms of resistance in sorghum to the sugarcane aphid, comprising germplasm accessions, parental lines (A/B and R lines), agronomic elite lines, hybrids, varieties, and locals, have been reported from different countries. Resistance to insects has been characterized into three components as antixenosis (nonpreference), antibiosis, and tolerance (Painter, 1951; Horber, 1980; Smith, 1989; Smith et al., 1994). However, antixenosis and antibiosis resistance to *M. sacchari* in sorghum do not vary with plant age (Teetes, 1980). Bothalate and apterous virginoparous adults showed a stronger tendency of preference to *S. bicolor* and *S. halepense* rather than *M. sinensis*. Tolerance of sugarcane aphid injury in sorghum increases greatly with slight increases in plant height, and has an inherent advantage over antibiosis and antixenosisin in that it does not impose selection pressure on aphid populations and thus may have greater permanence. Setokuchi (1988) reported that *M. sacchari* failed to establish on resistant lines under field conditions, although resistant sources served as suitable hosts in confinement studies, which indicate the involvement of both antixen osis and antibiosis mechanisms. Antixenosis for adult colonization was noticed in

TAM 428, IS 1144C, IS 1366C, IS 1598C, IS 6416C, IS 6426C, IS 12661C, and IS 12664C. Among them, IS 1144C and IS12664C were preferred less than the resistant check, TAM 428. High levels of antibiosis expressed in TAM 428, IS 1144C, IS 5188C, IS 12609C, and IS 12664C for the least number of days to reproduction; and in TAM 428, IS 12609C, and IS 12664C for greater mortality and shorter longevity of adults and production of fewer or no nymphs (Teetes et al., 1995).

Morpho-physiological traits: In sorghum, some morpho-physiological traits, such as genotypes with small, narrow, and fewer leaves and low leaf bending at the seedling stage (Mote and Kadam, 1984), greater plant height and greater distance between two leaves and the presence of waxy lamina (Mote and Shahane, 1994), and epicuticular wax on the ventral surface of the leaves, were associated with reduced susceptibility to the sugarcane aphid (Pi and Hsieh, 1982a,b). Tolerance of sugarcane aphid injury in sorghum increases greatly with slight increases in plant height and has an inherent advantage over antibiosis and antixenosis (Singh et al. 2004). Aphid density and plant damage under natural infestations have been used to select resistant sorghum genotypes in greenhouse and field conditions (Setokuchi, 1976; Pi and Hsieh, 1982a; Hagio and Ono, 1986). Seedling and mature plant evaluations displayed similar results (Pi and Hsieh, 1982a; Hagio and Ono, 1986); however, seedlings were preferred for easy handling and control of infestation levels (Teetes, 1980).

Biochemical factors: Host suitability to various phloem-feeding Homoptera has frequently been related to nitrogen levels in host plants. Hsieh (1988) stated that the presence of *p*-hydroxybenzaldehyde during HCN release from the sorghum leaves due to aphid biting, at the seedling stage, may be important in order to repel further attack. The mean HCN content of F1 hybrids produced from crosses between parents with high- and low-HCN content was intermediate or closer to the parent with a high-HCN content. The development of aphid populations and leaf sugary exudation was more pronounced in sorghum genotypes having higher nitrogen sugar and chlorophyll content of leaves (Mote and Jadhav, 1993; Mote and Shahane, 1994). The varieties ICSV 9, BTP 28, IS 1640, ICSV 148, and Swati (SPV 504), with higher content of phosphorus, potassium, and polyphenols, were less preferred by the sugarcane aphids and also showed less development of leaf sugary exudation (Mote and Shahane, 1994). There is a significant reduction in nitrogen, phosphorus, potash, total sugars, and chlorophyll content in sorghum due to infestation by the sugarcane aphid. In contrast, there is a significant reduction in polyphenols in resistant over the susceptible sorghum genotypes (Balikai, 2001). In India, planting in the third week of September caused a reduction in protein, total minerals, and fat content in the grain due to infestation by the sugarcane aphid. Similarly, there is also a decline in diastase activity in sorghum grain, while there is an increase in crude fiber and carbohydrate content when infested by *M. sacchari* (van den Berg et al., 2003). In sorghum fodder, there was a loss of 10% and 7.0% in crude protein and crude fiber content, respectively (Balikai, 2001).

Genetic basis of resistance: Greenhouse and field studies with the crosses between PI 257595 (highly resistant), 129-3A (moderately resistant), and RTx 430 (susceptible) have shown that resistance is monogenic and controlled by a single dominant gene (Hsieh and Pi, 1982; Pi and Hsieh, 1982b; Tan et al., 1985). Studies have also indicated that PI 257595 and 129-3A have the same gene for resistance, although the resistance gene of 129-3A has modifiers (Pi and Hsieh, 1982a). Although dominant and additive gene actions are involved, additive gene action accounts for the resistance expression (Hsieh and Pi, 1988).

FIGURE 7.13 Dead heart caused due to shoot fly infestation.

The cross between RTx 430_129-3A indicated the presence of complimentary gene action (Chang and Fang, 1984). Observations on aphid populations at various time intervals (51, 58, 65, 72, 86, and 93 DAE, and at maturity) revealed that inheritance of resistance was governed by two dominant genes with duplicate effects in the cross M35-1 (susceptible) × R354 (resistant; Deshpande et al., 2011).

7.4.2 Shoot Fly

Shoot fly attacks the sorghum at the seedling stage (18–30 DAE). Infestation rates are higher in late-sown rainy season and early-sown postrainy season sorghum crops. The levels of infestation may go up even to 90–100% (Usman, 1972). The losses due to this pest have been estimated to reach as high as 85.9% of grain and 44.9% of fodder yield (Sukhani and Jotwani, 1980). The annual losses in sorghum production due to shoot fly in India have been estimated at nearly US$200 million (ICRISAT, 1992). Shoot fly is an insect pest of seedling sorghum, primarily in Asia and Africa (Peterson et al., 1997). Since resistance to shoot fly in sorghum is apparently a complex trait, breeders for the last two decades have used glossiness as an indirect selection criteria for shoot fly resistance, although genetic resistance control of this trait is not well understood. However, nonavailability of absolute resistance sources in cultivated sorghum, coupled with the complex nature of the inheritance (as resistance is contributed by several component traits), has resulted in slow and inefficient progress in shoot fly resistance breeding through conventional approaches (Figure 7.13).

Inheritance of tolerance to shoot fly [*A. soccata* (Rondani): The majority of sources of resistance to shoot fly (*A. soccata* (Rondani)) in the world are reported to be in India (e.g., Rao et al., 1996). Resistance in sorghum to shoot fly was first reported by Ponnaiya (1951).

Subsequently, Rao and Rao (1956) and Jain and Bhatnagar (1962) selected a few promising resistant sources; however, no attempts have been made to utilize them.

Blum (1969) suggested that any genetic information about reaction of sorghum to shoot fly must be interpreted according to insect population factors, and resistance is primarily due to ovipositional nonpreference governed by additive genes. Further, Madhav Rao et al. (1970), based on genetic analysis of resistance to shoot fly in sorghum, reported polygenic inheritance of resistance governed by additive genes. Similarly, Rao et al. (1974), Sharma et al. (1977), Balakotaiah et al. (1975), Agrawal and House (1982), Nimbalkar and Bapat (1987), and Singh and Verma (1988) reported that resistance to shoot fly in terms of oviposition nonpreference was controlled by additive genetic components. Meanwhile, Borikar and Chopde (1981) reported that oviposition preference with its component traits (namely, dead heart percentage, eggs per plant, and plant recovery percentage) were controlled by additive gene action. Gene action for other traits (namely, plant height, number of tillers per 100 plants, and number of fertile tillers) was predominantly nonadditive. Similarly, Patil and Thombre (1985), with respect to inheritance of shoot fly, noted that resistance was conditioned by dominant gene– and recessive gene–controlled susceptibility for shoot fly. Singh and Jotwani (1980a,b) also worked on the mechanism of resistance to sorghum shoot fly. Results revealed that in addition to nonpreference to oviposition, antibiosis mechanism has been displayed in some varieties, while Rana et al. (1982a) found that the prominent mechanism governing resistance is nonpreference, and the nature of inheritance is generally additive for threshold characters and for insect resistance. Genetic analysis of generation means indicated that the additive gene effects were predominant for glossiness, trichome density, and percentage of dead hearts, while both additive and dominant gene effects were found to be important for the number of eggs per plant, percentage of plants having eggs, and grain yield per plant. Regarding nonallelic interactions, additive × additive was found to be important for most of the characters. The predominance of additive and additive × additive gene effects suggests that the selection would be more effective for improvement in shoot fly resistance, as well as other desirable attributes (Patil et al., 2005).

The importance of both additive and nonadditive gene effects for oviposition, expressed in terms of egg count per plant, dead heart percentage, and predominantly additive gene action for recovery resistance, was reported by Halalli et al. (1982), Biradar et al. (1984), Patel et al. (1985), Dabholkar et al. (1989), Elbadawi et al. (1997), Ravindrababu et al. (1997), and Ravindrababu and Pathak (2000a). Blum (1972) cited seedling resistance and recovery resistance as important components of tolerance of shoot fly. Gibson and Maiti (1983) suggested that presence of trichomes or leaf hairs (associated with reduced susceptibility) is recessive, controlled by a single locus, while Biradar et al. (1984) detected polygenic inheritance with additive gene effects for trichome density based on six generation mean analysis. Rana et al. (1985) highlighted that nonpreference for oviposition was a major mechanism of resistance, although antibiosis was also seen for shoot fly in sorghum. Singh and Rana (1986) thought that with respect to biophysical factors, glossy leaf surface and trichomes are associated with shoot fly resistance.

In another similar study, Patel and Sukhani (1990) suggested that long internodes, yellowish green leaves (glossy) with high trichome density at the seedling stage, and shoot peduncle should be considered when selecting resistant genotypes. Regarding the inheritance of shoot fly resistance, Nimbalkar and Bapat (1992) studied eight sorghum parents (three resistant and five susceptible) using a diallel model, and the results revealed that nonadditive gene action was operating for inheritance of egg count per plant. One or two dominant genes controlled the egg count per plant. Similarly, Sanjaykumar et al. (1996), based on a diallel study, stated that shoot fly resistance should be evaluated in terms of eggs per plant and dead heart percentage in order to explain oviposition nonpreference, tolerance, and antibiosis mechanisms. Nonadditive gene effects played an important role in governing glossiness, seedling vigor, and proportion of plants with dead hearts. For trichome density, both additive and nonadditive gene actions were important (Aruna and Padmaja, 2009). In another study, Jayanthi et al. (1999) worked out the genetics of glossy and trichome traits. They noted that the nonglossy trait associated with susceptibility was under the influence of dominant genes.

Regarding trichome density, the inheritance was complex and depended on the parents involved and season specificity. Further, Jayanthi et al. (2000), based on genetic analysis of shoot fly resistance in sorghum reported that least egg count (associated with resistance) was recessive over high egg count (associated with susceptibility). Susceptibility to shoot fly based on dead heart percentage was also dominant over resistance. Rao et al. (2000) assessed the genetic variability for shoot fly incidence in sorghum and suggested that resistance was mainly under genetic control, with only a moderate environmental influence. Additive gene effects were predominant for glossiness, trichome density, and percentage of dead hearts, while both additive and dominance gene effects were found important for number of eggs per plant, percentage of plants having eggs, and grain yield per plant. Regarding nonallelic interactions, additive × additive was found to be important for most of the characteristics. The predominance of additive and additive × additive gene effects suggests that the selection would be more effective for improvement in shoot fly resistance, as well as other desirable attributes (Patil et al., 2005).

The importance of both additive and nonadditive gene effects was observed for controlling the abovementioned threshold characters. Ravindrababu et al. (1997), with respect to host plant resistance against sorghum shoot fly, reported that although both additive and nonadditive gene action governed the number of eggs per plant and dead heart percentage, there was prominent additive gene action for both resistance parameters. Ravindrababu and Pathak (2000a) further studied the gene effects for shoot fly resistance using resistant (R) × R, R × susceptible (S) and S × S crosses. Individual crosses in the three categories (namely, R × R, R × S, and S × S) revealed gene effects in different ways. However, in the case of R × S crosses, all the three types of gene effects (namely, additive, dominance, and epistasis effects) were important for egg count per plant and dead heart percentage.

In another study, Ravindrababu and Pathak (2000b) obtained a higher magnitude of *gca* variance than *sca* variance, indicating the preponderance of additive and additive × additive gene effects for egg count per plant, dead heart percentage, and seedling vigor. On the other hand, with respect to seedling vigor, Jayanthi et al. (2002) found that low seedling vigor (susceptibility) was dominant over high seedling vigor (resistance) under low-temperature conditions against shoot fly.

In another study, Kulkarni (2002) revealed differences in gene action with respect to both parameters of resistance (namely, percentage of dead heart and number of shoot fly eggs per plant). For dead heart percentage, predominance of *gca* variance over *sca* variance was obtained, inferring that this trait was largely under the control of additive gene action, while inheritance of number of shoot fly eggs per plant was found to be under the control of nonadditive gene action, as revealed by high *sca* variance. Kamatar and Salimath (2003) studied morphological traits of sorghum associated with resistance to shoot fly and found high seedling vigor, glossiness, and taller seedlings were associated with resistance to shoot fly.

Jayanthi (1997) also observed significant negative correlation between seedling vigor and dead hearts in both rainy and postrainy seasons. The glossiness (pale green, shiny leaves) trait in sorghum is associated with shoot fly resistance (Blum, 1972; Maiti and Bidinger, 1979). Out of 17,536 genotypes screened, only 495 were glossy, and these showed multiple resistance to shoot fly, stem borer, drought salinity, and high temperature (Maiti, 1996). These are found in the *durra* race, and glossy leaves might possibly affect the quality of light reflected from leaves and influence the orientation of shoot flies toward their host plants. Glossy leaves might also influence host selection due to chemicals present in the surface wax of leaves (Jotwani et al., 1971; Maiti and Bidinger, 1979; Maiti, 1980; Bapat and Mote, 1982; Omori et al., 1988; Taneja and Leuschner, 1985). Vijayalakshmi (1993), Jeewad (1993), Darbha (1997), and Jayanthi (1997) also observed that glossiness was significantly correlated with oviposition and dead-heart percentage.

A_1 cytoplasm was more susceptible to shoot fly than the maintainer line cytoplasm (Reddy et al., 2003). The A_4 cytoplasm was found to be least susceptible to shoot fly, as it was comparatively less preferred for oviposition and had lower dead heart formation across seasons than the A1, A2, and A3 cytoplasms tested, and thus can be exploited for developing shoot fly–resistant hybrids (Umakanth et al., 2012). Oviposition and dead heart formation were significantly lower on the maintainer lines than on the corresponding male-sterile lines. Among the cytoplasms tested, A_4M cytoplasm showed antixenosis for oviposition and suffered less dead heart formation than the other cytoplasms tested. The A_4G_1 and A_4M cytoplasms suffered fewer dead hearts in tillers than the other cytoplasms. Recovery following shoot fly damage in A_4M, A_3, and A_2 cytoplasms was better than in the other cytoplasms tested. The larval and pupal periods were longer and male and female pupal weights lower in A_4M and A_4VzM CMS backgrounds compared to the other CMS systems. Fecundity and antibiosis indices on CMS lines were lower than on the B-lines. The A_4M cytoplasm was found to be relatively resistant to sorghum shoot fly, and can be exploited for developing shoot fly–resistant hybrids for sustainable crop production in the future (Dhillon et al., 2005a). The hybrids based on shoot fly–resistant CMS and restorer lines had significantly lower proportions of oviposition and dead hearts than the hybrids based on other cross-combinations. Leaf glossiness and trichome density revealed a high correlation with shoot fly resistance (Gomashe et al., 2012).

The levels of resistance in cultivated germplasm are low to moderate, and therefore it is important to identify sorghum genotypes with diverse mechanisms of resistance based on physicochemical and or molecular markers. Three resistant accessions (namely, IS 1054, IS 1057, and IS 4664) were found to be diverse to IS 18551, which is widely used as a shoot fly resistance donor. These diverse sources, after further characterization for resistance mechanisms, can be used in breeding programs for improving shoot fly resistance (Chamarthi et al., 2012). Incorporation of multiple resistant sources in crossing, as well as pyramiding of genes for resistance in subsequent generations, would help improve shoot fly and stem borer resistance in the elite lines. The biochemical constituents like ash, silica, phosphorus, calcium, fat, crude fiber, dry matter, acid detergent fiber, and NDF were observed more in resistant and moderately resistant varieties (Bangar et al., 2012).

Leaf glossiness and trichomes on leaf surfaces, two important component traits governing resistance to shoot fly in sorghum, are controlled by QTLs located on linkage group SBI-05 (Lakshmidevi et al., 2012) Mg, Zn, soluble sugars, tannins, fats, leaf glossiness, leaf sheath, plumule pigmentation, and trichome density explained 99.8% of the variation in shoot fly damage (Chamarthi et al., 2011a). The sorghum genotypes IS 1054, IS 1057, IS 2146, IS 4664, IS 2312, IS 2205, SFCR 125, SFCR 151, ICSV 700, and IS 18551 exhibited antixenosis for oviposition and suffered fewer dead hearts due to sorghum shoot fly. Compounds on leaf surface such as undecane 5-methyl, decane 4-methyl, hexane 2, 4-methyl, pentadecane 8-hexyl, and dodecane 2, 6, 11-trimethyl, which are present on the leaf surface of sorghum seedlings, were associated with susceptibility to shoot fly; while 4, 4-dimethyl cyclooctene was associated with resistance to shoot fly (Chamarthi et al., 2011b).

Proportional contributions of CMS lines for oviposition, dead hearts, leaf glossiness, and recovery resistance were greater than those of the restorer lines. The gca and sca estimates suggested that inheritance for oviposition nonpreference, dead hearts, recovery resistance, and the morphological traits associated with resistance or susceptibility to *A. soccata* were governed by the additive type of gene action. The sca effects and heterosis estimates indicated that heterosis breeding would not be rewarding in breeding for resistance to shoot fly (Dhillon et al., 2006).

Wild relatives of sorghum exhibited very high levels of antibiosis to *A. soccata*, while only low levels of antibiosis have been observed in the cultivated germplasm. Therefore, wild relatives with different mechanisms of resistance can be used as a source of alternate genes to increase the levels and diversify the basis of resistance to *A. soccata*. Accessions belonging to *S. exstans* (TRC 243601), *S. stipoideum* (TRC 243399), and *S. matarankense* (TRC 243576) showed an absolute nonpreference for oviposition under no-choice conditions. Accessions belonging to *Heterosorghum*, *Parasorghum*, and *Stiposorghum* were preferred for oviposition, but there was low dead heart formation. Larval mortality was recorded in main stems of the *Parasorghums*. Accessions belonging to *S. bicolor* ssp. *verticilliflorum* were highly susceptible to shoot fly, as were those of *S. halepense* (Kamala et al., 2009).

Predominance of nonadditive gene effects were recorded for plant height, number of leaves per plant, number of eggs per plant, trichomes on upper and lower surface of lamina, and dead heart percentage (Anandan et al., 2009).

Shoot fly–susceptible sorghum varieties emit attractive volatiles, but the compounds involved were unknown. Eight compounds that elicited an EAG response, which were identified by coupled GC-MS and GC peak enhancement on two GC columns of different polarity, were (Z)-3-hexen-1-yl acetate, (-)-a-pinene, (-)-(E)-caryophyllene, methyl salicylate, octanal, decanal, 6-methyl-5-hepten-2-one, and nonanal (Padmaja et al., 2010).

Physicochemical traits such as leaf glossiness, trichome density, and plumule and leaf sheath pigmentation were found to be associated with resistance, and chlorophyll content, leaf surface wetness (LSW), seedling vigor, and waxy bloom with susceptibility to shoot fly explained 88.5% of the total variation in dead hearts. Stepwise regression indicated that 90.4% of the total variation in dead hearts was due to leaf glossiness and trichome density. The direct and indirect effects, correlation coefficients, and multiple and stepwise regression analysis suggested that dead hearts, plants with eggs, leaf glossiness, trichomes on the abaxial surface of the leaf, and leaf sheath pigmentation can be used as marker traits to select for resistance (Dhillon et al., 2005b).

Antixenosis for oviposition was observed under multi- and dual-choice conditions in the case of IS 1054, IS 1057, IS 2146, IS 4664, IS 2312, IS 2205, SFCR 125, SFCR 151, ICSV 700, and IS 18551. Antibiosis (expressed in terms of IS 2312, SFCR 125, SFCR 151, ICSV 700, and IS 18551), which exhibited antixenosis, antibiosis, and tolerance components of resistance, may be used in sorghum improvement to develop sorghum cultivars with resistance to this pest. Under multi- and dual-choice conditions, antixenosis for oviposition was observed in the case of IS 1054, IS 1057, IS 2146, IS 4664, IS 2312, IS 2205, SFCR 125, SFCR 151, ICSV 700, and IS 18551. Antixenosis for oviposition was not apparent under no-choice conditions. Antibiosis (expressed in terms of prolonged larval and pupal development, and/or larval and pupal mortality,) was observed in the case of IS 2146, IS 4664, IS 2312, SFRC 125, ICSV 700, and IS 18551. The genotypes IS 1054, IS 1057, IS 2146, IS 2205, and IS 4664 showed lower percentages of tiller deadhearts than the susceptible check, while the genotypes IS 2312, SFCR 125, SFCR 151, ICSV 700, and IS 18551, exhibited antixenosis, antibiosis, and tolerance components of resistance. Such genotypes with diverse combination of characteristics associated with resistance to sorghum shoot fly can be used in breeding programs to broaden the genetic base and increase the levels of resistance to this pest (Siva Kumar et al., 2008).

The results revealed that, the lines 104A, 104B, RR 9817, RR 9818, RS 585, and RS 653 were found to be resistant to shoot fly by recording 18.7%, 17.0%, 14.9%, 16.9%, 13.4%, and 16.2% dead hearts, respectively. The parental lines (namely, AKR 150, C 43, R 354, RS 29, and Indore 12) were categorized as resistant to aphids, which scored less than 2 grade (Balikai and Biradar, 2007).

The landrace (namely, Madabhavi local) with 16.9% dead hearts was on a par with resistant check (IS 2312). The other landraces, like Kannolli local, Muddihali jola, Afzalpur local, Lakkadi, Chiitapur local, Yaranal local, and Khatizapur local, with 18.5%, 19.3%, 22.2%, 19.9%, 19.5%, 24.0%, and 24.9% dead hearts, which showed moderate level of resistance, were on a par with the popular *rabi* variety, M 35-1 (21.0%) (Balikai and Biradar, 2007).

The genotypes IS 2312 and IS 4664 showed stability of antixenosis for oviposition during postrainy season advanced-generation lines compared to the susceptible checks. Depletion in levels of reducing sugars and phosphorus in resistant genotypes played a significant role in dead heart formation in the test genotypes. The positive association of nitrogen and potassium with oviposition at early seedling stages indicated their role in releasing chemical cues for oviposition. Low levels of reducing sugars and total sugars seemed to enhance the degree of resistance to sorghum shoot fly. The total chlorophyll content had no relationship with antixenosis for oviposition. No relationship was observed between moisture content of sorghum seedlings and shoot fly resistance. Low concentrations of reducing sugars, total sugars, nitrogen, phosphorus, and potassium in sorghum seedlings greatly enhanced the degree of antixenosis for oviposition/feeding and dead heart formation, and they can be used as selection criteria for resistance to shoot fly (Singh et al., 2004).

The varietal approach to increase production and combat insect attack appeared to be a futile exercise in sorghum. The obvious choice was to shift to hybrid development with the sole aim of exploiting hybrid vigor, but there too, the situation is not very encouraging except for the performance of a few hybrids (such as CSH 9 and CSH 12R). Others belied the hope and did not show promise. Perhaps the level of resistance transferred in these genotypes was unable to cope with the intensity of insect attack, especially when the sorghum plant is attacked by a number of key pests at different GSs. The situation became even worse due to overlapping generations of these pests. In addition, resistant × resistant and resistant × susceptible crosses did not exhibit an improvement over the parents, indicating no diversity among resistant lines. All these shortcomings suggested the futility of the single-cross (temperate × tropical) approach and opened new avenues to develop a different strategy to meet the ensuing challenge and possibly improve the level of resistance to shoot fly and stem borer, which included crossing intermediate × intermediate resistant crosses and subsequently selecting the resistant lines using the pedigree method (Prem Kishore, 2005).

The susceptibility of sorghum to the shoot fly *A. soccata* Rondani, (Diptera: Muscidae) is affected by seedling age and is highest when seedlings are 8–12 days old. This corresponds with high moisture accumulation on the central leaf, which is the path of newly hatched larva as it moves downward from the oviposition site and toward the growing apex. Studies showed that LSW of the central shoot leaf was higher in 10-day-old seedlings than in seedlings of other ages. Similarly, LSW was much higher in the susceptible sorghum genotype CSH 1 than in the resistant genotype IS 2146. Larvae moved faster toward the growing point and produced dead hearts much earlier in CSH 1 than in IS 2146. They also moved faster in 10-day-old seedlings than in seedlings of other ages. It was also shown that the LSW of the central shoot leaf is a more reliable parameter of resistance than the glossy leaf trait or trichome density (Nwanze et al., 1990).

Resistance, as indicated by lower dead heart percentage, is governed by recessive genes. Both additive and nonadditive gene actions were important for resistance, and this trait is influenced by environment (Aruna et al., 2011a). Environment had the greatest effect (69.2%) followed by G × E interactions (24.6%) and genotype (6.2%). Low heritability and high environmental influence for dead heart percentage suggested that shoot fly resistance is a highly complex characteristic, emphasizing the need for marker-assisted selection. Transgressive variation was observed in the RIL population for all the traits, indicating the contribution of alleles to resistance from both resistant and susceptible parents. Since the alleles for shoot fly resistance are contributed by both resistant and susceptible parents, efforts should be made to capture favorable alleles from resistant and susceptible genotypes (Aruna et al., 2011b). One QTL that is common for glossiness, oviposition, and dead hearts was detected following composite interval mapping (CIM) on linkage group A (Apotikar et al., 2011).

Sajjanar (2002) identified eight QLTs for shoot fly resistance components. One major QLT for glossiness was detected on linkage group J, with phenotypic expression ranging from 34.3% to 46.5% in the three screening environments with the highest expression in postrainy season. The largest consistent effect for glossiness due to this QTL on linkage group J comapped with genomic regions associated with dead heart percentage under high shoot fly pressure. This QTL may be a useful target for marker-assisted selection (MAS) for shoot fly resistance in sorghum.

FIGURE 7.14 Damage caused due to shoot bug infestation. *Source: Courtesy: Gawali, IIMR.*

A total of 29 QTLs were detected by multiple QTL mapping; namely, 4 each for leaf glossiness and seedling vigor, 7 for oviposition, 6 for dead hearts, 2 for adaxial trichome density, and 6 for abaxial trichome density. For most of the QTLs, IS18551 contributed resistance alleles; however, at six QTLs, alleles from 296B also contributed to resistance. QTLs of the related component traits were colocalized, suggesting pleiotropy or tight linkage of genes. QTLs identified in this study correspond to QTLs/genes for insect resistance at the syntenic maize genomic regions, suggesting the conservation of insect resistance loci between these crops (Satish et al., 2009).

7.4.3 Shoot Bug/Corn Planthopper

The corn plant hopper, *P. maidis* (Ashmead) (Homoptera: Delphacidae), was first recorded on corn in Hawaii by Perkins in 1892. It thrives mostly in low, humid elevations, but with a sharp decline at greater than 800-m elevations in the tropics and coastal areas of subtropical and temperate regions of the world (Napompeth, 1973).

Nature of damage: P. maidis pierces the vascular tissues in the vessels of sorghum by sucking sap from the leaves, leaf sheaths, and stems during exploratory feeding. Adults and nymphs are massed inside the leaf whorl and on the inner side of the leaf sheath, causing reduced plant vigor, stunting, yellowing of leaves, and predisposition in the plant to moisture stress. Severe infestations result in withering of the leaves from the top of the plant, inhibition of panicle formation, and sometimes death of the plant (Chelliah and Basheer, 1965) through girdling of the stems (Singh, 1997). Photosynthetic flow to the root system is disrupted, leading to leaf senescence. Tissue surrounding eggs sometimes becomes septic and turns reddish (Napompeth, 1973). In India, the leaf sugary exudation (*chikta*) due to oviposition and feeding punctures, as well as excretion of honeydew by *P. maidis*, is a serious menace in sorghum, and more so in soils of low fertility and in bunded areas (Borade et al., 1993) (Figure 7.14).

In India, it has been estimated to cause a loss of 10–15% due to leaf sugar exudation (Mote and Shahane, 1993), 10–18% loss of plant stand (Managoli, 1973), and 30% of grain sorghum yield (Mote et al., 1985). An economic injury level of 3.7 nymphs per plant has been determined in sorghum (Rajasekhar, 1996).

Seasonal interactions: In India, the seasonal fluctuations of *P. maidis* on sorghum was found from July–December, with a peak in September (Agarwal et al., 1978); July–March, with a peak during August–October in Madhya Pradesh (Rawat and Saxena, 1967); September–January, with a population decline in February in Tamil Nadu (Chelliah and Basheer, 1965); and June–October, with a peak in August and decline in October in Andhra Pradesh (Rajasekhar, 1989). The prevalence of hot weather during March–June was unfavorable for *P. maidis*, either due to the absence of cultivated crops or by rendering inadequate nutrient supply in the alternate hosts/weeds.

Sources of resistance: Few sources of resistance have been reported in sorghum and are linked to virus diseases transmitted by *P. maidis*. The genotypes of Kafir Suma and Dwarf Hegari (Khan and Rao, 1956); and I 753, H 109, GIB, 3677B, and BP 53 (IS 1055) in sorghum are free of infestation (Agarwal et al., 1978). The genotypes IS 19349 (Chandra Shekar, 1991; Chandra Shekar et al., 1993a,b) and IS 18657, IS 18677, and PJ 8K(R) were stable in resistance across different stages of crop growth in sorghum (Singh and Rana, 1992; Chandra Shekar, 1991; Chandra Shekar et al., 1992, 1993a,b). In contrast, the sorghum genotypes M 35-1, Bilinchigan, BS 12-2-11, Farm Aispuri, Fulgar White, Fulgar Yellow, GAR 2, Gundinni, Hegari 1, Improved Ramkel, Improved Saoner, Khadi BK 1-1, M 47-3, Nandyal, ND 15, NJ 16K, NJ 156, PJ 16K, PJ 24K, PJR 4, PS 13, Sampgaon, Shanali 4-2, Shanali 4-5, and Striga-1 are highly susceptible to *P. maidis* (Capoor et al., 1968).

Behavioral responses vis-à-vis mechanisms of resistance: Oviposition is greater during the vegetative crop stage and greatly declines after flowering. Behavioral responses of adults and nymphs indicate that antixenosis for colonization, oviposition, feeding, or any combination is one of the predominant mechanisms of resistance in sorghum (Chandra Shekar, 1991; Chandra Shekar et al., 1992,1993a). In sorghum, the orientation and settling behavior of nymphs and brachypterous adults is influenced by olfactory and visual responses, but feeding is not sustained, suggesting that a lack of gustatory stimuli plays a significant role in determining the degree of antixenosis for feeding (Chandra Shekar, 1991; Chandra Shekar et al., 1992, 1993a). Accumulations of *P. maidis* nymphs and adults on leaves near the top of plants indicate a greater preferential response to feeding on tender shoots over those in the middle or basal portion of plants. Low preferential response to colonization was observed on sorghum cultivars such as SPV 736 and MSH 65 (Rajasekhar et al., 1995). *P. maidis* normally oviposit on the upper surface of the leaf midrib of corn and sorghum, but on the commercial sorghum hybrid, CSH 1 oviposition occurs on both the upper and lower surfaces of the midrib. Resistant genotypes of sorghum receive 5–10 times fewer eggs, and those deposited are arranged in a disorderly manner compared to the susceptible hybrid, CSH 1, even in no-choice conditions (Chandra Shekar, 1991; Chandra Shekar et al., 1993a). There is also a marked preference for oviposition on mature leaves of older plants compared to very few eggs on young leaves (Napompeth, 1973), even in no-choice conditions (Fisk, 1978a; Singh and Rana, 1992). Poor growth of *P. maidis* on smaller sorghum plants is associated with less time spent in feeding and a corresponding increase in probing activity (Fisk, 1978a,b). Antibiosis is expressed as increased *P. maidis* mortality, prolonged nymphal development, and reduced fecundity in sorghum (Chandra Shekar, 1991; Shekar et al., 1993b). Fewer offspring of *P. maidis* were noted to be produced on the younger plants (Fisk, 1978a,b), and some sorghum germplasm accessions are relatively less susceptible to leaf sugary exudates (Mote and Shahane, 1993).

Morphological traits, physiological mechanisms, and biochemical factors associated with resistance: The invasion of macropterous adults of *P. maidis* on 20-cm-high sorghum plants has been attributed to physical factors rather than chemical cues, as compact and tightly wrapped whorl leaves around the stem of some sorghum genotypes impart resistance (Agarwal et al., 1978). High levels of nitrogen, sugar, and total chlorophyll content have been shown to be strongly associated with susceptibility to *P. maidis* in sorghum; conversely, genotypes with high phosphorus, potash, and polyphenol content are less preferred by *P. maidis* (Mote and Shahane, 1994). Fisk (1980) observed that mixtures of phenolic acids and esters or their proportions interfere with settling behavior and reduce feeding by *P. maidis* in a complex manner, not only in plants, but also on assayed diets with sorghum. However, higher concentration of phenolic acids was also correlated with a reduced ability of *P. maidis* to locate the phloem tissues. These results indicate that phenolic acids and esters play a dual role in promoting probing activity, but also deterring feeding. Generally, *P. maidis* feeds on the xylem and phloem sap of resistant and susceptible sorghum genotypes, respectively. Xylem fluid is poor in amino acids, but it contains monosaccharides, organic acids, potassium ions, and other minerals, but phloem is rich in sucrose and relatively poor in amino acids and minerals. During the process of feeding, the salivary stylets of *P. maidis* block the phloem, resulting in an accumulation of nitrogen, eventually turning into ammonia and causing chlorosis; and also changing the carbohydrate partitioning patterns affecting the source–sink relationships in the plant. Plant moisture stress causes a conversion of starch to sugars and the accumulation of high concentrations of proline, which acts as a phagostimulant and increases the nitrogen titers. Heavy feeding generally induces proteolysis and a dramatic increase in amino acids, which results in promoting higher fecundity.

Population dynamics: The success of *P. maidis* as a key pest largely stems from its adaptability to continual changes of the environment in both quality and quantity. Continuous cropping, reduced genetic variability in short-duration and high-yielding cultivars, and the application of high levels of nitrogenous fertilizers have compounded the problem. In India, the growing severity of outbreaks of *P. maidis* in the sorghum agroecosystem have been reported from different states such as Maharashtra (AICSIP, 1982, 1987), Karnataka (AICSIP, 1989, 1990), Tamil Nadu (AICSIP, 1987, 1989, 1990), Andhra Pradesh (AICSIP, 1988, 1989, 1990, 1991), and Madhya Pradesh (AICSIP, 1980, 1985). During the last decade, *P. maidis* has established as a key pest in the sorghum agroecosystem. The quantitative analysis of the reaction of a maize/sorghum agroecosystem to infestation by *P. maidis* can be based on the initial wave of immigrant macropterous populations. Continuous cultivation of corn and sorghum stabilizes the spread of infestation of *P. maidis* from one crop or field to another within a localized area. When the crop is harvested, residual populations leave the field to settle on alternate weedy hosts. The macropterous adults are capable of moving short distances from their wild host plants and may take advantage of long-range dispersal by wind and weather frontal systems in search of suitable habitats. By the time that volunteer crop plants become available, the population decreases considerably but is still large enough to colonize and settle on the plants. However, when volunteer crop plants are unavailable or destroyed, macropterous adults survive on alternate hosts as refugia, either in the field or in surrounding areas.

The population dynamics of *P. maidis* have fundamentally different characteristics depending on the strength and form of density-dependent and density-independent interactions. However, it is generally agreed that some degree of negative density dependence is required for population persistence. But density-independent interactions exert their effects and play an important role in determining the components of fitness for *P. maidis*.

Density-dependent interactions: The effects of abiotic factors are indirectly involved in density-dependent interactions on the components of fitness of *P. maidis*, partly by the innate capacity to increase, changes in the quality and quantity of food associated with aging, and the condition of the corn/sorghum plants. The negative effects of density on fitness are intensified by poor host quality (aging, senescing, or wilting) in the production of macropters, from where they disperse to new habitats. However, the levels of dispersal are further influenced by a variety of interactions, including the host canopy, habitat persistence, planting succession, and resource isolation (Napompeth, 1973). With high population densities, intraspecific competition can be intense and adversely affect many components of fitness. However, in crowded conditions, the rates of survival are reduced, individuals develop more slowly, and fecundity is lower because of competition for oviposition sites. Thus, density-dependent interactions result in decreased reproductive rate (R0) and population growth. Reductions in fitness may be associated not only with crowded conditions, but also very low densities. When individuals are very rare and scattered, the searching ability for mates may be limiting. Above all, factors such as host plant

TABLE 7.4 Screening Techniques

Pest	Screening technique
Shoot fly	Selection of "hot spots" and interlards + fishmeal + September last week planting
Corn plant hopper/ shoot bug	Selection of "hot spots" and glass house screening with artificial infestation under multiple and no-choice conditions
Sugarcane aphid	Selection of "hot spots" and laboratory testing with leaf disc and intact leaf techniques under free-choice and no-choice conditions

TABLE 7.5 Mechanisms of Resistance

Pest	Component of resistance
Shoot fly	**1.** Antixenosis for oviposition **2.** Antibiosis for larvae **3.** Recovery resistance in terms of productive tillers
Corn plant hopper/shoot bug	**1.** Antixenosis for adult colonization and oviposition **2.** Antibiosis for nymphal population buildup **3.** Tolerance for plant damage/ transmission of viruses
Sugarcane aphid	**1.** Antixenosis for adult colonization **2.** Antibiosis for population buildup **3.** Tolerance to plant damage

TABLE 7.6 Inheritance of Resistance

Pest/parameter	Inheritance
Shoot fly	Polygenic
Eggs/seedling	Nonadditive gene action (two dominant genes)
Dead hearts	Predominantly additive gene effects, non-additive gene action and high broad sense heritability
Seedling vigor	Tall dominant to short
LSW	Nonadditive gene action and low to moderate heritability
Glossiness	High heritability, recessive (gl)
Trichomes	Low heritability, presense is recessive (tr) on abaxial leaf surface
Epicuticular wax	Waxy bloom dominant (Bm) to sparse bloom
Sugarcane aphid	Single dominant gene, additive gene action and incomplete dominance

TABLE 7.7 Selection Criteria for Resistance to Key Pests in *Rabi* Sorghum

Pest	Selection criterion
Shoot fly	1. Eggs at seedling stage 2. Dead hearts at 14, 21 and 28 DAE
Corn plant hopper	1. Brachypterous and Macropterous adults/plant 2. Total eggs/plant 3. Nymphal population/ plant 4. Plant damage % at 45, 60, 75 DAE
Sugarcane aphid	1. Population density rating on 1–9 scale 2. Plant damage rating on 1–9 scale

TABLE 7.8 Morpho Physiological Traits Associated with Resistance

Pest	Morphophysiological trait
Shoot fly	*Seedling and leaf characteristics*
	High seedling vigor, purple plant pigment, long, erect and narrow leaves, tight and prickle hairs on leaf sheath, glossy trait and high trichome density
	Anatomical characteristics
	Lignification and thickness of cell walls and silicon bodies
	Physiological traits
	Low LSW, low moisture content in seedlings, low stomata number on the dorsal leaf surface, smooth, amorphous epicuticular wax layer on adaxial surface of the basal leaves
Corn plant hopper	Tan plant pigment and compactly wrapped leaves
Sugarcane aphid	Tan plant pigment, small, narrow and erect leaves and waxy leaf lamina

TABLE 7.9 Biochemical Factors Associated with Resistance

Pest	Biochemical factor
Shoot fly	Absence of lysine in the seedling leaves, high levels of calcium, phenols, polyphenol oxidase, silicon and total amino acids and low levels of chlorophyll a and b, dhurrin, magnesium, nitrogen, phosphorus, reducing sugars and total sugars
Corn plant hopper	High levels of phosphorus, potash and polyphenols
Sugarcane aphid	Epicuticular wax on the abaxial surface of the leaves and high levels of HCN content

TABLE 7.10 Off-Season Survival and Reproductive Potential

Pest	Offseason survival	Reproductive potential
Shoot fly	Overwinter on ratoon sorghum	75 eggs/female
Corn plant hopper	*Sorghum verticilliflorum*	60 eggs/female
Sugarcane aphid	*S. halepense*, *Panicum maximum* and *Setaria* spp.	96 nymphs/female

TABLE 7.11 Resistant Genotypes for the Key Pests of *Rabi* Sorghum

Pest	Resistant genotypes	Reference
Shoot fly	RSV 175, RSV 176, RSV 182, RSV 290, IS 18551, IS 2312, ICSV 705, 104A/B, RSE 03	Narkhede et al. (2002)
	ICSR 170, SPV 1156, M 148-138, SPV 1173, IS 33742, A 1, SPV 462, IS 33859, SPV 570, SPV 489, KSV 18R, IS 33843, 5-4-1 (Muguti), IS 18366, IS 12611, SPV 655, SPV 1155, Afzalpur local, SPV 839, IS 4657, IS 33751, DRC 1000, BRJ 17, Selection-3, DRV 20, IS 188758, M 35-1	Balikai and Biradar (2003)
	IS 2191, IS 4481, IS 4516, IS 17596, IS 18366, IS 33714, IS 33717, IS 33722, IS 33740, IS 33742, IS 33756, IS 33761, IS 33764, IS 33810, IS 33820, IS 33839, IS 33843 and IS 33889	Balikai and Biradar (2004)
Shoot bug	IS 18676, IS 19677, IS 19349, IS 18657, IS 18677, IS 19349, PJ 8K	Shekhar et al. (1993b)
	SPV 736, MSH 65	Sekhar et al. (1995)
	CK 60B, Swati and RS 29	Anaji (2005)
Aphid	IS 12664C, IS 12609C, IS 12158C, and IS 12661C, IS 12610, IS 12664, IS 12609, IS 12158, IS 12661, TAM 428, IS 12608, SC 108-3, C 43, RS 29	Teetes et al. (1995)

nutrition, natural enemies, dispersal competition, and physical stresses also influence the population growth and determine the spatial and temporal variation. Diverse vegetation texture has a great influence on the functional and numerical responses of natural enemies by affecting the population abundance.

Density-independent interactions include cultural practices, and microenvironmental and climatic factors, which concurrently operate, but their effects are obscured by endogenous factors. Soil moisture and relative humidity changes in the agroecosystem may exert immense effects on the nutritional quality and quantity of crop/weed hosts. The populations of *P. maidis* fluctuate dramatically within and among seasons in the same patch, as well as spatially among the patches. Population growth can be exponential during different growing seasons, and particularly rapid in local patches, where aggregations of brachypterous adults occur.

Summaries of insect pest screening and resistance sources are given in Tables 7.4–7.11.

7.4.4 Conventional Breeding Methodology for Resistance to Biotic and Abiotic Stresses

Drought resistance breeding methodology: An approach for breeding for drought resistance and yield potential has been established (Reddy, 1986) by selecting breeding materials for specific traits such as emergence under crust, seedling drought recovery, and grain yield under drought-prone and yield-potential areas for early stage drought, for drought recovery, and for grain yield under drought-prone and yield-potential areas alternatively for midseason drought; and for stay-green and nonlodging and grain yield under drought-prone and yield-potential areas alternatively for terminal drought.

Landrace hybrid approach for *rabi* season: *Rabi*-adapted landrace sorghum possesses excellent adaptive characteristics for the prevalent moisture-limiting conditions. It was demonstrated that landrace hybrids would have almost all the characteristics of the landraces preferred by farmers and a 15% superiority in grain yield over cultivated landraces (ICRISAT, 1995). Therefore, restorers of the *rabi* landrace hybrids for breaking the yield plateau in the *rabi* season in India is proposed.

Combining earliness and productivity: It was shown that earliness, grain yield productivity, and biomass can be simultaneously improved by following S_1 family selection in development of a gene pool by incorporating the selected landraces in ms_3 bulk.

Breeding methods for specific purposes: Most materials derived from the pedigree program were released globally than those derived from the population improvement program. Thus, it was confirmed that for a short specific adaptation, pedigree selection appears to be more appropriate than population improvement. Thus, it is evident that the targeted gene pool approach is appropriate for a program that aims at a broad geographic mandate (Reddy et al., 2003).

A moving mean to evaluate large number of progenies: A checkerboard design was used to account for local variation and increase the precision of the experiment for screening for *striga* resistance (Rao, 1985). But this design requires a lot of land and resources, and the fixed checkerboard arrangement is cumbersome to execute under field conditions. Therefore, the moving mean concept for use in screening for resistance to shoot fly, stem borer, striga, and other problems was developed (Reddy, 1993).

$$\text{For example, striga resistance index (\%)} = \frac{[S - (C1 + C2)/2]}{[1 + (C1 + C2)/2]} * 100,$$

where S is the number of striga plants in the test entry, C1 and C2 are the number of striga plants in the two adjacent resistant control plants.

Simultaneous selection and conversion method: A breeding scheme involving simultaneous selection for resistance and grain yield and converting the maintainer selections into male-sterile lines was used effectively to develop male-sterile lines for resistance to pests and diseases in the shortest possible period of 4 years (Reddy et al., 2003).

7.5 BREEDING FOR GRAIN, FODDER, AND NUTRITIONAL QUALITY

Sorghum grains are consumed as food mostly by the rural poor in the semiarid developing countries. The quality, amount, and availability of nutrients are important to the nutritional status of these populations. Sorghum also assumes importance as feed for animals in developed nations, and in the future, it may be utilized as feed in developing countries as well. In general, the nutritional quality of grain as human food is somewhat lower than other cereals (Hamaker and Axtell, 1997). Starch content in the whole grain is about 70%. Crude protein content is about 11% (flour weight basis; 12% moisture). Lysine is the limiting amino acid in sorghum than normal maize and is about 2.0% of total protein or about 0.25% of flour weight (Axtell and Ejeta, 1990), due in part to sorghum's approximately 10–15% higher prolamin content (Hamakar et al., 1995) and lower amounts of high-lysine-containing nonprolamin proteins. Two sources of high-lysine sorghum exist. The first is a naturally occurring Ethiopian mutant identified by Singh and Axtell (1973), with lysine levels of 3.1% and a total crude protein content of 15–17% (Axtell et al., 1974). On a flour weight basis, the lysine content is 0.5%. A second high-lysine gene mutation was induced by chemical mutagenesis, and the mutant was named P-721 opaque (Mohan, 1975). The P721 opaque mutant (designated as P721Q) resulted in a 60% increase in lysine content. The mutant is controlled by a single gene that is simply inherited as a partially dominant factor. The high-lysine lines show high digestibility as well (Mohan, 1975).

Several researchers have evaluated the grain processing and food quality traits of sorghum cultivars, particularly those of recent origin, in comparison with traditionally grown cultivars (Rao et al., 1964; Viraktamath et al., 1972; Pattanayak, 1977, 1978; Juarez, 1979; Scheuring et al., 1982; Obilana, 1982; Khan et al., 1980). Large differences were reported among genotypes for various food quality traits. Consumer cooking trials, laboratory taste panels, and standardized *roti*-making procedures have consistently shown that sorghum cultivars produce *rotis* with vastly different acceptabilities (Rao et al., 1964; Rao, 1965; Anantharaman, 1968; Viraktamath et al., 1972; Waniska, 1976; Murty and House, 1980; Murty et al., 1982a). In general, *roti* made from grains with a pale yellow-white color, with an intermediate endosperm texture, and without a subcoat and with a thin pericarp had acceptable organoleptic quality. The presence of a tough, leathery pericarp produced *rotis* with inferior texture and flavor. Floury grains produced a poor-quality dough, while waxy grains produced a sticky dough and gummy *rotis*. The physical and chemical properties of sorghum that significantly affect *roti* quality are only partially understood (Murty et al., 1982a; Subramanian and Jambunathan, 1982).

Grain quality traits such as bold grain with thin pericarp and luster with semicorneous endosperm are important for making *roti,* while hard grain (small grain size) contributes to lessening grain mold damage. Grain nutritional trait selection programs, such as for high protein or lysine, are not successful in sorghum. Heritabilities of fodder digestibility, high protein, and less fiber content are reasonably high, and it is possible to breed for high grain yield and high fodder quality. Landraces of sorghum exhibit higher protein content than improved varieties. Protein content was higher in landraces like Dood Moghra (12.42%), Yennigar Jola (11.55%), and SPV 1155 (11.37%).

7.5.1 Tests to Predict the Quality of Sorghum

Most sorghum-breeding programs have been focused on agronomic performance in order to ensure food security. However, grain quality is also an essential requirement for the development and the use of improved cultivars. Many quality criteria can be considered regarding the wide range of culinary dishes prepared with sorghum grains. This multiplicity of uses and the difficulty of designing rapid simple methods to evaluate complex parameters have delayed the development of improved cultivars with acceptable grain quality (Deu et al., 2000). It would be useful to identify simple physicochemical tests that could predict the quality of sorghum varieties for use in foods. Such tests have been used effectively for evaluating wheat and rice quality (Heyne and Barmore, 1965; Juliano, 1979). A major problem limiting the development of quick tests to predict sorghum quality was the lack of clearly identified cultivars with good and poor quality.

The discovery that the *Opaque-2* gene in maize improves protein quality has stimulated great interest among breeders, nutritionists, and biochemists, and considerable progress has been made toward genetic improvement of plant protein quality in other cereals. Grain quality can perhaps be considered to consist of two main parts: (i) evident quality based on appearance, flavor, and cooking quality characteristics; and (ii) cryptic quality based on nutritional value.

The amylose content of sorghum does not vary as much as that of rice and has not been clearly shown to be related to food quality (Waniska, 1976; Akingbala et al., 1982; Subramanian and Jambunathan, 1982). The amylose content of 495 nonwaxy sorghum genotypes varied from 20% to 30% of the starch in the endosperm. Additional information to determine the potential value of amylase content is needed; but the amylose content of sorghum does not appear to be as important as it is for rice.

A test to predict the color of tortillas and alkali *to* is to soak five kernels of sorghum overnight or by boiling for 2 h has been used by Khan et al. (1980). The color of the cooked kernels or steeped kernels was evaluated subjectively by comparing with known standards. A variation of this method was used by Iruegas et al. (1982) in Mexico. Waniska (1976) modified the alkali spread test that has been used successfully with rice by applying the alkali to milled kernels of sorghum. The method clearly distinguished waxy from nonwaxy kernels but could not make clear distinctions among the International Sorghum Food Quality Trial (ISFQT) samples. A major problem may be variability in milling damage to the decorticated sorghum kernels, which causes variation in the rate of alkali absorption.

A number of tests based on gelatinization of flour water dispersions followed by measuring the consistency of the gel appear promising but must be evaluated more carefully (Da et al., 1982; Murty et al., 1982b). A significant association of swelling power and starch solubility with the cooking properties of boiled sorghum was reported (Subramanian et al., 1982). However, the correlations are low, and more information is required to evaluate the potential of these tests.

The amylograph cooking characteristics of sorghum starches and flour have been tentatively related to the food quality of sorghum (Waniska, 1976). The setback viscosity of the sorghum starch and flour was high for sorghums with acceptable thick porridge making quality and was low for sorghums with acceptable *roti*-making properties. Similar observations were recorded by Desikachar and Chandrasekhar (1982).

The particle size distribution and starch damage in a flour affect the quality of the flour significantly. Murty and House (1980) studied the flour particle size index (PSI) of several cultivars using the method of Waniska (1976) and found a range of 25–80 PSI among genotypes. PSI values were affected by grinding and sieving methods and were subject to considerable error. However, the PSI was correlated consistently with the texture of the endosperm. Particle size measurements are important and should be given high priority in future research into sorghum quality testing procedures.

The amount of water absorbed by the grains after soaking them in water for 5 h at room temperature has been expressed as a percentage of water absorption (Murty and House, 1980). This parameter showed a broad range of variation among various grain types and was negatively correlated with *roti* quality. Desikachar and Chandrasekhar (1982) found that water uptake of flour was related to dough and *roti* quality.

The single most critical property of sorghum foods that affects their acceptance is related to texture. Simple objective methods to measure texture are needed and are not readily available. Maintaining quality is a critical factor that relates to texture measurements. The Instron universal testing machine has been used to measure the texture of *to*, tortillas, and *roti*, as well as the hardness of individual sorghum kernels (Waniska, 1976; Johnson et al., 1979; Da et al., 1982). However, the Instron is an expensive, sophisticated instrument that requires considerable expertise to operate. It is not practical in routine plant-breeding programs, but it is extremely useful to determine basic information on texture. Then the basic information can be used to develop "quick-and-dirty" tests that can provide screening. A few simple tests have been applied to sorghum *to* (Waniska, 1976; Da et al., 1982), such as stickiness measured using double pan balance and softness using a penetrometer. Both techniques can be used to distinguish between *to* samples prepared from a single head of sorghum. The penetrometer provides a relatively low-cost objective method that can improve upon the use of subjective methods.

The Hunter Colorimeter, Agtron, and other instruments can be used to measure color objectively in terms of reflectance and a and b values that measure the intensity of the primary colors. The instruments are expensive, require sophisticated maintenance and constant voltage, and are generally impractical in routine breeding programs. An effective

TABLE 7.12 Sorghum Parents with Significant gca Effect for Protein Content and Other Important Chemical Constituents and Grain Yield

Parents	Protein (%)	Starch (%)	Soluble/free sugars (%)	Grain yield (%)
Barshi Joot	1.17**	−0.10**	0.05	2.09**
Dagdi Solapur	1.13**	−1.59**	0.10**	−6.87**
Yennigar Jala	0.48**	−0.49**	−0.08**	5.81**
Ramkhe	0.45**	−0.05	0.28**	3.75**
SPV 1457	0.28**	0.38**	0.11**	−0.88**

***Significant at $P < 0.01$.*
Source: Adapted from Deshpande et al. (2003).

and inexpensive method is to compare the color of the product with that of standard color charts. Murthy et al. (1979) have used the Munsell soil color charts to describe the colors of *roti*. A standard set of colors representing the range observed for the particular food product can be purchased inexpensively and easily. The correct Munsell plates can be selected by using an instrument to determine the range in color values for an array of specific foods, or the soil color charts can be compared until the appropriate color match is obtained (Rooney and Murty, 1982).

The proportion of floury versus corneous endosperm in the kernel is called *endosperm texture*. Endosperm texture is related to the hardness, milling properties, and cooking characteristics of the flour. The most common method to evaluate endosperm texture is to cut 10–20 individual, sound representative kernels with a pocket knife. Then the relative proportion of corneous to floury endosperm is rated subjectively on a scale of 1–5, where 1 means 81–100%; 2 means 61–80%; 3 means 41–60%; 4 means 21–40%, and 5 means 0–20%. The texture of the endosperm is subject to environmental effects; variation among individual grains within a sample is common. In some samples, 20 or more kernels are sampled to secure an average value. Sophisticated laboratory facilities are not needed to do this, and considerable progress can be made by using it in selection programs. More accurate measurements of texture have been made by Munck et al. (1982) and Kirleis and Crosby (1982), who measured the relative proportion of corneous to floury endosperm in individual kernels. In Munck's procedure, highly sophisticated equipment is required, which limits its application only to basic research. The Vicker's hardness tester can measure the hardness of individual endosperm cells (Munck et al., 1982). Endosperm texture is related to various indices of grain hardness that have been developed using standard milling and sifting procedures (Maxson et al., 1971), or, alternatively, by recording the time required to dehull a standard quantity of grain to a specified level and recording the extent of breakage in the recovered endosperm (Shepherd, 1979; Oomah et al., 1981). Although these measurements are subject to error due to the interaction of grain shape with the abrasive mechanism, they seem to be quite reliable and are related to endosperm texture scores or breaking strength measurements taken with the Kiya rice hardness tester. Kernel shape affects the measurements taken with the kiya tester; flat and turtle-beaked sorghum kernels frequently give erroneously high values (Murty and House, 1980).

7.5.2 Genetics of Grain Quality Traits

Genetic studies revealed that the landraces were good general combiners for higher protein content, low starch, and high soluble/free sugar content (Table 7.12). Rao et al. (1982) also observed that a positive gca effect for protein content was accompanied by a negative gca effect for starch content. Correlation studies indicated that for protein content improvement in terms of quantity, the starch content must decrease or compensate for biochemical levels in the grain during the grain development stage (Deshpande et al., 2003).

7.5.3 Selection Criteria

Crop improvement programs generate a range of segregating material by making crosses between lines possessing good agronomic characteristics, disease and pest resistance, drought resistance, and other traits. These programs are confronted with the problem of choosing and advancing families that combine several economic characters, including food quality. Currently, there are no clear-cut methods to help sorghum breeders select for good food quality, as there are in wheat- and rice-breeding programs. Breeders in national or regional programs may select cultivars suitable for a particular product, while those in international programs may find it necessary to identify cultivars that are suitable to make a range of foods. Obviously, from the review discussed earlier, for most sorghum foods, there is no clear identification of the physicochemical properties of the grain that can be used to predict preferred quality, although several tests of possible significance have been reported. Simple tests tailored for laboratory use are urgently required to permit rapid progress in breeding for food quality in the developing countries.

For quality testing in a breeding program, parents with sources of good food quality, grain mold and weathering resistance, and other yield limiting factors are crossed to generate F_1 hybrids. In the F_2 generation, selection is done for absence of testa, colorless thin pericarp, appropriate endosperm texture, medium size, round shape, tan-colored plant, grain mold resistance, and other good agronomic characteristics, including straw-colored glumes in the field. In F_3 generation, desired grain and glume characteristics are selected in the field. Then selection for a light color reaction of grains with KOH is done in the laboratory. The percentage of water absorption of grain is checked. In the F_4 generation, selection is done for grain mold resistance in the field. The hardness and milling quality with small grain samples is done in the laboratory, and gel viscosity is checked. In the F_5 generation, selection is done for milling quality, flour particle size, and gel viscosity. In the F_5 generation, multilocational tests are done in the field. The laboratory taste panel studies on the selected entries for the appropriate food system are done. In the F_7 generation, laboratory taste panel studies are carried out on the food product, and in the F_8 generation, advanced yield tests and consumer tests are carried out on a few selected entries.

Considerable progress in quality breeding is possible by an empirical selection of the precise endosperm texture, while the food technologists and chemists continue research into the development of objective physicochemical quality tests. Selection in the F2 generation should be for those grain characters that are controlled by major genes (Rooney and Miller, 1982), such as colorless (rryy or R-yy?) and thin pericarp (Z-), absence of testa (b1 b1 or b2 b2), endosperm texture, and tan plant color (pp-). These characters could be selected by subjective methods in the field. A laboratory is not required. Where the sorghum crop is expected to mature toward the end of the rainy season, grain mold resistance is an important selection criterion. Grain quality characters, associated with the preferred food quality traits and mold resistance, may not necessarily be the same, and the best recombinants that combine these two characters should be chosen. Grain of individual F3 selections from the off-season crop could be used for laboratory testing of KOH color reaction. Since grain quantities might be limiting, samples from selections in the F4 and F5 generations might be used either for the study of gel viscosity or milling and flour quality. The evaluation of the qualities of the product per se needs to be done only on those entries that are selected for improved yield, adaptation, and other traits. It is important that assessments on the food product be conducted with grain harvested in the main crop season for which the variety is intended. Consumer tests at the farmer's level should use only the most promising cultivars from multilocation tests of yield and adaptation. The most preferred local varieties should always be included in these tests for comparison. This scheme could be modified to suit the major objectives of any breeding program involved in the improvement of yield-limiting factors like disease and pest resistance.

Association analysis of 300 accessions between 333 single-nucleotide polymorphisms (SNPs) in candidate genes and loci and grain quality traits resulted in eight significant marker–trait associations. An SNP in starch synthase *IIa* (*SSIIa*) gene was associated with kernel hardness (KH) with a likelihood ratio–based R^2 ($R_{LR}{}^2$) value of 0.08, a SNP in starch synthase (*SSIIb*) gene was associated with starch content with an $R_{LR}{}^2$ value of 0.10, and a SNP in loci *pSB1120* was associated with starch content with an $R_{LR}{}^2$ value of 0.09 (Sukumaran et al., 2012).

A chromosomal segment located on LG F was found to play a major role in grain quality (Rami et al., 1998). In a RIL, four QTLs for flouriness, dehulling yield, amylose content, and mold resistance during germination were detected, which were very closely linked with each other. For the same linkage group, four important QTLs were detected on another RIL for flouriness, kernel friability, KH, and amylose content (Deu et al., 2000). These results are consistent with the close correlation found between amylose content and endosperm texture. The *B2/b2* gene controlling the presence of a high-tannin testa layer in the grain has been phenotypically mapped in this region. This explains the visual quality criteria used by breeders (a quality grain has a vitreous and hard endosperm and does not have testa). QTLs for kernel friability, KH, dehulling yield, and protein content were detected on LG A for RIL249. This is consistent with the close correlations found between these traits; the *guinea* allele conferred a higher mechanical resistance to the kernel. Colocalizations of QTLs for protein content and kernel physical properties were also found in other segments of the genome: for example, QTLs for protein content and flouriness had a same map position on LG C. Major QTLs for amylose content, dehulling yield, and kernel texture are not linked to productivity traits, while they are colocated with major QTLs for the mold resistance during germination and the tannin content. No genetic obstacle was observed for the recombination of genetic components of both productivity and

TABLE 7.13 Average Grain and Fodder Yield of New Varieties in Comparison to M35-1 over 3 Years Across All Locations from 1999–2000 to 2001–2002

Entry	No. of trials	Maturity (days)	Grain yield (t/ha)	Increase over M35-1 (%)	Fodder yield (t/ha)	Increase over M35-1 (%)
Phule Yashoda	585	125	1.58	25	3.99	26
Parbhani Moti	176	120	1.86	47	4.39	39
Mauli	196	117	1.21	−5 (20%)[a]	3.80	20
M35-1 (Check)	59	120	1.27	–	3.16	–

[a]*The overall average is based on trials in all type of soils. However, the variety is released for shallow to medium soils where normally M35-1 does not perform well. It gives about 20% more yield than M35-1 in shallow soils.*

TABLE 7.14 Traits Other Than Yield for Which New Varieties Excel M35-1 (Farmer Perceptions)

Variety	Traits other than grain yield
P. Yashoda	Tall plants, bold grain with easy threshability, better flour recovery with good keeping quality of flour
P. Moti	Very bold and attractive lustrous grains which fetch a higher market price; fodder preferred by animals; very good roti quality
Mauli	Fodder preferred by cattle; tolerant to aphids and leaf sugary disease; tolerant to terminal drought.

FIGURE 7.15 A woman farmer selecting high-yielding earheads of sorghum as a part of participatory varietal selection.

grain quality in *caudatum* × *guinea* crosses (Deu et al., 2000). A large number of QTLs are involved in the regulation of the quantity of storage proteins in the grain, especially albumins and prolamins, and several QTLs for albumin quantity have the same map position as QTLs for grain hardness, flouriness, and dehulling yield (Rami, 1999).

7.6 PARTICIPATORY VARIETAL SELECTION

Participatory varietal selection: Growers of *rabi* sorghum in the project states did not have any alternative to the popular but very old local varieties Maldandi and M35-1, a selection from Maldandi. Maldandi was released in 1930 in Maharashtra, and M35-1 was released for Maharashtra, Karnataka, and AP in 1984. This low genetic diversity made the *rabi* sorghum crop potentially highly vulnerable to disease. It also meant that farmers were not benefiting from decades of plant breeding research because they were not adopting new varieties that had been released, largely because plant breeders had concentrated on yield rather than providing varieties that at least matched M35-1 for grain quality (Witcombe et al., 1998). The participatory varietal selection (PVS) approach removed the major limitation of the public-sector breeding program by selecting varieties in farmers' fields under their own management, and thus minimized genotype × environment interaction that resulted from the high input and management conditions on research stations.

The market price of sorghum varieties is generally based on grain size, grain color/luster, and whether the variety was a composite or a hybrid. Participatory trials were conducted by farmers who were provided 1- or 2-kg seed of each new variety. The farm trials were conducted on 27 varieties from 1999–2000 to 2001–2002. They were baby trials, where a new variety was grown alongside the local check; or mother trials, where all entries were grown together in a single replicate (Witcombe, 2002). Each farmer's field was considered a replication. All trials were conducted under the farmer's normal management. The farm trials were conducted by the six centers of the All India Coordinated Sorghum Improvement Project (AICSIP) located at the Natural Resource Conservation Service (NRCS), the state agricultural universities in Maharashtra and Karnataka, and at ICRISAT with the help of seven nongovernmental organizations (NGOs). A total of 720, 838, and 1026 farm trials were conducted in 1999–2000, 2000–2001, and 2001–2002, respectively, and were jointly monitored by researchers and farmers. Both preharvest and postharvest traits were evaluated by both participating and nonparticipating farmers. M35-1 is the most popular variety of *rabi* sorghum and was used as a check to compare new varieties. Overall, in trials between 1999–2000 and 2001–2002, the new varieties gave 19–47% more grain yield and 14–43% more fodder yield (Tables 7.13 and 7.14) (Figure 7.15).

In addition to higher yield, farmers recorded a number of other preferred traits in the new varieties. The most important was the drought tolerance of Mauli under rainfed conditions in shallow soils with very poor water-holding capacity.

References

Abdulai, A.L., Parzies, H., Kouressy, M., Vaksmann, M., Asch, F., Brueck, H., 2012. Yield stability of photoperiod sensitive sorghum (*Sorghum bicolor* L. Moench) accessions under diverse climatic environments. Int. J. Agric. Res. 7, 17–32.

Abu, A.A., Joseph, D., Suprasanna, P., Choudhury, R.K., Saxena, A., Bapat, V.A., 2002. Study of trace element correlations with drought tolerance in different sorghum genotypes using energy dispersive X-ray fluorescence technique, In En. Biol. Trace. Elem. Res. 85 (3), 255–267.

Acevedo, E., Ceccarelli, S., 1989. Role of physiologist-breeder in a breeding program for drought resistance. In: Baker, F.W.G. (Ed.), Drought Resistance in Cereals. CAB International, UK, pp. 117–139.

Ackerson, R.C., 1981. Osmoregulation in cotton in response to water stress. Plant Physiol. 67, 489–493.

Ackerson, R.C., Krieg, D.R., Sung, S.J.M., 1980. Leaf conductance and osmoregulation of field grown sorghum. Crop Sci. 20, 10.

Agarwal, R.K., Verma, R.S., Bharaj, G.S., 1978. Screening of sorghum lines for resistance against shoot bug, *Peregrinus maidis* Ashmead (Homoptera: Delphacidae). JNKVV Res. J. 12, 116.

Aggarwal, P.K., Sinha, S.K., 1984. Differences in water relations and physiological characteristics in leaves of wheat associated with leaf position on the plant. Plant Physiol. 74, 1041–1045.

Agrawal, B.L., House, L.R., 1982. Breeding for pest resistance in sorghum. Sorghum in the Eighties. Proceedings of the International Symposium on Sorghum. 2–7 November 1981, ICRISAT, Patancheru, India, pp. 435–446.

Agrawal, V.D., Sharma, D., 1972. Estimates of combining ability of maize lines having different intensities of selfing. Indian J. Agric. Sci. 42, 565–568.

AGRHYMET, 1992. Atlas of Agroclimatic Zone Countries of CILSS. AGRHYMET, Niamey, Rep. Niger, 162 p.

Ahmed, M., Fayyaz-ul-Hassen, Qadeer, U., Aslam, M.A., 2011. Silicon application and drought tolerance mechanism of sorghum. Afr. J. Agric. Res. 6 (3), 594–607.

AICSIP (All India Coordinated Sorghum Improvement Project), 1979–2003. Annual Progress Reports for Each Year. All India Coordinated Sorghum Improvement Project. ICAR & Cooperative Agencies, Indian Council of Agricultural ResearchAll India Coordinated Sorghum Improvement Project), 1979–2003, New Delhi.

Akingbala, J.O., Rooney, L.W., Palacios, L.G., Sweat, V.E., 1982. Thermal properties of sorghum starches. In: Rooney, L.W, Murty, D.S, Mertin, J.V (Eds.), Proceedings of the International Symposium on Sorghum Grain Quality. ICRISAT Center, Patancheru, India, pp. 251–261.

Alam, S., Asghar, A., Qamar, I.A., Arshad, M., Sheikh, S., 2001. Correlation of economically important traits in sorghum bicolor varieties. Online J. Biol. Sci. 1 (5), 330–331.

Ali, M.G., Naylor, R.E.L., Mathews, S., 2006. Distinguishing the effects of genotype and seed physiological age on low temperature tolerance of rice (*Oryza sativa* L.). Exp. Agric. 42, 337–349. Available from: http://dx.doi.org/10.1017/S0014479706003619.

Ali, M.A., Abbas, A., Niaz, S., Zulkiffal, M., Ali, S., 2009a. Morpho-physiological criteria for drought tolerance in sorghum (*Sorghum bicolor*) at seedling and post-anthesis stages. Int. J. Agric. Biol. 11, 674–680.

Ali, M.A., Niaz, S., Abbas, A., Sabir, W., Jabran, K., 2009b. Genetic diversity and assessment of drought tolerant sorghum landraces based on morph-physiological traits at different growth stages. Plant Omics J. 2, 214–227.

Ali, M.A., Hussain, M., Khan, M.I., Ali, Z., Zulkiffal, M., Anwar, J., et al., 2010. Source-sink relationship between photosynthetic organs and grain yield attributes during grain filling stage in spring wheat (*Triticum aestivum*). Int. J. Agric. Biol. 12, 509–515.

Ali, M.A., Jabran, K., Awan, S.I., Abbas, A., Ehsanullah, Zulfikar, M., Acet, T., et al., 2011. Morphophysiological diversity and its implications for improving drought tolerance in grain sorghum at different growth stages. Aust. J. Crop Sci. 5 (3), 308–317.

Allard, R.W., 1960. Principles of Plant Breeding. Wiley, New York, NY.

Allen, D.J., Ort, D.R., 2001. Impacts of chilling temperatures on photosynthesis in warm-climate plants. Trends Plant Sci. 6, 36–42.

Al-Naggar, A.M.M., El-Kadi, D.A., Abo-Zaid, Z.S.H., 2007. Genetic analysis of drought tolerance traits in sorghum. Egypt. J. Plant Breed. 11 (3), 207–232.

Amsalu, A.A., Bapat, D.R., 1990. Diallel analysis of combining ability in sorghum. J. Maharashtra Agric. Univ. 15 (3), 302–305.

Anahosur, K.H., Rao, M.V.H., 1977. A note on epidemics of sorghum in regional research station. Dharwad. Sorghum Newsl. 20, 22.

Anahosur, K.H., Naik, S.T., 1985. Relationship of sugars and phenols on root and stalk rot of sorghum with charcoal rot. Indian Phytopathol. 38, 131–134.

Anahosur, K.H., Patil, S.H., 1983. Assessment of losses in sorghum seed weight due to charcoal rot. Indian Phytopathol. 36, 85–88.

Anahosur, K.H., Rao, M.V.H., Patil, S.H., Hegde, R.K., 1974. Assessment of losses in sorghum seed eight due to charcoal rot. Indian Phytopathol. 24, 85–88.

Anahosur, K.H., Patil, S.H., Hedge, R.K., 1985. Relationship of salt, sugars and phenols with charcoal rot of sorghum. Indian Phytopathol. 38, 335–337.

Anahosur, K.H., Naik, S.T., Nadaf, S.K., 1987. Correlation and path coefficient analysis of loss in seed weight due to charcoal rot in sorghum. Indian Phytopathol. 40, 478–481.

Anaji, R., 2005. Studies on Crop Loss Estimation and Management of Shoot Bug, *Peregrinus maidis* (Ashmead) in *Rabi* Sorghum (M.Sc. (Agri.) thesis). Univ. Agric. Sci. Dharwad (India).

Anandan, A., Huliraj, H., Veerabadhiran, P., 2009. Analysis of resistance mechanism to *Atherigona soccata* in crosses of sorghum. Plant Breed. 128 (5), 443–450.

Ananthraraman, P.V., 1968. Grain quality in sorghum. Sorghum Newsl. 11, 49–53.

Andrews, D.J., Reddy, B.V.S., Talukdar, B.S., Reddy, L.J., Saxena, N.P., Saxena, K.B., 1983. Breeding for drought resistance. House Symposium on Drought Resistance. ICRISAT, Patancheru, India, pp. 1-14.

Andrew, K.B., Hammer, G.L., Henzell, R.G., 2000. Does maintaining green leaf area in sorghum improve yield under drought II. Dry mater production and yield. Crop Sci. 40, 1037–1048.

Annandale, J.E., Hammers, P.S., Nel, P.C., 1984. The effect of soil fertility on the vegetative growth, yield and water use of wheat. Crop Prod. 13, 30.

Apotikar, D.B., Venkateswarlu, D., Ghorade, R.B., Wadaskar, R.M., Patil, J.V., Kulwal, P.L., 2011. Mapping of shoot fly tolerance loci in sorghum using SSR markers. J. Genet. 90 (1), 59–66.

Arjomand, A., Siadat, S.A., Dezfuli, A.H., Rahnama, A., 2000. Study of some physiological indices of drought tolerance of forage sorghum (*Sorghum bicolor* var. Speed Feed) in the presence of potassium ions. J. Agric. Sci. 6 (2), 113–124.

Arkin, G.F., Monk, R.L., 1979. Seedling photosynthetic efficiency of a grain sorghum hybrid and its parents. Crop Sci. 19, 128–130.

Aruna, C., Padmaja, P.G., 2009. Evaluation of genetic potential of shoot fly resistant sources in sorghum *(Sorghum bicolor* (L.) Moench). J. Agric. Sci. 147 (1).

Aruna, C., Bhagwat, V.R., Sharma, V., Hussain, T., Ghorade, R.B., Khandalkar, H.G., et al., 2011a. Genotype x environment interactions for shoot fly resistance in sorghum (*Sorghum bicolor*(L.) Moench): response of recombinant inbred lines. Crop Prot. 30, 623–630. Available from: http://dx.doi.org/10.1016/j.cropro.2011.02.007.

Aruna, C., Bhagwat, V.R., Madhusudhana, R., Sharma, V., Hussain, T., Ghorade, R.B., et al., 2011b. Identification and validation of genomic regions that affect shoot fly resistance in sorghum [*Sorghum bicolor* (L.) Moench]. Theor. Appl. Genet. 122, 1617–1630. Available from: http://dx.doi.org/10.1007/s00122-011-1559-y.

Ashby, E., 1937. Physiological, biochemical and genetic basis of heterosis. Ann. Bot. 1, 11–41.

Assar, A.H.A., Joseph, D., Suprasanna, P., Choudhury, R.K., Saxena, A., Bapat, V.A., 2002. Study of trace element correlations with drought tolerance in different sorghum genotypes using energy-dispersive X-ray fluorescence technique. Biol. Trace Elem. Res. 85 (3), 255–267.

Assay, K.H., Johnson, D.A., 1983. Breeding for drought resistance in range grasses. IOWA St. J. Res. 57, 441–444.

Asthana, O.P., Asthana, N., Sharma, R.L., Shukla, K.C., 1996. Path analysis for immediate components of grain yield in exotic sorghum II 100-grain weight. Adv. Plant Sci. 9 (2), 29–32.

Atkins, R.E., 1979. Yield comparisons among a short statured seed parents of grain sorghum. IOWA State J. Res. 53, 269–272.

Atkins, R.E., 1988. Registration of 19 sorghum parental lines. Crop Sci. 28, 387.

Audilakshmi, S., Aruna, C., 2005. Genetic analysis of physical grain quality characters in sorghum. J. Agric. Sci. 143 (3), 267–273.

Avadhani, K.K., Ramesh, K.V., 1979. Charcoal rot incidence in some released and pre-released varieties and hybrids (Bijapur). Sorghum Newsl. 27, 37.

Awala, S.K., Wilson, J.P., 2005. Expression and segregation of stay-green in pearl millet. Int. Sorghum Millets Newsl. 46, 87–100.

Awari, V.R., Gadakh, S.R., Shinde, M.S., Kusalkar, D.V., 2003. Correlation study of morpho-physiological and yield contributing characters with grain yield in sorghum. Ann. Plant Physiol. 17 (1), 50–52.

Axtell, J.D., Ejeta, G., 1990. Improving sorghum grain protein quality by breeding. In: Ejeta, G. (Ed.), Proceedings of the International Conference on Sorghum Nutritional Quality. Purdue University. West Lafayette, IN, pp. 117–125.

Axtell, J., Kapran, I., Ibrahim, Y., Ejeta, G., Andrews, D.J., Coors, J.G., et al., 1999. Heterosis in sorghum and pearl millet. The Genetics and Exploitation of Heterosis in Crops in Proceedings of an International Symposium. CIMMYT, Mexico, pp. 375–386.

Axtell, J.D., Mohan, D., Cummings, D., 1974. Genetic improvement of biological efficiency and protein quality in sorghum. Proceedings 29th Annual Corn and Sorghum Research Conference, Chicago. American Seed Trade Association, Washington, DC, pp. 29–39.

Azhar, F.M., Hussain, S.S., Mahamood, I., 1998. Heterotic response of F1 sorghum hybrids to NaCl salinity at early stage of plant growth. Pak. J. Sci. Ind. Res. 41, 50–53.

Badhe, P.L., Patil, H.S., 1997. Line × tester analysis in sorghum. Ann. Agric. Res. 8 (3), 281–284.

Bakheit, B.R., 1989. Variability and correlation in grain sorghum genotypes (*Sorghum biocolor* (L.) Moench) under drought conditions at different stages of growth. Asian J. Agric. Sci. 20 (4), 227–237.

Balakotaiah, K., Rana, B.S., Tripathi, D.P., Rao, N.G.P., 1975. Genetic analysis of some exotic and Indian crosses in sorghum: inheritance of resistance to sorghum shoot fly. Indian J. Genet. Plant Breed. 35, 344–347.

Baldwa, P.L., Desai, K.B., SBS, T., 1979. Estimation of combining ability in F2 population of sorghum. Sorghum Newsl. 22, 14–15.

Baldy, C., Konate, J.M., Olufayo, Ruelle, O., Fernandes, A., 1993. Drought resistance of grain sorghum in Mediterranean climate and optimum management of a limited water supply. Sci. et Chang. Planet. Secheresse. 4 (2), 85–93.

Balikai, R.A., 2001. Bioecology and Management of Aphid, *Melanaphis sacchari* (Zehntner) on Rabi Sorghum (Ph.D. thesis). University of Agricultural Sciences, Dharwad.

Balikai, R.A., 2004. Chemical control of sugarcane leaf aphid, *Melanaphis sacchari* (Zehntner) on rabi sorghum. Agric. Sci. Dig. 24 (2).

Balikai, R.A., Biradar, B.D., 2003. Field screening of sorghum entries against shoot fly, *Atherigona soccata* (Rondani). Agric. Sci. Dig. 23 (4), 282–284.

Balikai, R.A., Biradar, B.D., 2004. Performance of sorghum germplasm lines against shoot fly in *rabi* sorghum. Agric. Sci. Dig. 24 (1), 63–64.

Balikai, R.A., Biradar, B.D., 2007. Field evaluation of sorghum parental lines for resistance to shoot fly and aphid. Agric. Sci. Dig. 27 (4), 291–292.

Balikai, R.A., Lingappa, S., 2002a. Avoidable loss in different genotypes of sorghum due to aphid, *Melanaphis sacchari* (Zehntner). J. Aphidology. 16, 125–128.

Balikai, R.A., Lingappa, S., 2002b. Biophysical and biochemical basis of resistance in sorghum genotypes to aphid, *Melanaphis sacchari* (Zehntner) (Homoptera: Aphididae). J. Aphidology. 16 (1 and 2), 59–66.

Balikai, R.A., Lingappa, S., 2003. Management of aphid, *Melanaphis sacchari* (Zehntner) on *rabi* sorghum. J. Aphidology. 17 (1&2), 67–72.

Bangar, M.S., Patel, C.C., Kher, H.R., Parmar, H.P., 2012. Biochemical basis of resistance in forage sorghum to shoot fly *Atherigona soccata* (Rondani). Indian J. Entomol. 74 (2).

Bapat, D.R., Mote, U.N., 1982. Sources of shootfly resistance in sorghum. J. Maharashtra Agric. Univ. 7, 238–240.

Bapat, D.R., Gujar, A.R., 1990. Effect of soil moisture stress on the yield and yield attributes of different sorghum genotypes. J. Maharashtra Agric. Univ. 15 (2), 185–188.

Barbara George-Jaeggli. 2009. Physiology and Genetics of Height-Yield Associations in Sorghum (Ph.D. thesis). School of Land. Crop and Food Sciences. The University of Queensland.

Barrs, H.D., Weatherly, P.E., 1962. A re-examination of relative turgidity for estimating water deficits in leaves. Aust. J. Biol. Sci. 15, 413–428.

Basnayake, J., Cooper, M., Ludlow, M.M., Henzell, R.G., Snell, P.J., 1995. Inheritance of osmotic adjustment to water stress in three-grain sorghum crosses. Theor. Appl. Genet. 90 (5), 675–682.

Basnayake, J., Sihathep, V., Sonekham, P.S., Senthonghae, M., Sibounheuang, V., Phamixay, V., et al., 2003. Effects of time of planting on agronomic and yield performance of several rice cultivars under various temperature conditions in Lao PDR. Proceedings 11th Agron. Conf. Melbourne. 1–6 February 2003, Available from: <http://www.regional.org.au/au/asa/2003/>.

Bechoux, N., Bernier, G., Lejeune, P., 2000. Environmental effects on the early stages of tassel morphogenesis in maize (*Zea mays* L.). Plant Cell Environ. 23, 91–98. Available from: http://dx.doi.org/10.1046/j.1365-3040.2000.00515.x.

Behura, B.K., Bohidar, K., 1983. Effect of temperature on the fecundity of five species of aphids. Pranikee. 4, 23–27.

Belawatagi, S.F., 1997. Genetic Studies on Sweet Stalk Based Sorghum [*Sorghum bicolor* (L.) Moench] Hybrids (M.Sc.(Agriculture) thesis). University of Agricultural Sciences, Dharwad.

Beltrano, J., Ronco, M.G., 2008. Improved tolerance of wheat plants (*Triticum aestivum* L.) to drought stress and rewatering by the arbuscular mycorrhizal fungus *Glomus claroideum*: effect on growth and cell membrane stability. Braz. J. Plant Physiol. 20, 29–37.

Berenji, J., 1988. Evaluation of combining ability and heterosis and analysis of yield components in grain sorghum. Sirak i Lekovito Bilje. 20, 47–49.

Bhadouriya, N.S., Saxena, M.K., 1997. Combining ability studies in sorghum through diallel analysis. Crop Res. 14, 253–256.

Bhagmal, B., Mishra, U.S., 1985. Hybrid vigour for culm and leaf characters in forage sorghum. Sorghum Newsl. 28, 6–7.

Bhagwat, V.Y., 1975. Charcoal rot of sorghum. Sorghum Newsl. 18, 45–46.

Bhale, N.L., Borikar, S.T., 1982. Combining ability for yield and yield components in *rabi* sorghum. J. Maharashtra Agric. Univ. 7, 247–249.

Bhale, N.L., Khidse, S.R., Borikar, S.T., 1982. Heterosis for proline accumulation in sorghum. Indian J. Genet. Plant Breed. 42, 101.

Bharud, R.W., Durgudge, A.G., Takate, A.S., 1995. Effect of soil [profile depth on physiological parameters and yield of rabi sorghum genotypes under dryland conditions]. Mysore J. Agric. Univ. 29, 315–319.

Bhavsar, V.V., Borikar, S.T., 2002. Combining ability studies in sorghum involving diverse CMS steriles. J. Maharashtra Agric. Univ. 27, 35–38.

Bhutta, W.M., 2007. The effect of cultivar on the variation of spring wheat grain quality under drought conditions. Cereal Res. Commun. 35, 1609–1619.

Bichkar, R.P., 2005. Study of Combining Ability and Inheritance of Physiological Traits and Grain Yield in *Rabi* Sorghum (Thesis abst. M.Sc.(Agri), thesis submitted to M.P.K.V). Rahuri, (Maharashtra).

Bidinger, F.R., Mahalakshmi, V., Rao, G.D.P., 1987a. Assessment of drought resistance in pearl millet (*Pennisetum americanum* L. Leeke) I. Factors affecting yields under stress. Aust. J. Agric. Res. 38, 37–48.

Bidinger, F.R., Mahalakshmi, V., Rao, G.D.P., 1987b. Assessment of drought resistance in pearl millet (*Pennisetum americanum* L. Leeke) II. Estimation of genotype response to stress. Aust. J. Agric. Res. 38, 49–59.

Biradar, B.D., 1995. Genetic Studies Involving Diverse Sources of Cytoplasmic-Genetic Male Sterility in Sorghum [*Sorghum bicolor* (L.) Moench] (Ph.D. thesis). University of Agricultural Sciences, Dharwad.

Biradar, S.G., Borikar, S.T., Chundurwar, R.D., 1984. Genetic analysis of shoot fly resistance in sorghum. Sorghum Newsl. 27, 109.

Biradar, B.D., Parameswarappa, R., Patil, S.S., Parameswargoud, P., 1996. Inheritance of seed size in sorghum (*Sorghum bicolor* (L.) Moench). Crop Res. 11, 331–337.

Biradar, B.D., Vastrad, S.M., Balikai, R.A., Nidagundi, J.M., 2000. Combining ability studies in *rabi* sorghum. Karnataka J. Agric. Sci. 13 (2), 721–723.

Bishonoi, L.K., 1983. Canopy Development, Light Interception, Leaf Photosynthesis and Respiration in Maize (Ph. D. thesis). Haryana Agricultural University, Hissar, India.

Blum, A., 1969. Oviposition preference by the sorghum shoot fly (*Atherigona varia soccata*) in progenies of susceptible x resistant sorghum crosses. Crop Sci. 9, 695–696.

Blum, A., 1970. Effect of plant density and growth duration on sorghum yield under limited water supply. Agron. J. 62, 333–336.

Blum, A., 1972. Sorghum breeding for shoot fly resistance. In: Jotwani, M.G., Young, W.R (Eds.), Control of Sorghum Shoot Fly. Oxford and IBH, New Delhi, pp. 180–191.

Blum, A., 1979a. Genetic improvement of drought resistance in crop plants: a case for sorghum. In: Mussell, H., Staples, R.C. (Eds.), Stress Physiology in Crop Plants. John Wiley, New York, NY, pp. 429–445.

Blum, A., 1979b. Principles and methodology of selecting for drought resistance in sorghum. Monogra-fie di Genet. Agrano. 4, 205–215.

Blum, A., 1983. Genetic and physiological relationships in plant breeding for drought resistance. Agri. Water Manage. 7, 195–202.

Blum, A., 1987. Genetic and environmental considerations in the improvement of drought stress avoidance in sorghum. In: Food grain production in semi-arid Africa. Proceedings of International Drought Symposium, 19–23 May 1983, Nairobi, Kenya, pp. 91–99.

Blum, A., 1988. Plant Breeding for Stress Environments. CRC Press, Boca Raton, FL.

Blum, A., 1990. Productivity and drought resistance of genetically improved cultivars as compared with native landless of sorghum. Sorghum Newsl. 32, 41.

Blum, A., 1991. The comparative productivity and drought response of semitropical hybrids and open pollinated varieties of sorghum. Sorghum Newsl. 32, 55.

Blum, A., 2004. Sorghum physiology. In: Nguyen, HT, Blum, A (Eds.), Physiology and biotechnology integration for plant breeding. Marcel Dekker, New York, pp. 141–223.

Blum, A., 2005. Drought resistance, water-use efficiency, and yield potential—are they compatible, dissonant, or mutually exclusive? Aust. J. Agric. Res. 56, 1159–1168.

Blum, A., 2011. Chapter 2. Plant Water Relations, Plant Stress and plant production pages 11-52 in plant breeding for water limited environments. first ed. Springer, New York, NY, 258pp.

Blum, A., Arkin, G.F., 1984. Sorghum root growth and water-use as affected by water supply and growth duration. Field Crops Res. 9, 131–142. Available from: http://dx.doi.org/10.1016/0378-4290(84)90019-4.

Blum, A., Sullivan, C.Y., 1986. The comparative drought resistance of landraces of sorghum and millet from dry and humid regions. Ann. Bot. 57 (6), 835–846.

Blum, A., Arkin, F.G., Jordan, W.R., 1977. Sorghum root morphogenesis and growth II, manifestation of heterosis. Crop Sci. 17, 149–157.

Blum, A., Bojarcova, H., Golan, G., Mayer, J., 1983a. Chemical dessication of wheat plants as a simulator of post-anthesis stress. I. Effects on translocation and kernel growth. Field Crops Res. 6, 51–58.

Blum, A., Mayer, J., Golan, G., 1983b. Chemical dessication of wheat plants as a simulator of post-anthesis stress. I. Relations to drought stress. Field Crops Res. 6, 149–155.

Blum, A., Mayer, J., Golan, G., 1989a. Agronomical and physiological assessments of genotypic variation for drought resistance in sorghum. Aust. J. Agric. Res. 16 (1), 49–61.

Blum, A., Golan, G., Mayer, J., Simmena, B., Shpiler, L., Burra, J., 1989b. The drought response of landraces of wheat from the northern Negev desert in Israel. Euphytica. 43, 87–96.

Blum, A., Ramaiah, S., Kanemaso, E.T., Pavlsen, G.M., 1990. A physiology of heterosis in sorghum with respect to environmental stress. Ann. Bot. 65, 149–158.

Blum, A., Mayer, J., Sinmena, B., Oblina, T., 1992. Comparative productivity and drought response of semi-tropical hybrids and open pollinated varieties of sorghum. J. Agric. Sci. 118 (1), 29–36.

Blum, A., Golan, G., Mayerm, J., Sinmena, B., 1997. The effect of dwarfing genes of sorghum grain filling from remobilized stem reserves under stress. Field Crops Res. 52, 43–54.

Bohnert, H.J., Nelson, D.E., Jensen, R.G., 1995. Adaptations to environmental stresses. Plant Cell. 7, 1099–1111.

Borade, B.V., Pokharkar, R.N., Salunkhe, G.N., Gandhale, D.N., 1993. Effects of dates of sowing on leaf sugar malady on *rabi* sorghum. J. Maharashtra Agric. Univ. 18, 124–125.

Borikar, S.T., Chopde, P.R., 1981. Inheritance of shoot fly resistance in sorghum. J. Maharashtra Agric. Univ. 6, 47–48.

Borole, D.N., 2002. Heterosis and Combining Ability for Yield and Yield Components and Genesis of Seed Setting in *Rabi* Sorghum (*Sorghum bicolor* (L.) Moench) (Thesis submitted to M.P.K.V). Rahuri, Maharashtra.

Borrell, A.K., Douglas, A.C.L., 1997. Maintaining green leaf area in grain sorghum increases nitrogen uptake under post-anthesis drought. Int. Sorghum Millets Newsl. 38, 89–92.

Borrell, A.K., Hammer, G.L., 2000. Nitrogen dynamics and the physiological basis of stay-green in sorghum. Crop Sci. 40, 1295–1307.

Borrell, A.K., Bidinger, F.R., Sunitha, K., 1999. Stay-green associated with yield in recombinant inbred sorghum lines varying in rate of leaf senescence. Int. Sorghum Millets Newsl. 40, 31–33.

Borrell, A.K., Hammer, G.L., Douglas, A.C.L., 2000a. Does maintaining green leaf area in sorghum improve yield under drought? I. Leaf growth and senescence. Crop Sci. 40, 1026–1037.

Borrell, A.K., Hammer, G.L., Henzell, R.O., 2000b. Does maintaining green leaf area in sorghum improve yield under drought? II. Dry matter production and yield. Crop Sci. 40, 1037–1048.

Borrell, A.K., Hammer, G.L., van Oosterom, E.J., 2001. Staygreen: a consequence of the balance between supply and demand of nitrogen during grain filling? Ann. Appl. Biol. 138, 91–95.

Bos, I., 1977. More arguments against intermating F_2 plants of a self-fertilizing crop. Euphytica. 26, 33–46.

Boyer, J.S., 1982. Plant productivity and environment. Science. 218, 443–448.

Boyle, M.G., Boyer, J.S., Morgan, P.W., 1991. Stem infusion of liquid culture medium prevents reproductive failure of maize at low water potentials. Crop Sci. 31, 1248–1252.

Bramel Cox, P.J., Claflin, L.E., 1989. Selection for resistance to *Macrophomina phaseolina* and *Fusarium moniliforme* in sorghum. Crop Sci. 29, 1468–1472.

Bramel Cox, P.J., Stein, I.S., Rodgers, D.M., Claflin, L.E., 1988. Inheritance of resistance to *Macrophomina phaseolina* (Tassi) Gold and *Fusarium moniliform* Sheldon in sorghum. Crop Sci. 28, 37–40.

Bueno, A., Atkins, R.C., 1982. Growth analysis of grain sorghum hybrids. IOWA State J. Res. 56, 367–381.

Burke, J.J., Franks, C.D., Burow, G., Xin, Z., 2010. Selection system for the stay-green drought tolerance trait in sorghum germplasm. Agron. J. 102 (4), 1118–1122.

Burow, G.B., Burke, J.J., Franks, C., Xin, Z., 2011. Genetic enhancement of cold tolerance to overcome a major limitation in sorghum. American Seed Trade Association Conference Proceedings.

Buttery, B.R., Tan, C.S., Buzzell, R.I., Gaynor, J.D., MacTavish, D.C., 1993. Stomatal numbers of soybean and response to water stress. Plant Soil. 149, 283–288.

Capoor, S.P., Rao, D.G., Varma, P.M., 1968. Chlorosis of sorghum. Indian J. Agric. Sci. 38, 198–207.

Ceccarelli, S., Acevedo, E., Grando, S., 1991. Breeding for yield stability in unpredictable environments: single traits, interaction between traits, and architecture of genotypes. Euphytica. 56, 169–185.

Chamarthi, S.K., Sharma, H.C., Sahrawat, K.L., Narasu, L.M., Dhillon, M.K., 2011a. Physico-chemical mechanisms of resistance to shoot fly, *Atherigona soccata* in sorghum, *Sorghum bicolor*. J. Appl. Entomol. 135 (6), 446–455.

Chamarthi, S.K., Sharma, H.C., Vijay, P.M., Lakshmi Narasu, M., 2011b. Leaf surface chemistry of sorghum seedlings influencing expression of resistance to sorghum shoot fly, *Atherigona soccata*. J. Plant Biochem. Biotechnol. 20 (2), 11–216.

Chamarthi, S.K., Sharma, H.C., Deshpande, S.P., Hash, C.T., Rajaram, V., Ramu, P., et al., 2012. Genomic diversity among sorghum genotypes with resistance to sorghum shoot fly, *Atherigona soccata*. J. Plant Biochem. Biotechnol. 21 (2), 242–251.

Chan, B.S.P., Eccleston, J.A., 2003. On the construction of nearest neighbor balanced row column designs. Biometrics. 45, 97–106.

Chand, P., Kumar Singh, P., 2003. Evaluation of sorghum (*Sorghum bicolor* L.) genotypes for drought tolerance. Indian J. Dryland Agric. Res. Dev. 18 (1), 92–94.

Chandak, R.R., Nandanwanvar, K.G., 1993. Gene effects in *kharif* sorghum. J. Maharashtra Agric. Univ. 18 (3), 355–358.

Chandra Shekar, B.M., 1991. Mechanisms of Resistance in Sorghum to Shoot Bug, *Peregrinus maidis* (Ashmead) (Homoptera: Delphacidae) (M.Sc. thesis). Andhra Pradesh Agricultural University, Hyderabad, India, 106pp.

Chandra Shekar, B.M., Dharma Reddy, K., Singh, B.U., Reddy, D.D.R., 1992. Components of resistance to corn planthopper, *Peregrinus maidis* (Ashmead), in sorghum. Resist. Pest Manage. Newsl. 4, 25.

Chandra Shekar, B.M., Reddy, K.D., Singh, B.U., Reddy, D.D.R., 1993a. Antixenosis component of resistance to corn planthopper, *Peregrinus maidis* (Ashmead) in sorghum. Insect Sci. Appl. 14, 77–84.

Chang, C.P., Fang, M.N., 1984. Studies on the resistance of sorghum variety to sorghum aphid, *Melanaphis sacchari* (Zehntner). China J. Entomol. 4, 97–105.

Chang, N.T., 1981a. Resistance of some grain sorghum cultivars to sorghum aphid injury. Plant Prot. Bull. (Taiwan). 23, 35–41.

Chang, S.C., 1981b. Sources of resistance in sorghum to sugarcane aphid, *Melanaphis sacchari* (Zehntner). Rep. Corn Res. Center. Taiwan DAIS. 15, 10–14.

Channappagoudar, B.B., Biradar, N.R., Bharamagoudar, T.D., Rokhade, C.J., 2007. Morpho-physiological characters contributing for drought tolerance in sorghum under receding soil moisture conditions. Karnataka J. Agric. Sci. 20 (4), 719–723.

Chapman, S.C., Barreto, H.J., 1997. Using a chlorophyll meter to estimate specific leaf nitrogen of tropical maize during vegetative growth. Agron. J. 89, 557–562.

Chapman, S.C., Ludlow, M.M., Blamey, F.P.C., Fisher, K.S., 1993. Effect of drought during early reproductive development on growth of cultivars of groundnut (*Arachis hypogaea* L.). I. Utilization of radiation and water during drought. Field Crops Res. 32, 193–210.

Chaudhary, S.D., 1992. Heterosis in high energy sorghum. J. Maharashtra Agric. Univ. 17, 28–29.

Chaudhary, S.B., Narkhede, B.N., Pawar, S.V., 2003. Hybrid vigour involving diverse cytosteriles sorghum [*Sorghum bicolor* (L.) Moench]. J. Soils Crops. 13, 162–164.

Chaugle, D.S., Birmani, S.P., Jamdagni, B.M., 1982. Harvest index, biological yield and other characters in rabi. J. Maharashtra Agric. Univ. 17, 52.

Chavan, U.D., Patil, J.V., Shinde, M.S., 2009. Nutritional and roti quality of sorghum genotypes. Indones. J. Agric. Sci. 10 (2), 80–87.

Chavda, D.H., Drolsom, D.N., 1970. Heterosis among crosses of eight selected parental strains in sorghum. Indian J. Agric. Sci. 40, 967–973.

Chelliah, S., Basheer, M., 1965. Biological studies of *Peregrinus maidis* (Ashmead) on sorghum. Indian J. Entomol. 27, 466–471.

Chen, X.Q., 1988. Study on the genetic effect of cytoplasmic nuclear interaction of sorghum. I. Hybrid vigor of sorghum nucleus substitution line. Acta Agric. Univ. Jilinensis. 10, 1–5.

Chen, X., 1994. Study on genetic effect of yield heterosis in sorghum nucleo-cytoplasm hybrids. J. Jilin Agric. Univ. 16, 7–11.

Cheng, B.C., Liu, Q.Y., Jiang, H., 1989. Analysis of grain weight of sorghum in diallel crosses. Hereditas Beijing. 11, 12–14.

Chen-Huiming, L.-X., 2001. Inheritance of seed colour and luster in mungbean (*Vigna radiata*). Human Agric. Sc. Techn. News Lett. 2 (1), 8–12.

Chenu, K., Chapman, S.C., Hammer, G.L., McLean, G., Tardieu, F., 2008. Short-term responses of leaf growth rate to water deficit scale up to whole plant and crop levels. An integrated modelling approach in maize. Plant, Cell Environ. 31, 378–391.

Childs, K.L., Frederick, R.M., Pratt, M.M.C., Pratt, L.H., Morgan, P.W., Mullet, J.E., 1997. The sorghum photoperiod sensitivity gene, Ma3 encodes a phytochrome B. Plant Physiol. 113, 611–619.

Chimmad V.P., Kamatar M.Y., 2003. Physiological characterization of sorghum parents of kharif and rabi sorghum. II International Congress of Plant Physiology. 8–12 January 2003, New Delhi, pp. 77.

Chinna, B.S., Paul, P.S., 1988. Heterosis and combining ability studies in grain sorghum under irrigated and moisture stress environment. Crop Improv. 15, 151–155.

Choudhary, S.I., Wardlaw, I.F., 1978. The effect of temperature on kernel development in cereals. Aust. J. Agric. Res. 29, 205–223.

Christiansen, M.N., Lewis, C.F., 1982. Breeding Plants for Less Favourable Environments. John Wiley, New York, NY.

Cisse, N., Ejeta, G., 2003. Genetic variation and relationships among seedling vigor traits in sorghum. Crop Sci. 43, 824–828. Available from: http://dx.doi.org/10.2135/cropsci2003.0824.

Clarke, J.M., Richards, R.A., Condon, A.G., 1991. Effect of drought on residual transpiration and its relationship with water use of wheat. Can. J. Plant Sci. 71, 695–702.

Clerget, B., Dingkuhn, M., Chantereau, J., Hemberger, J., Louarn, G., Vaksmann, M., 2004. Does panicle initiation in tropical sorghum depend on day-to-day change in photoperiod? Field Crops Res. 88, 21–37.

Colom, M.R., Vazzana, C., 2003. Photosynthesis and PSII functionality of drought-resistant and drought sensitive weeping lovegrass plants. Environ. Exp. Bot. 49, 135–144.

Conner, A.B., Karper, R.E., 1927. Hybrid vigor in sorghum. Texas Experiment Stations Bulletin no 359, Texas A&M University, Texas, USA.

Cook, G.E., Boosalis, M.G., Dunkle, L.D., Ovody, N.G., 1973. Survival of *Macrophomina phaseolina* in corn stalk and sorghum residue. Plant Dis. Rep. 57, 873–875.

Crasta, O.R., Xu, W.W., Rosenow, D.T., Mullet, J.E., Nguyen, H.T., 1999. Mapping of post-flowering drought resistance traits in grain sorghum: association between QTLs influencing premature senescence and maturity. Mol. General Genet. 262, 579–588.

Craufurd, P.Q., 1993. Effect of heat and drought stress on sorghum grain yield. Exp. Agric. 29, 77–86.

Craufurd, P.Q., Peacock, J.M., 1993. Effect of heat and drought stress on sorghum II grain yield. Exp. Agric. 29 (1), 77–86.

Craufurd, P.Q., Qi, A., 2001. Photothermal adaptation of sorghum (*Sorghum bicolor*) in Nigeria. Agric. For. Meteorol. 108, 199–211.

Craufurd, P.Q., Flower, D.J., Peacock, J.M., 1993. Effect of heat and drought stress on sorghum (*Sorghum bicolor*). I. Panicle development and leaf appearance. Aust. J. Exp. Agric. 29, 61–76.

Craufurd, P.Q., Wheeler, T.R., Ellis, R.H., Summerfield, R.J., Williams, J.H., 1999. Effect of temperature and water deficit on water-use efficiency, carbon isotope discrimination and specific leaf area in peanut. Crop Sci. 39, 136–142.

Da, S., Akingbala, J.O., Scheuring, J.F., Rooney, L.W., 1982. Evaluation of quality in a sorghum breeding program. Proceedings, International Symposium on Sorghum Grain Quality, ICRISAT. 28–31 October 1981, Patancheru, India.

Dabholkar, A.R., Baghel, S.S., 1980. Combining ability analysis of yield and yield components in sorghum. JNKVV-Res. J. 14, 53–59.

Dabholkar, A.R., Bhadouriya, N.S., Mishra, V.K., 1970. Combining ability analysis of some physiological attributes of sorghum. J. Maharashtra Agric. Univ. 20, 435–437.

Dabholkar, A.R., Lal, G.S., Mishra, R.C., 1984. Genetic analysis of grain size and other characteristics of sorghum. Madras Agric. J. 71, 750–753.

Dabholkar, A.R., Lal, G.S., Mishra, R.C., Barche, N.B., 1989. Combining ability analysis of resistance of sorghum to shoot fly. Indian J. Genet. Plant Breed. 49, 325–330.

Dabholkar, A.R., Bhadouriya, N.S., Mishra, V.K., 1995. Combining ability analysis of some physiological attributes of sorghum. J. Maharashtra Agric. Univ. 20, 435–437.

Damodar, R., Subbarao, I.V., Rao, N.G.P., 1978. Heterosis for root activity in grain sorghum. Indian J. Genet. Plant Breed. 38, 431–436.

Dankov, T., 1965. An experiment in early testing of the general combining ability of maize inbred lines. Resten. Naukpl. Grow. Safija. 2, 3–10.

Darbha, S., 1997. Componential analysis of plant morphological factors associated with sorghum resistance to shoot fly *Atherigona soccata Rondani*. Masters thesis, Acharya N G Ranga Agricultural University.

Desai, S.G.M., 1991. Studies on Recombinational Variability for Combining Ability in B x R Crosses of Sorghum [*Sorghum bicolor* (L.) Moench] (M.Sc. (Agri) thesis). University of Agricultural Sciences, Dharwad.

Desai, 1998.

Desai, M.S., Shukla, P.T., 1997. Estimation of genetic parameters of yield characters in grain sorghum (*Sorghum bicolor* (L.) Moench), GAU. Res. J. 22 (2), 22–27.

Desai, K.B., Tikka, S.B.S., Patil, D.V., Desai, D.T., Kukadia, M.V., 1980. Effect of hybridity on panicle characters under low management in grain sorghum. Sorghum Newsl. 23, 24–25.

Desai, M.S., Desai, K.B., Kukadia, M.V., 1983. Heterobeltiosis for grain and its components in sorghum. Sorghum Newsl. 26, 88.

Desai, M.S., Desai, K.B., Kukadia, M.V., 1985. Heterosis in grain sorghum. Sorghum Newsl. 28, 14.

Desai, S.A., Singh, P., Shrotria, P.K., Singh, R., 2000. Variability and heterosis for forage yield and its components in interspecific crosses of forage sorghum. Karnataka J. Agric. Sci. 13, 315–320.

Deshpande, S.K., 2005. Genetic Studies Involving Maintainer and Restorer Lines for Improving Yield and Tolerance to Biotic Stresses in Rabi Sorghum (Ph.D. thesis). UAS, Dharwad.

Deshpande, S.P., Borikar, S.T., Ismail, S., Ambekar, S.S., 2003. Genetic studies for improvement of quality characters in *rabi* sorghum using landraces. Int. Sorghum Millets Newsl. 44, 6–8.

Deshpande, S.K., Biradar, B.D., Salimath, P.M., 2011. Remove from marked records studies on inheritance of charcoal rot resistance and aphid resistance in *rabi* sorghum [*Sorghum bicolor* (L.) Moench]. Plant Arch. 11 (2), 635–643.

Desikachar, H.S.R., Chandrasekhar, A., 1982. Quality of sorghum for use in Indian foods. In: Proceedings, International Symposium on Sorghum Grain Quality, ICRISAT. 28–31 October 1981, Patancheru, India: ICRISAT.

Deu, M., Grivet, L., Trouche, G., Barro, C., Ratnadass, A., Diabaté, M., et al., 2000. Use of molecular markers in the sorghum breeding program at CIRAD. In: Haussmann, BIG, Greiger, HH, Hess, DE, Hash, CT, Bramel-Cox, P (Eds.), Application of Molecular Markers in Plant Breeding, Training Manual on Seminar Held at IITA. Ibadan, Nigeria, August 16–17, 1999.

Dhanda, S.S., Sethi, G.S., Behl, R.K., 2004. Indices of drought tolerance in wheat genotypes at early stages of plant growth. J. Agron. Crop Sci. 190, 6–12.

Dhillon, M.K., Sharma, H.C., Reddy, B.V.S., Singh, R., Naresh, J.S., 2006. Nature of gene action for resistance to sorghum shoot fly, *Atherigona soccata*. Crop Sci. 46, 1377–1383.

Dhillon, M.K., Sharma, H.C., Reddy, B.V.S., Singh, R., Naresh, J.S., Kai, Z., 2005a. Relative susceptibility of different male-sterile cytoplasms in sorghum to shoot fly, *Atherigona soccata*. Euphytica. 144 (3), 275–283.

Dhillon, M.K., Sharma, H.C., Singh, R., Naresh, J.S., 2005b. Mechanisms of resistance to shoot fly, *Atherigona soccata* in sorghum. Euphytica. 144 (3), 301–312.

Dhingra, O.D., Sinclair, J.B., 1977. An Annotated Bibliography of *Macrophomina phaseolina* 1905-1975. lmprensia Universitaria, Universidade Federal de Vicosa, Vicosa, Brazil, p. 244.

Dhingra, O.D., Sinclair, J.B., 1978. Biology and Pathology of *Macrophomina phaseolina*. lmprensia Universitaria, Universidade Federal de Vicosa, Vicosa, Brazil, p. 166.

Dhole, V.J., 2004. Genetic Analysis and Multiple Trait Selection Indices Based on Selection Methods in Rabi Sorghum (*Sorghum bicolor* (L.) Moench) (Thesis abst., thesis submitted to M.P.K.V. Rahuri). Maharashtra.

Dhoble, M.V., Kale, U.V., 1988. Recovery of drought resistance in sorghum genotypes. J. Maharashtra Agric. Univ. 13 (1), 118.

Dinakar, B.L., 1985. Genetic Analysis of Some Quantitative Characters in Sorghum (*Sorghum bicolor* L. Moench) (M.Sc. (Agri.) thesis). University of Agricultural Science, Bangalore, Karnataka (India).

Dinakar, B.L., Rao, T.S., Goud, J.V., 1988. Genetic analysis of yield and its attributes in *rabi* grain sorghum. Mysore J. Agric. Sci. 22, 456–459.

Dodd, J.L., 1977. A photosynthetic stress-translocation balance concept of corn stalk rot. In: Loden, H.D., Wilkinson, D. (Eds.), Proceedings of the 32nd Annual Corn and Sorghum Research Conference. American Seed Trade Association, Washington, DC, pp. 122–130.

Doggett, E., 1988. Sorghum. John Wiley & Sons, New York, NY.

Donald, C.M., 1962. In search of yield. J. Aust. Inst. Agric. Sci. 28, 171–178.

Donald, C.M., 1968. The breeding of crop ideotypes. Euphytica. 17, 385–403.

Donatellli, M., Hammer, G.L., Vanderlip, R.L., 1992. Genotypic and water limitation effects on phenology growth and transpiration efficiency in grain sorghum. Crop Sci. 32 (3), 781–786.

Downes, R.W., 1972. Effect of temperature on the phenology and grain yield of sorghum bicolor. J. Agric. Res. 23, 385–394.

Downes, R.W., Marshall, D.R., 1971. Low temperature induced male sterility in *Sorghum bicolor*. Aust. J. Exp. Anim. Husbandary. 11, 352–356. Available from: http://dx.doi.org/10.1071/EA9710352.

Dremlyuk, G.K., 1980. Some methods of selecting parent forms for breeding grain sorghum. Nauch. Tekhn. Byul. Vses. Selekts. Genet. In-ta. 4, 44–48.

Du Plessis, J., 2008. Sorghum Production. ARC Grain Crops Institute, Department of Agriculture, Republic of South Africa.

Duncan, W.G., McCloud, D.E., McGraw, R., Boote, K.J., 1978. Physiological aspects of peanut yield improvement. Crop Sci. 18, 1015–1020.

Duncan, R.R., Bockholt, A.J., Miller, F.R., 1981. Descriptive comparison of senescent and non-senescent sorghum genotypes. Crop Sci. 21, 849.

Eastin, J.D., 1983. Sorghum. Proceedings of the Symposium of Potential Productivity of Field Crops Under Different Environments. IRRI, Los Bonas Manilla, Philliphines, pp. 181–204.

Eberhart, S.A., Russell, w.A., 1966. Stability parameters for comparing varieties. Crop Sci. 6, 36–40.

Edmunds, L.K., 1964. Combined relation of plant maturity, temperature and soil moisture to charcoal rot development in grain sorghum. Phytopathology. 54, 514–517.

Ejeta, G., 1987. Breeding sorghum hybrids for irrigated and rainfed conditions in Sudan. Food Grain Production in Semi-Arid Africa, Proceedings of International Drought Symposium. 19–23 May 1986, Nairobi, Kenya, 121.

Ejeta, Q., Rosenow, D.T., 1993. Registration of P89001 to P89010, 10 fertility restorer parental lines of sorghum. Crop Sci. 33, 222–223.

Ejeta, G., Goldsbrough, M.R., Tunistra, E.M., Grote, A., Menkir, Y., Ibrahim, Y., et al., 2000. Molecular marker applications in sorghum. In: Haussmann, B.I.G., Geiger, H.H, Hess, D.E., Hash, C.T., Bramel-cox, P. (Eds.), Application of Molecular Markers in plant breeding, Training manual for a seminar held at IITA. Ibadan, Nigeria, pp. 16–17. August 1999. International Crops Research Institute for the Semi-Arid Tropics, Patancheru 502 324, Andhra Pradesh, India, p. 81–89.

Ekanayaka, I.J., O'Toole, J.C., Garrity, D.P., Masajo, T.M., 1985. Inheritance of root characteristics and their relations to drought resistance in rice. Crop Sci. 25, 927–933.

Elbadawi, A.A., Nour, A.M., Pathak, R.S., 1997. Combining ability of resistance to the sorghum shoot fly, *Atherigona soccata* and stem borer *Chilo partellus* in grain sorghum, *Sorghum bicolor*. Insect Sci. Appl. 17, 329–333.

Ellis, E.B., 1975. The effects of endosperm characteristics on seed and grain quality of Sorghum bicolor (L.) Moench. Ph. D. Dissertation. Texas A&M University, College Station, TX.

Ellis, R.H., Guaufurd, G.A., Summerfield, R.J., Robert, E.H., 1997. Effect of photoperiod, temperature and asynchrony between thermoperiod and photoperiod on development to panicle initiation in sorghum. Ann. Bot. 79 (2), 169–178.

Erick, H.V., Hanway, J.J., 1965. Some factors affecting development of longevity of leaf of corn. Agron. J. 57, 1–12.

Erick, H.V., Musick, J.T., 1979. Plant water stress effects on irrigated sorghum. I. Effect on yield. Crop Sci. 19, 589–592.

Esha, C.S., 2001. Genetic Evaluation and Stability of New Experimental Hybrids in Rabi Sorghum (M.Sc. (Agri) thesis). University of Agricultural Sciences, Dharwad.

Falconer, D.S., 1986. Introduction to Quantitative Genetics. Oliver and Boyd, Edinburgh, UK.

Fan, L., Zheng, S., Wang, X., 1997. Antisense suppression of phospholipase D retards abscisic acid– and ethylene-promoted senescence of postharvest *Arabidopsis* leaves. Plant Cell. 9, 2183–2196.

Fang, M.N., 1990. Population fluctuation and timing for control of sorghum aphid on variety, Taichung 5. Bull. Taichung Dist. Agric. Improv. Stn. 28, 59–71.

Farrell, T.C., Fox, K.M., Williams, R.L., Fukai, S., Reinke, R.F., Lewin, L.G., 2001. Temperature constraints to rice production in Australia and Lao PDR: a shared problem. In: Fukai, S., Basnayake, J. (Eds.), ACIAR Proc. 101: Increased Lowland Rice Production in the Mekong Region. Australian Centre for International Agricultural Research, Canberra, ACT, pp. 129–137.

Fawzi, A.L., 1962. Effectiveness of early testing method of breeding in isolating better inbred lines of maize. Agric. Res. Rev. Egypt. 40, 85–96.

Fayed, M.F.S., 1975. Evaluation of newly developed cytoplasmic male sterile lines for their combining ability in Sorghum. Agric. Res. Rev. Egypt. 53 (8), 5–26.

Feltus, F.A., Hart, G.E., Schertz, K.F., et al., 2006a. Genetic map alignment and QTL correspondence between inter- and intra-specific sorghum populations. Theor. Appl. Genet. 112 (7), 1295.

Feltus, F.A., Singh, H.P., Lohithaswa, H.C., Schulze, S.R., Silva, T., Paterson, A.H., 2006b. Conserved intron scanning primers: targeted sampling of orthologous DNA sequence diversity in orphan crops. Plant Physiol. 140, 1183–1191.

Ferraris, R., 1981. Early assessment of sweet sorghum as an agro–industrial crop. 1 Varietal evaluation. Aust. J. Exp. Agric. Anim. Husbandry. 21, 75–82.

Finlay, K.W., Wilkinson, G.N., 1963. The analysis of adaptation in a plant breeding programme. Aust. J. Agric. Res. 14, 742–754.

Fischer, K.S., Wilson, G.L., 1971. Studies on grain production in (*Sorghum vulgare* L. Moench) I the contribution of pre-flowering photosynthates to grain yield. Aust. J. Agric. Sci. 22, 33–37.

Fischer, R.S., Wilson, J.T., 1975. Drought resistance in spring wheat cultivars. Aust. J. Agric. Sci. 26, 31–42.

Fischer, R.A., Maurer, R., 1978. Drought resistance in spring wheat cultivars. I. Grain yield responses. Aust. J. Agric. Res. 29, 897–912.

Fischer, R.A., Wood, J.T., 1979. Drought resistance in spring wheat cultivars. III. Yield association with morpho-physiological traits. Aust. J. Agric. Res. 30, 1001–1020.

Fischer, K.S., Johnson, E.C., Edmeades, G.O., 1982. Breeding and selection for drought resistance in tropical maize. Drought Resistance in Crops with Emphasis on Rice. IRRI, Philippines, pp. 377–399.

Fisk, J., 1978a. A field study of *Peregrinus maidis* on sorghum in India. COPR/ICRISAT Collaborative Project on the Planthopper, *Peregrinus maidis*. Centre for Overseas Pest Research, London, UK, 2ik5pp.

Fisk, J., 1978b. Resistance of *Sorghum bicolor* to *Rhopalosiphum maidis* and *Peregrinus maidis* as affected by differences in the growth of the host. Entomol. Exp. Appl. 23, 227–236.

Fisk, J., 1980. Effect of HCN, phenolic acids and related compounds in *Sorghum bicolor* on the feeding behaviour of the planthopper, *Peregrinus maidis*. Entomol. Exp. Appl. 27, 211–222.

Flattery, K.E., 1982. An assessment of pest damage of grain sorghum in Botswana. Exp. Agric. 18, 319–328.

Flower, D.J., Rani, A.U., Peacock, J.M., 1990. Influence of osmotic adjustment on the growth, stomatal conductance and light interception of contrasting sorghum lines in a harsh environment. Aust. J. Pl. Physio. 17, 91–105.

Fokar, M., Blum, A., Nguyen, H.T., 1998. Heat tolerance in springwheat. II. Grain filling. Euphytica. 104, 9–15. Available from: http://dx.doi.org/10.1023/A:1018322502271.

Folliard, A., Traore, P.C.S., Vaksmann, M., Kouressy, M., 2004. Modelling of sorghum response to photoperiod: a threshold-hyperbolic approach. Field Crops Res. 89, 59–70.

Franca, J.G., Murthy, D.S., Nicodemos, D., House, L.R., 1986. Genetic analysis of some agronomic traits in grain sorghum. Heterosis Rev. Bras. Genet. 9, 659–678.

Franks, C.D., Burrow, G.B., Burke, J.J., 2006. A comparison of U.S. and Chinese sorghum germination for early season cold tolerance. Crop Sci. 46, 1371–1376. Available from: http://dx.doi.org/10.2135/cropsci2005.08-0279.

Fukumoto, G., Mau, R.F.L., 1989. The yellow sugarcane aphid in Hawaiian pastures. HNFAS Animal Sciences Program, Collaboration on Tropical Agriculture and Human Resources. University of Hawaii, Manoa, Hawaii, 3pp.

Ganesh, S., Fazullah Khan, A.K., Senthil, N., 1996. Heterosis studies for grain yield characters in sweet sorghum. Madras Agric. J. 83, 655–657.

Ganesh, S., Fazullah Khan, A.K., Senthil, N., 1997. *Per se* performance of parents and hybrids in sweet sorghum for grain yield characters. Madras Agric. J. 84, 323–325.

Gangadhar Rao, D., Sinha, S.K., 1988. Leaf water relations of sorghum hybrids and their parents. Proceedings of International Congress of Plant Physiology. New Delhi, pp. 885–889.

Gao, S.J., 1993. Analysis of gene effects on yield characteristics in sorghum. Heriditas (Beijing). 15 (2), 25–27, L.

Gaosegelwe, P.L., Kirkhan, M.B., 1990. Evaluation of wild, primitive and adapted sorghums for drought resistance (En.) Pages 224-226. In: Challenges of Dryland Agriculture: A Global Perspective Proceedings of the International Conference and Dry Land Farming. 15–19 August 1988, Amarillo, Bushland, TX.

Garrity, D.P., Sullivan, C.Y., Ross, W.M., 1982. Alternative approaches to improving grain sorghum productivity under drought stress. Drought Resistance in Crops with Emphasis in Rice. International Rice Research Institute, Manila, Philippines, pp. 339–356.

Garrity, D.P., Sullian, C.Y., Watts, D.G., 1983. Moisture deficits and grain sorghum performance: drought stress conditioning. Agron. J. 75 (6), 997–1004.

Garrity, D.P., Sullian, C.Y., Watts, D.G., 1984. Changes in grain sorghum stomatal and photosynthetic response to moisture stress growth stages. Crop Sci. 24 (1), 441–446.

Garud, T.B., Borikar, S.T., 1985. Genetics of charcoal rot resistance in sorghum. Sorghum Newsl. 28, 87.

Garud, T.B., Seetharama, N., Deshpande, S.P., Ismail, S., Dadke, M.S., 2002. Usefulness of non-senescent parents for charcoal rot resistance breeding in sorghum. Int. Sorghum Millets Newsl. 43, 63–65.

Geeta, S., Rana, B.S., 1988. Genetic change over six generations in a pedigree breeding programme in sorghum. Indian J. Genet. Plant Breed. 48, 61–66.

Gepstein, S., 2004. Leaf senescence—not just a "wear and tear" phenomenon. Genome Biol. 5, 212–215.

Gerasenkov, B.I., Goncharova, L.P., 1972. Evaluation of combining ability in maize inbred lines in early stages of inbreeding. Referativnyl Zhurnal. 3, 27–39.

Ghanashyam, C., Jain, M., 2009. Role of auxin-responsive genes in biotic stress responses. Plant Signal Behav. 4, 846–848.

Gholipoor, M., Prasad, P.V.V., Mutava, R.N., Sinclair, T.R., 2010. Genetic variability of transpiration response to vapor pressure deficit among sorghum genotypes. Field Crops Res. 119 (1), 85–90.

Ghorade, R.B., Gite, B.D., Sakhare, B.A., Thorat, A., 1997. Analysis of heterosis and heterobeltiosis for commercial exploitation of sorghum hybrids. J. Soil Crops. 7, 185–189.

Ghorade, R.B., Kalpande, V.V., Pallavi, B., Bhongle, S.A., Nemade, S., Barabde, N., 2014. Character association for grain yield and some of the growth parameters associated with drought tolerance in *rabi* sorghum. Agrotechnology. 2, 4.

Gibson, S., 2005. Control of plant development and gene expression by sugar signaling. Curr. Opin. Plant Biol. 8, 93–102.

Gibson, P.T., Schertz, K.F., 1977. Growth analysis of sorghum hybrid and its parents. Crop Sci. 17, 387–391.

Gibson, P.T., Maiti, R.K., 1983. Trichomes in segregating generation of sorghum matings I. Inheritance of trichome presence and density. Crop Sci. 23, 73–78.

Giorda, L., Martinez, M.J., Chulze, S., 1995. Fusarium root and stalk rot in Argentina. In: Leslie, J.F., Frederiksen, R.A. (Eds.), Disease Analysis Through Genetics and Molecular Biology: Interdisciplinary Bridges to Improved Sorghum and Millet Crops. Iowa State University Press, Ames, IA, USA, pp. 185–193.

Giriraj, K., 1983. Studies on correlation, path analysis, genetic divergence, heterosis, combining ability and nature of gene action in eight parent diallel cross of sorghum [*Sorghum bicolor* (L.) Moench]. Thesis Abstr. 9, 287–288.

Giriraj, K., Goud, J.V., 1981. Heterosis for vegetative characters in grain sorghum. Indian J. Hered. 13, 9–13.

Giriraj, K., Goud, J.V., 1983. Inheritance study of vegetative characters in grain sorghum [*Sorghum bicolor* (L.) Moench]. Mysore J. Agric. Sci. 16, 142–146.

Giriraj, K., Goud, J.V., 1985. Studies on combining ability for grain yield and panicle components in sorghum. Mysore J. Agric. Sci. 19, 162–170.

Girma, F.S., Krieg, D.R., 1992. Osmotic adjustment in sorghum. I. Mechanisms of diurnal osmotic potential changes. Plant Physiol. 99, 577–582.

Gite, B.D., Khorgade, P.W., Ghorade, R.B., Sakhare, B.A., 1997. Combining ability of some newly developed male sterile and restorer lines in sorghum [*Sorghum bicolor* (L.) Moench]. J. Soils Crops. 7, 80–82.

Gomashe, S.S., Misal, M.B., Mehtre, S.P., Rakshit, S., Ganapathy, K.N., 2012. Assessment of parental lines and crosses for shoot fly resistance mechanism in sorghum (*Sorghum bicolor* (L.) Moench). Indian J. Genet. Plant Breed. 72 (1), 31–37.

Gomez, F., Chanterau, J. 1997. Breeding photoperiod sensitive sorghums. Proceedings of the International Conference on Genetic Improvement of Sorghum and Pearl Millet, INTSORMIL and ICRISAT. 22–27 September 1996, Lubbock, TX, pp. 66–70.

Gomez, L.D., Vanacker, H., Buchner, P., Noctor, G., Foyer, C.H., 2004. Intercellular distribution of glutathione synthesis in maize leaves and its response to short-term chilling. Plant Physiol. 134, 1662–1671. Available from: http://dx.doi.org/10.1104/pp.103.033027.

Gonalez-Hernadez, 1985. Growth efficiency of four sorghum genotypes. Sorghum Newsl. 28, 118.

Gonalez-Hernandez, Manjarrez-Sandoval, P., Mandoza-Onofre, L.E., Engleman, E.M., 1986. Leaf elongation rate stomatal diffusive resistance of sorghum plants under water stress at different development stages. Sorghum Newsl. 29, 92.

Gonalez-Hernadez, Manjarrez-Sandoval, P., Mandoza-Onofre, L.E., 1992. Drought stress effect on dry matter production and distribution in sorghum plants. Sorghum Newsl. 33, 56.

Gorham, J., Wyn Jones, R.G., McDonnell, E.M., 1985. Some mechanisms of salt tolerance in crop plants. Plant Soil. 89, 15–40.

Govil, J.N., Murty, B.R., 1979. A comparative study on diallel and partial diallel analysis. Indian J. Genet. Plant Breed. 39, 298–304.

Govil, J.N., Murty, B.R., Mukherjee, B.K., 1979. Studies on nature and magnitude of genetic parameters estimated in the original and advanced generations of certain sorghum hybrids. Zeitschrift-fur-Pflanzenzuchtung. 82, 340–348.

Gowda, B.T.S., Anahosur, K.H., Parameshwarappa, R., 1981. Breeding charcoal rot resistant sorghum varieties. Mysore J. Agric. Sci. 15, 503–506.

Griffin, R.H., 1966. Effect of water management and surface applied utilization of grain sorghum in the southern great plains. Agron. J. 58, 449–452.

Griffing, B., 1956. Concept of general and specific combining ability in relation to diallel crossing systems. Aust. J. Biol. Sci. 9, 463–493.

Grumet, R., Albrechtsen, R.S., Hanson, A.D., 1987. Growth and yield of barley isopopulations differing in solute potential. Crop Sci. 27, 991–995.

Gunawardena, T.A., Fukai, S., Blamey, P., 2003a. Low temperature induced spikelet sterility in rice. I. Nitrogen fertilization and sensitive reproductive period. Aust. J. Agric. Res. 54, 937–946. Available from: http://dx.doi.org/10.1071/AR03075.

Gunawardena, T.A., Fukai, S., Blamey, P., 2003b. Low temperature induced spikelet sterility in rice II. Effects of panicle and root temperature. Aust. J. Agric. Res. 54, 947–956. Available from: http://dx.doi.org/10.1071/AR03076.

Guo, Y., Cai, Z., Gan, S., 2004. Transcriptome of *Arabidopsis* leaf senescence. Plant Cell Environ. 27, 521–549.

Guohua, L., Yaohu, K., Lan, L., Shuqin, W., 2010. Effect of irrigation methods on root development and profile soil water uptake in winter wheat. Irrig. Sci. 28, 387–398.

Gupta, J.P., 1986. Moisture and thermal regimes of the desert soils of Rajasthan, India and their management for higher plant production. J. Hydro. Sci. 31, 347–359.

Gurumurthy, B.R., 1982. Desirable Growth Characteristics in Relation to Productivity in Finger Millet (*Eleusinea coracana*) (M.Sc. (Agri.) thesis). University of Agricultural Sciences, Bangalore.

Gururaj Rao, M.R., Patil, S.J., Parameshwarappa, R., 1993. Heterosis for development of panicle characters in *rabi* sorghum. Mysore J. Agric. Sci. 27, 223–228.

Gururaja Rao, M.R., Patil, S.J., Giriraj, K., 1994. Generation mean analysis of grain yield and some panicle characters in two exotic x Indian *rabi* sorghum crosses. Indian J. Genet. 54 (2), 158–163.

Guzhov, U.L., Malyuzhenets, N.S., 1984. Heterosis and the inheritance of economically important characteristics in grain sorghum. Izvestiya-Akademii-Nauk-SSSR-Biologicheskaya. 6, 811–822.

Hagio, T., Ono, S., 1986. Varietal reactions of sorghum to sugarcane aphid in the seedling test and field assessment. Sorghum Newsl. 29, 72.

Halalli, M.S., Gowda, B.T.S., Kulkarni, K.A., Goud, J.V., 1982. Inheritance of resistance to shoot fly (*Atherigona soccata* Rond.) in sorghum (*Sorghum bicolor*(L.) Moench). Sabrao J. 14, 165–170.

Hamaker, B.R., Axtell, D., 1997. Nutritional quality of sorghum. Proceedings of the International Conference on Genetic Enhancement of Sorghum and Pearl Millet. 22–27 September 1996, INTSORMIL and ICRISAT, Lubbock, TX, 531.

Hamakar, B.R., Mohamed, A.A., Habben, J.E., Huang, C.P., Larkins, B.A., 1995. Efficient procedure for extracting maize and sorghum kernel proteins reveals higher prolamin contents than the conventional method. Cereal Chem. 72, 583–588.

Hammer, G.L., Vanderlip, R.L., Gipson, G., Wade, L.J., Henzell, R.G., Founger, D.R., et al., 1989. Genotype-by-environment interaction in grain sorghum II. Effects of temperature and photoperiod on ontogeny. Crop Sci. 29 (2), 376–384.

Hammer, G.L., Farquhar, G.D., Broad, I.J., 1997. On the extent of genetic variation for transpiration efficiency in sorghum. Aust. J. Agric. Res. 48, 649–655.

Hammer, G.L., Dong, Z.S., McLean, G., Doherty, A., Messina, C., Schusler, J., et al., 2009. Can changes in canopy and/or root system architecture explain historical maize yield trends in the US corn belt? Crop Sci. 49, 299–312.

Hammer, G.L., van Oosterom, E., McLean, G., Chapman, S.C., Broad, I., Harland, P., et al., 2010. Adapting APSIM to model the physiology and genetics of complex adaptive traits in field crops. J. Exp. Bot. 61 (8), 2185–2202.

Hanks, J.R., Keller, J., Rasmusson, V.P., Wilson, G.D., 1976. Line source sprinkler for continuous variable irrigation: crop production studies. Soil Sci. Soc. Am. Proc. 40, 426–429.

Harden, M.L., Krieg, D.R., 1983. Contribution of preanthesis to grain fill in sorghum. Cereal-food-World. 28 (9), 562–563.

Hardwick, R.C., 1988. Critical physiological traits in pulse crops. In: Summerfield, R.J. (Ed.), World Crops: Cool Season Food Legumes. Kluwer Academic Publishers, Dordrecht, The Netherlands, pp. 885–896.

Harris, K., Subudhi, P.K., Borrell, A., Jordan, D., Rosenow, D., Nguyen, H., et al., 2007. Sorghum stay-green QTL individually reduce post-flowering drought-induced leaf senescence. J. Exp. Bot. 58 (2), 327–338.

Hash, C.T., Bhaskar Raj, A.G., Lindup, S., Sharma, A., Beniwal, C.R., Folkertsma, R.T., et al., 2003. Opportunities for marker assisted selection (MAS) to improve the feed quality of crop residues in pearl millet and sorghum. Field Crops Res. 84, 79–88.

Hattori, T., Inanaga, S., Araki, H., Ping, A., Morita, S., Luxová, M., et al., 2005. Application of silicon enhanced drought tolerance in Sorghum bicolor. Physiol. Plant. 123 (4), 459–466.

Hausmann, B.I.G., Mahalakshmi, V., Reddy, B.V.S., Seetharama, N., Hash, C.T., Geiger, H.H., 2002. QTL mapping of stay green in two sorghum recombinant inbred populations. Theor. Appl. Genet. 106, 143–148.

Haussmann, B.I.G., Obilana, A.B., Blum, A., Ayiecho, P.O., Schipprack, W., Geiger, H.H., 2000. Yield and yield stability of four population types of grain sorghum in a semi-arid area of Kenya. Crop Sci. 40, 319–329.

Hayman, B.I., 1954. The theory and analysis of diallel crosses. Genetics. 39, 789–809.

Hedhly, A., Hormaza, J.I., Herrero, M., 2008. Global warming and plant sexual reproduction. Trends Plant Sci. 14, 30–36.

Heinrich, G.M., Francis, G.A., Eastin, J.D., Saeed, M., 1985. Mechanisms of yield stability in sorghum. Crop Sci. 25, 1109–1113.

Heinrich, G.M., Fracis, C.A., Eastin, J.D., Saeed, M., 1993. Mechanism of yield stability in sorghum. Crop Sci. 25 (6), 1109–1113.

Henderson, G.R., 1952. Specific and general combining ability. In: Gowen, J.W. (Ed.), Heterosis. Iowa State College Press, Ames, IA.

Henrich, 1983. Correlation between carbon isotype discrimination and transpiration efficiency in liens of the C4 species of *Sorghum bicolor* in the glasshouse and the field. Aust. J. Plant Physiol. 25, 111–123.

Henzell, R.G., Gillerion, W., 1973. Effects of partial and complete panicle removal on the rate of death of some Sorghum bicolor genotypes under moisture stress. Qld. J. Agric. Anim. Sci. 30, 291–299.

Henzell, R.G., Hare, B.W., 1996. Sorghum breeding in Australia: public and private endeavors. Proceedings of the Third Australian Conference. 20–22 February 1996, Tamworth, NSW Australian Institute of Agricultural Science, Melbourne, Occasional Publication No. 93, pp. 159–171.

Henzell, R.G., Dodman, R.L., Done, A.A., Brengman, R.L., Meyers, R.E., 1984. Lodging, stalk rot and root rot in sorghum in Australia. In: Mughogho, L.K. (Ed.), Sorghum Root and Stalk Rots: A Critical Review, Proceedings of Consultative Group Discussion of Research Needs and Strategies for Control of Sorghum Root and Stalk Rot Diseases, Bellagio, Italy. 27 November–2 December 1983, ICIRISAT, Patancheru, India, pp. 225–236.

Henzell, R.G., Brengman, R.L., Fletcher, D.S., McCosker, A.N., 1992. Relationship between yield and non-senescence (stay green) in some grain sorghum hybrids grown under terminal drought stress. Proceedings of the Second Australian Sorghum Conference. 4–6 February 1992, Gatton, QLD Australian Institute of Agricultural Science, Melbourne, Occasional Publication No. 68, pp. 355–358.

Henzell, R.G., Hammar, G.I., Borrell, A.K., McIntyre, C.L., Chapman, S.C., 1997. Research on drought resistance in grain sorghum in Australia. Int. Sorghum Millets Newsl. 38, 1–8.

Hermus, R.C., Fukal, S., Wilson, G.L., 1982. Quantitative studies of water stress in sorghum. Sorghum Newsl. 25, 125.

Hewenan, 2001. An analysis of the heterosis and yield component factors of middle short stalked sorghum hybrids. J. Jilin Agric. Univ. 23, 11–14.

Heyne, E.R., Barmore, M.A., 1965. Breeding wheat for quality. Adv. Agron. 17, 85–114.

Hiremath, S.M., Parvatikar, S.R., 1985. Growth and yield analysis in sorghum identification of genotypes with low leaf area and high dry matter production. Sorghum Newsl. 28, 108.

Hiremath, R.V., Palakshappa, 1991. Evaluation of high yielding cultivars and hybrids of sorghum against diseases. Sorghum Newsl. 32, 39.

Hoekstra, F.A., Buitink, J., 2001. Mechanisms of plant dessiccation tolerance. Trends Plant Sci. 8 (9), 431–438.

Hoogmoed, W.B., Klaij, M.C., 1990. Soil management per crop production in the West African Sahel. I. Soil and climate parameters. Soil Tillage Res. 16, 85–103.

Hookstra, G.H., Ross, W.M., Mumm, R.F., 1983. Simultaneous evaluation of grain sorghum A lines and random mating population with top crosses. Crop Sci. 23, 977–981.

Horber, E., 1980. Types and classification of resistance. In: Maxwell, F.G., Jennings, P.R. (Eds.), Breeding Plant Resistant to Insects. Wiley, New York, NY, pp. 15–21.

Horsfall, J.G., Diamond, A.E., 1957. The diseased plant. In: Horsfall, J.G., Diamond, A.F. (Eds.), Plant Pathology—An Advanced Treatise. Academic Press, New York, NY, pp. 1–16.

House, L.R., 1980. A Guide to Sorghum Breeding. International Crops Research Institute for the Semi-Arid Tropics, Patancheru, India.

House, L.R., Verma, B.N., Ejeta, G., Rana, B.S., Kapran, I., Obilana, A.B., et al., 1997. Developing countries breeding and potential of hybrid sorghum. INTSORMIL Publication. Proceedings of the International Conference on Genetic Improvement of Sorghum and Pearl Millet, Lubbock, TX. 22–27 September 1996, Lincoln, NE. Collaborative Research Support Program on Sorghum and Pearl Millet. 97–5, pp. 84–96.

Hovny, M.R.A., El-Noguiy, O.O., Hassaballa, E.A., 2000. Combining ability and heterosis in grain sorghum (*Sorghum bicolor* (L.) Moench). Assiut J. Agric. Sci. 31 (3), 1–16.

Hovny, M.R.A., Menshawi, M.M., Nagouly, O., 2001. Combining ability and heterosis in grain sorghum [*Sorghum bicolor* (L.) Moench]. Bull. Fac. Agric., Cairo Univ. 52, 47–60.

Hsi, D.C.H., 1961. An effective technique for screening sorghum for resistance to charcoal rot. Phytopathology. 51, 340–341.

Hsieh, J.S., 1988. Cyanogenesis and aphid resistance in sorghum. Guoli Taiwan Daxue Nongxue Yuan Baogao. 28, 81–87.

Hsieh, J., Pi, C.P., 1982. Genetic study on aphid resistance in sorghum. J. Agric. Assoc. China New Ser. 117, 6–14.

Hsieh, J., Pi, C.P., 1988. Diallel analysis of resistance to aphid in sorghum. J. Agric. Assoc. China. 142, 67–84.

Hukkeri, S.B., Shukla, N.P., 1983. Effect of soil moisture stress at different stages of growth on the fodder yield of MP Chari sorghum. Indian J. Agric. Sci. 53 (1), 44–48.

Humphreys, M.O., 1994. Variation in carbohydrate and protein content of rye grasses: potential for genetic manipulations. In: Reheul, D., Ghesquiere, A. (Eds.), Breeding for Quality: Proceedings of 19th Eucarpia Fodder Crops Section Meeting. Merelbeke, Belgium, Rijksstation plantenveredeling.

Hunt, L.A., 1993. Designing improved plant types: a breeder's viewpoint. In: Penning de Vries, F.W.T., Tang, P., Metselaar, K. (Eds.), Systems Approaches for Sustainable Agricultural Development. Kluwer Academic Publishers, Dordrecht, The Netherlands, pp. 3–17.

Hussaini, S.H., Rao, P.V., 1964. A note on the spontaneous occurrence of cytoplasmic male sterility in Indian sorghum. Sorghum Newsl. 7, 27–28.

ICRISAT, 1992. The Medium Term Plan, vol. 2. International Crops Research Institute for the Semi-Arid Tropics (ICRISAT), Patancheru, India.

ICRISAT (International Crops Research Institute for the Semi-Arid Tropics), 1995. Restorers. ICRISAT Asia Region Annual Report 1992. ICRISAT, Patancheru, India, 75.

Indi, S.K., 1978. Genetic Analysis of Different Quantitative Characters in Inter Varietal Crosses of Sorghum (M.Sc. (Agri) thesis). University of Agricultural Sciences, Bangalore.

Indi, S.K., Goud, J.V., 1981. Gene effects in sorghum. Indian J. Genet. Plant Breed. 41, 25–29.

Indira, S., Rana, B.S., 1983. Host plant resistance to *Macrophomina phaseolina* in F1 varietal crosses of Sorghum. Int. J. Trop. Plant Dis. 1, 187–192.

IPCC, 2007. Climate Change 2007: impacts, adaptation and vulnerability. In: Parry, M.L., Canziani, O.F., Palutikof, J. P., van der Linden, P.J., Hanson, C.E. (Eds.), Contribution of Working Group II to the Fourth Assessment Report of the Intergovernmental Panel on Climate Change. Cambridge University Press, Cambridge, UK, p. 976.

Iruegas, A., Cejudo, H., Guiragossian, V., 1982. Screening and evaluation of *tortilla* from sorghum and sorghum-maize mixtures. Proceedings, International Symposium on Sorghum Grain Quality. 28–31 October 1981, ICRISAT, Patancheru, India.

Irvine, J.E., 1975. Relation of photosynthetic rate of leaf and canopy characters on sugarcane yield. Crop Sci. 15, 1671–1676.

Iyanar, K., Gopalan, A., Ramasamy, P., 2001. Combining ability analysis in sorghum [*Sorghum bicolor* (L.) Moench]. Ann. Agric. Res. 22, 341–345.

Jagadeshwar, K., Shinde, V.K., 1992. Combining ability in *rabi* sorghum (*Sorghum bicolor* L. Moench). Indian J. Genet. 52 (1), 22–25.

Jahagirdar, S., Ravikumar, M.R., Jamadar, M.M., Pawar, K.N., 2002. Field screening of local genotypes against charcoal rot of sorghum caused by *Macrophomina phaseolina* (Tassi) Goid. Agric. Sci. Dig. 22 (2), 87–89.

Jahagirdar, S., Kulkarni, S., Biradar, B.D., 2006. Identification of sources of resistance against *M. phaseolina* an incitant of charcoal rot of *rabi* sorghum. J. Basic Appl. Mycol. 5 (I & II), 32–33.

Jain, K.K., Bhatnagar, M.P., 1962. Studies on varietal resistance to the Jowar shoot fly. Indian J. Genet. Plant Breed. 22, 224–229.

Jain, K.K., Kulashreshta, V.P., 1976. Dwarfing genes and breeding for yield in bread wheat. Pflangenziecht. 76, 102–112.

Jain, M., Khurana, J.P., 2009. Transcript profiling reveals diverse roles of auxin-responsive genes during reproductive development and abiotic stress in rice. FEBS J. 276, 3148–3162.

Jayanthi, P.D.K., 1997. Genetics of Shoot Fly Resistance in Sorghum Hybrids of Cytoplasmic Male Sterile Lines (Ph.D. thesis). Acharya N. G. Ranga Agricultural University, Hyderabad, India.

Jayanthi, P.D.K., Reddy, B.V.S., Gour, T.B., Reddy, D.D.R., 1999. Genetics of glossy and trichome characters in sorghum hybrids of cytoplasmic male sterile lines. J. Maharashtra Agric. Univ. 24, 251–256.

Jayanthi, P.D.K., Reddy, B.V.S., Reddy, D.D.R., Gour, T.B., 2000. Genetic analysis of shoot fly resistance in sorghum. PKV Res. J. 24 (1), 35–41.

Jayanthi, P.D.K., Reddy, B.V.S., Gour, T.B., Reddy, D.D.R., 2002. Early seedling vigour in sorghum and its relationship with resistance to shoot fly, *Atherigona soccata* Rondani. J. Entomol. Res. 26, 93–100.

Jayaprakash, P., Das, L.D.V., 1994. Effect of heterosis in sorghum of leaf area and dry fodder yield. Int. Sorghum Millets Newsl. 35, 78.

Jebaraj, S., Sundaram, M.K., Sundarapandian, G., 1988. A comparative study of heterosis in hybrids in MS lines with indigenous sorghum cultivation. Madras Agric. J. 75, 161–163.

Jeewad, A.S.B., 1993. Studies on the Inheritance of Resistance to Shoot Fly (*Atherigona soccata* Rondani) of Sorghum (Masters thesis). Andhra Pradesh Agricultural University.

Jefferson, P.G., Johnson, D.A., Runbaugh, M.D., Asay, K.H., 1989. Water stress and genotypic effects on epicuticular wax production of alfalfa and crested wheat grass in relation to yield and excised leaf water loss rate. Can. J. Plant Sci. 69, 481–490.

Jenkins, M.T., 1935. The effect of inbreeding and of selection within inbred lines of maize upon the hybrids made after successive generations of selfing. Iowa State J. Sc. 3, 429–750.

Jenkins, M.T., 1940. The segregation of genes affecting yield of grain in maize. J. Am. Soc. Agron. 32, 55–63.

Jeyaprakash, P., Ganapathy, S., Pillai, A., 1997. Correlation and analysis in sorghum. Ann. Agric. Res. 18 (3), 309–312.

Jhansi Rani, K., Rana, B.S., Kaul, S., Rao, S.S., Ganesh, M., 2007. Genetic analysis of certain morpho-physiological characters in *rabi* sorghum. Indian J. Genet. 67 (3), 281–283.

Jiban Mitra, 2001. Genetics and genetic improvement of drought resistance in crop plants. Curr. Sci. 80 (25), 758–763.

Jirali, D.I., Biradar, B.D., Rao, S.S., 2007. Performance of rabi sorghum genotypes under receding soil moisture conditions in different soil types. Karnataka J. Agric. Sci. 20 (3), 603–604.

Johnson, B.A., Rooney, L.W., Khan, M.N., 1979. *Tortilla* making characteristics of micronized sorghum and corn flours. J. Food Sci. 45, 671–674.

Johnson, S.S., Geadelmann, J.L., 1989. Influence of water stress on grain yield response to recurrent selection in maize. Crop Sci. 29, 558–565.

Jones, H.G., 1980. Interaction and integration of adaptive responses to water stress: the implications of an unpredictable environment. In: Mussell, H., Staples, R.C. (Eds.), Stress Physiology in Crop Plants. Wiley, New York, NY, pp. 408–428.

Jones, M.M., Osmond, C.B., Turner, N.C., 1980. Accumulation of solutes in leaves of sorghum and sunflower in response to water deficits. Aust. J. Plant Physiol. 7, 193–205.

Jordan, W.R., Miller, F.R., 1980. Genetic variability in sorghum root systems: implications for drought tolerance. In: Turner, N.C., Kramer, P.J. (Eds.), Adaptation of Plants to Water and High Temperature Stress. Wiley, New York, NY, pp. 383–399.

Jordan, W.R., Monk, R.L., 1980. Enhancement of the drought resistance of sorghum: progress and limitations. Proceedings, Thirty-fifth Corn and Sorghum Research Conference. American Seed Trade Association, Chicago, IL.

Jordan, W.R., Sullivan, C.Y., 1982. Reaction and resistance of grain sorghum to heat and drought. Sorghum in the Eighties, ICRISAT, Proceedings of International Sorghum Workshop. 2–7 November 1981, Hyderabad, India, pp. 131–142.

Jordan, W.R., Dugas, W.A., Shouse, P.J., 1983. Strategies for crop improvement for drought prone regions. Agric. Water Manage. 7, 281–299.

Joshi, M.W., Jamadagni, B.M., 1990. Physiological characters in relation to yield in *rabi* sorghum. J. Maharashtra Agric. Univ. 15, 228–229.

Jotwani, M.G.I., Sharma, G.C., Srivastava, B.G., Marwaha, K.K., 1971. Ovipositional response of shootfly, *Atherigona varia soccata* (Rondani) on some promising resistant lines of sorghum. Investigations on Insect Pests of Sorghum and Millets (1965-70) (ed S Pradhan) Final Technical Report. Division of Entomology, IARI, New Delhi, pp. 119–122.

Juarez, M.A., 1979. Estudio de adopcion variedad mejoradas de sorgo. Memoral XXV (Volume 3). Proceedings, PCCMA. Teguigalpa, Hondura. Secretoria de Recuersos Naturales.

Juliano, B.O., 1979. The chemical basis of rice grain quality. Chemical Aspects of Rice Grain Quality. IRRI, Los Banos, Philippines.

Kachave, K.S., Nandanwankar, K.G., 1980. Estimates of gene effects for grain yield in sorghum. Sorghum News Lett. 23, 15–16.

Kadam, D.E., Kulkarni, V.M., Katule, B.K., Patil, S.V., 2000. Combining ability studies in Rabi sorghum *Sorghum bicolor* L Moench under rainfed condition. Adv. Plant Sci. 20 (1), 39–41.

Kadam, G.N., Gadakh, S.R., Awari, V.R., 2002. Physiological analysis of *rabi* sorghum genotypes for shallow soil. J. Maharashtra Agric. Univ. 27 (3), 274–276.

Kadioglu, A., Terzi, R., 2007. A dehydration avoidance mechanism: leaf rolling. Bot. Rev. 73 (4), 290–302.

Kajjari, N.B., Chavan, V.M., 1953. A male-sterile jowar. Indian J. Genet. Plant Breed. 13, 48–89.

Kakani, V.G., Prasad, P.V.V., Craufurd, P.Q., Wheeler, T.R., 2002. Response of *in vitro* pollen germination and pollen tube growth of groundnut (*Arachis hypogaea* L.) genotypes to temperature. Plant, Cell Environ. 25, 1651–1661.

Kakani, V.G., Reddy, K.R., Koti, S., Wallace, T.P., Prasad, P.V.V., Reddy, V.R., et al., 2005. Differences in *in vitro* pollen germination and pollen tube growth of cotton cultivars in response to high temperature. Ann. Bot. 96, 59–67.

Kamala, V., Sharma, H.C., ManoharRao, D., Varaprasad, K.S., Bramel, P.J., 2009. Wild relatives of Sorghum as sources of resistance to Sorghum shoot fly, *Atherigona soccata*. Plant Breed. 128 (2), 137–142.

Kamatar, M.Y., Salimath, P.M., 2003. Morphological traits of sorghum associated with resistance to shoot fly, *Atherigona soccata* Rondani. Indian J. Plant Prot. 31, 73–77.

Kambal, A.E., Webster, O.J., 1965. Estimates of general and specific combining ability in sorghum. Crop Sci. 5, 521–523.

Kamoshita, A., Muchow, R.C., Cooper, M., Fukai, S., 1996. Influence of genotype and environment on the utilization of nitrogen by grain sorghum. In: Foale, M.A., Henzell, R.G., Kneipp, J.F. (Eds.), Proceedings Third Australian Sorghum Conference, Tamworth, 20-22 February 1996. Australian institute of Agricultural Science, Melbourne, Occasional Publication no. 93, pp. 338–391.

Kanaka, S.K., 1979. Genetic Analysis of Ten Quantitative Characters in Sorghum (M.Sc. (Agri) thesis). University of Agricultural Sciences, Bangalore.

Kanaka, S.K., 1982. Genetic Analysis of Ten Quantitative Characters in Sorghum (Thesis abstracts). 8, 72–73.

Kandekar, 2008. Genetics of Some Physiological Parameters Related to Drought Tolerant in *Rabi* Sorghum (*Sorghum bicolor* (L.) Moench) (M.Sc. (Agri.) thesis submitted to M.P.K.V). Rahuri, Maharashtra.

Kannangara, T., Seetharama, N., Durley, R., Simpson, G.M., 1983. Drought resistance of *Sorghum bicolor* (L.) changes in endogenous growth regulators. Can. J. Plant Sci. 63 (1), 147–157.

Karamanos, A.J., Papatheohari, A.Y., 1999. Assessment of drought resistance of crop genotypes by means of the water potential index. Crop Sci. 39, 1792–1797.

Karthik, V.T., 2004. Heterosis and Stability in *Rabi* Sorghum (*Sorghum bicolor* (L.) Moench (M.Sc.(Agri.) thesis). University of Agricultural Sciences, Dharwad.

Karunakar, R.I., Pande, S., Satyaprasad, K., Ramrao, P., 1993. Varietal reaction of non-senescent sorghums to the pathogens causing root and stalk rot of sorghums in India. Int. J. Pest Manage. 39 (3), 343–346.

Kasenko, G.A., 1986. Reproductive heterosis in hybrids of grain sorghum. Selektsiya-isemenovodstvo zernovykh-kultur, pp. 92–95.

Kashiwagi, J., Krishnamurthy, L., Upadhyaya, H.D., Krishna, H., Chandra, S., Vadez, V., et al., 2004. Genetic variability of drought avoidance root traits in the mini-core germplasm collection of chickpea (*Cicer arietinum* L.). Euphytica. 146, 213–222.

Kebede, H., Subudhi, P.K., Rosenow, D.T., Nguyen, H.T., 2001. Quantitative trait influencing drought tolerance in grain sorghum (*Sorghum bicolor* L. Moench). Theor. Appl. Genet. 103, 266–276.

Kempthorne, O., 1957. An Introduction to Genetic Statistics. first ed. John Wiley & Sons, New York, NY, pp. 456–471.

Ketring, D.L., 1986. Physiological response of groundnut to temperature and water deficits-breeding implications. In: Sivakumar, M.V.K., Virmani, S.M. (Eds.), Agrometeorology of Groundnut. ICRISAT, Patancheru, India, pp. 135–143.

Khalfaoui, J.L.B., Havard, M., 1993. Screening peanut cultivars in the field for root growth: a test by herbicide injection in the soil. Field Crops Res. 32, 173–179.

Khan, D.R., Mackill, D.J., Vergara, B.S., 1986. Selection for tolerance to low temperature induced spikelet at anthesis in rice. Crop. Sci. 26, 694–698.

Khan, M.Q., Rao, A.S., 1956. The influence of the black ant (*Camponotus compressus* F.) on the incidence of two homopterous crop pests. Indian J. Entomol. Soc. 18, 199–200.

Khan, M.N., Rooney, L.W., Rosenow, D.T., Miller, F.R., 1980. Sorghum with improved *tortilla* making characteristics. J. Food Sci. 45, 720.

Khanna, C.R., Sinha, S.K., 1988. Enhancement of drought induced senescence by the reproductive sink in fertile lines of wheat and sorghum. Ann. Bot. 61 (6), 649–653.

Khanure, S.K., Hemalatha, B., Anahosur, K.H., 1997. Charcoal rot disease resistance in sorghum. Karnataka J. Agric. Sci. 10, 1215–1216.

Khidse, S.R., Bhale, N.L., Borikar, S.T., 1983. Combining ability for seedling vigour in sorghum. J. Maharashtra Agric. Univ. 8, 59–60.

Khizzah, B.W., Miller, F.R., 1992. Correlations between sorghum components of drought resistance and various agronomic characters at four locations. Sorghum Newsl. 33, 54.

Khizzah, B.W., Miller, F.R., Newton, R.J., 1995. Genetic and physiological components of post-flowering drought tolerance in sorghum. Afr. Crop Sci. J. 3 (1), 15–21.

Kholova, J., Hash, C.T., Kakkera, A., Kocova, K., Vadez, V., 2009. Constitutive water-conserving mechanisms are correlated with the terminal drought tolerance of pearl millet [*Pennisetum glaucum* (L.) R. Br.]. J. Exp. Bot.1–9. Available from: http://dx.doi.org/10.1093/jxb/erp314 (http://jxb.oxfordjournals.org/).

Kholová, J., Hash, C.T., Kakkera, A., Kočová, M., Vadez, V., 2010a. Constitutive water conserving mechanisms are correlated with the terminal drought tolerance of pearl millet [*Pennisetum glaucum* (L.) R. Br.]. J. Exp. Bot. 61, 369–377.

Kholová, J., Hash, C.T., Lava Kumar, P., Yadav, R.S., Kocova, M., Vadez, V., 2010b. Terminal drought tolerant pearl millet (*Pennisetum glaucum* (L.) R. Br) have high leaf ABA and limit transpiration at high vapour pressure deficit. J. Exp. Bot. 61 (5), 1431–1440.

Kholová, J., Hash, C.T., Kocova, M., Vadez, V., 2011. Does a terminal drought tolerance QTL contribute to differences in ROS scavenging enzymes and photosynthetic pigments in pearl millet exposed to drought? Environ. Exp. Bot. 71 (1), 99–106.

Khot, K.B.,2008. Generation Mean Analysis and Molecular Assay of Yield and Yield Components for Drought Tolerance in *Rabi* Sorghum (Ph.D. thesis submitted to M.P.K.V). Rahuri, Maharashtra.

Khotyleva, L.V., Neshina, L.P., 1983. Methods of establishing useful parent forms of grain sorghum. Selektsiyz-i-semenovodstvo USSR. 4, 22–23.

Khusnmetdinova, T.G., Elkonin, L.A., 1989. Inheritance of some qualitative characters in sorghum. Selektisiya, Agroteckhniva i Prozvodstra Sorgo Shornik Nauchnyvh Trudor Zernograd. USSR, pp. 96–108.

Kide, B.R., Bhale, N.L., Borikar, S.T., 1985. Study of heterosis in single and three way crosses in sorghum [*Sorghum bicolor* (L.) Moench]. Indian J. Genet. Plant Breed. 45, 203–208.

Kim, J.G., Lee, S.B., Han, M.S., 1987. A study on dry matter accumulation and net energy values of sorghum plants. I physiological analysis of yield components and chemical composition of sorghum plants. Korean J. Anim. Sci. 29, 269–272.

Kim, J.S., Klein, P.E., Klein, R.R., Price, H.J., Mullet, J.E., Stelly, D.M., 2005. Chromosome identification and nomenclature of *Sorghum bicolor*. Genetics. 169 (2), 1169.

Kim, H.K., Luquet, D., van Oosterom, E., Dingkuhn, M., Hammer, G., 2010. Regulation of tillering in sorghum: genotypic effects. Ann. Bot. 106, 69–78.

Kirleis, A.W., Crosby K.D., 1982. Sorghum hardness: comparison of methods for its evaluation. Proceedings, International Symposium on Sorghum Grain Quality. October 1981, ICRISAT, Patancheru, India, pp. 28–31.

Kishan, A.G., Borikar, S.T., 1988. Heterosis and combining ability in relation to cytoplasmic diversity in sorghum (*Sorghum bicolor*). Indian J. Agric. Sci. 58, 715–717.

Kishan, A.G., Borikar, S.T., 1989. Line x tester analysis involving diverse cytoplasmic systems in sorghum (*Sorghum bicolor*). Plant Breed. 102, 153–157.

Kishore, P., 2005. Breeding for resistance to shoot fly, *Atherigona soccata* Rondani and stem borer, *Chilo partellus* (swinhoe) in sorghum. J. Entomol. Res. 29 (1), 1–8.

Klein, P.E., Klein, R.R., Cartinhour, S.W., et al., 2000. A high-throughput AFLP-based method for constructing integrated genetic and physical maps: progress toward a sorghum genome map. Genome Res. 10 (6), 789.

Klein, R.R., Klein, P.E., Chhabra, A.K., et al., 2001. Molecular mapping of the rf1 gene for pollen fertility restoration in sorghum (*Sorghum bicolor* L.). Theor. Appl. Genet. 102 (8), 1206.

Klein, R.R., Klein, P.E., Mullet, J.E., Minx, P., Rooney, W.L., Schertz, K.F., 2005. Fertility restorer locus Rf1 of sorghum (*Sorghum bicolor* L.) encodes a pentatricopeptide repeat protein not present in the colinear region of rice chromosome 12. Theor. Appl. Genet. 111 (6), 994.

Knoll, J.E., Ejeta, G., 2008. Marker-assisted selection for early season cold tolerance in sorghum: QTL validation across populations and environments. Theor. Appl. Genet. 116, 541–553.

Knoll, J., Gunaratna, N., Ejeta, G., 2008. QTL analysis of early-season cold tolerance in sorghum. Theor. Appl. Genet. 116, 577–587. Available from: http://dx.doi.org/10.1007/s00122-007-0692-0.

Kouressy, M., Dingkuhn, M., Vaksmann, M., Heinemann, A.B., 2008. Adaptation to diverse semi-arid environments of sorghum genotypes having different plant type and sensitivity to photoperiod. Agric. For. Meteorol. 148, 357–371.

Kramer, P.J., Boyer, J.S., 1995. In: Teare, I.D., Peet, M.M. (Eds.), Water Relations. John Wiley & Sons, New York, NY, pp. 352–380.

Kramer, N.W., 1959. Combining value in sorghums. Agron. Abstr.1959–1961.

Kramer, P.J., 1980. Drought, stress and the origin of adaptations. In: Turner, N.C., Karmer, P.J. (Eds.), Adaptation of Plants to Water and High Temperature Stress. Wiley–Interscience, New York, NY, pp. 7–20.

Kresovich, S., Barbazuk, B., Bedell, J.A., et al., 2005. Toward sequencing the sorghum genome. A U.S. National Science Foundation-Sponsored Workshop Report. Plant Physiol. 138 (4), 1898–1902.

Krieg, D.R., 1983. Sorghum. In: Teare, E.D., Peet, M.M. (Eds.), Crop-Water Relations. Wiley, New York, NY, pp. 351–388.

Krieg, D.R., 1988. Water use efficiency of grain sorghum. Proceedings of the Forty-Third Annual Corn and Sorghum Industry Research Conference. 8–9 December 1988, Chicago, IL, Organized by American Seed Trade Association, Washington, pp. 27–41.

Krieg, D.R., 1993. Stress tolerance mechanisms in above ground organs. Proceedings of a Workshop on Adaptation of Plants to Soil Stresses. 1–4 August 1993, Lincoln, NE, INTSORMIL Publication No. 94-2, pp. 65–79.

Krieg, D.R., Dalton., 1990. Analysis of genetic and cultural improvements in crop production: sorghum. In: Sinha, S.K., Sane, P.V., Bhargava, S.C., Agarwal, P.K. (Eds.), Proceedings of the International Congress of Plant Physiology Society of Plant Physiology and Biochemistry. 20 February 1988, New Delhi, pp. 142–151.

Krishnamurthy, K., Bommegouda, G., Raghynath, G., Rajashekar, B.G., Venugopal, N., Jagannath, M.K., et al., 1973. Structure in grain yield in sorghum genotypes under comparable leaf area index at flowering. Investigation on Structure of Yield Cereals (Maize and Sorghum) Technical Report (PL 480). Department of Agronomy, UAS, Bangalore.

Krishnamurthy, K., Bommegowda, G., Venugopal, N., 1974. Growth and yield differences in transplanted sorghum. Mysore J. Agric. Sci. 8, 60–68.

Krishna Murthy, S.L., Ambekar, S.S., Dushyanta kumar, B.M., Satish, R.G., Shashidhara, N., Sunil Kumar, S.V., 2010. Heterosis studies for yield and its attributing traits in sorghum (*Sorghum bicolor* (L.) Moench). Gregor Mendel Found. J. 1, 73–81.

Kudasomannavar, B.T., 1974. Effect of Nitrogen and Plant Population on Growth and Yield of Sorghum, CSH1 (M.Sc. (Agri.) thesis). University of Agricultural Sciences, Dharwad.

Kulkarni, V.V., 2002. Combining Ability and Heterosis Studies for Grain Yield, its Components and Shoot Fly Tolerance in *Rabi* Sorghum (M.Sc. (Agri) thesis). University of Agricultural Sciences, Dharwad.

Kulkarni, N., Shinde, V.K., 1987. Genetic analysis of yield components in *rabi* sorghum. J. Maharashtra Agric. Univ. 12, 378–379.

Kulkarni, L.P., Narayana, R., Krishna Sastry, K.S., 1981. Photosynthetic efficiency and translocation in relation to leaf characters and productivity in sorghum genotypes. Sorghum Newsl. 24, 124–125.

Kulkarni, L.P., Choudhari, S.B., Titbotkar, A.B., Kalyankar, S.P., 1983. Relationship of physiological parameter with grain sorghum grain yield in sorghum under *rabi* season. Sorghum Newsl. 26, 234.

Kusalkar, D.V., Awari, V.R., Pawar, V.Y., Shinde, M.S., 2003. Physiological parameters in relation to grain yield in *rabi* sorghum on medium soil. Adv. Plant Sci. 16 (1), 119–122.

Lakshmidevi, G., Janila, P., Anuradha, G., Shivashankar, A., 2012. Validation of foreground and background SSR markers for introgression of QTL governing leaf glossiness into a sorghum variety-NTJ 2. Indian J. Genet. Plant Breed. 72 (3).

Lamani, B.B., 1996. Influence of Plant Densities on Portion in of Assimilates During Post Anthesis Period in *Rabi* Sorghum (*Sorghum bicolor* (L.) Moench) Genotypes (M.Sc. (Agri.) thesis). University of Agricultural Sciences, Dharwad.

Lara, M.E.B., Garcia, M.C.G., Fatima, T., et al., 2004. Extracellular invertase is an essential component of cytokinin-mediated delay of senescence. The Plant Cell. 16, 1276–1287.

Lasavio, N., Mastrorilli, M., Venezian Scarascia, M.A., 1982. The water stress and air temperature influence on growth and development of grain sorghum. Annali bell Ishtiuto—Sperimentale Agronomico. 13 (2), 249–262.

Lawn, R.J., Imrie, B.C., 1991. Crop improvement for tropical and subtropical Australia: designing plants for different climates. Field Crops Res. 26, 113–139.

Laxman, S., 2001. Fertility restoration and magnitude of heterosis in cross between A1 restores and A2 CMS line in sorghum [*Sorghum bicolor* (L.) Moench]. J. Res. 28, 104–105, Acharana N.G. Ranga Agricultural University, Hyderabad.

Laxman, S., Rao, K.V.K., 1995. Effect of low temperature on anther dehiscence and seed setting in CSH-9 hybrid sorghum (*Sorghum bicolar* (L.) Moench) during rabi season. J. Res. APAU. 23 (3/4), 34–35.

Lazanyi, J., Bajai, S., 1986. Combining ability for yield and forage component in diallel crosses of some male sterile and maintainer lines of sorghum. Acta Agron. Hung. 35, 3–4.

Lee, M.H., 2001. Low temperature tolerance in rice: The Korean experience. In: Fukai, S., Basnayake, J. (Eds.), ACIAR Proc. 101: Increased Lowland Rice Production in the Mekong Region. Australian Centre for International Agricultural Research, Canberra, pp. 138–146.

Leopold, A.C., Musgrave, M.E., Williams, K.M., 1981. Solute leakage resulting from leaf desiccation. Plant Physiol. 68, 1222–1225.

Levitt, J., 1992. Responses of Plants to Environmental Stresses. Academic Press, New York, NY.

Lia, X.H., Dai, L.Y., Zhang, Z.H., 1998. Factors related to cold tolerance at booting stage in *Oryza sativa* L. Chin. J. Rice Sci. 12, 6–10.

Lin, J.L., Yeh, M.S., 1990. Variation in the growth dynamics of sorghum cultivars. J. Agric. Forestry. 39 (2), 63–67.

Lin, Y.R., Schertz, K.F., Paterson, A.H., 1995. Comparative analysis of QTLs affecting plant height and maturity across the Poaceae, in reference to an interspecific sorghum population. Genetics. 141, 391–411.

Lodhi, G.P., Paroda, R.S., Ram, H., 1978. Heterosis and combining ability in forage sorghum. Indian J. Agric. Sci. 48, 205–210.

Lokapur, R.G., 1997. Heterosis and Combining Ability Studies in Sorghum [*Sorghum bicolor* (L.) Moench] (M.Sc. (Agri) thesis). University of Agricultural Sciences, Dharwad.

Lonnquist, J.H., 1950. The effect of selecting for combining ability within segregating lines of corn. Argon J. 42, 503–508.

Lu, X.L., Niu, A.L., Cai, H.Y., Zhao, Y., Liu, J.W., Zhu, Y.G., et al., 2007. Genetic dissection of seedling and early vigor in a recombinant inbred line population of rice. Plant Sci. 172, 212–220.

Ludlow, M.M., 1980a. Stress physiology of tropical pasture plants. Trop. Grasslands. 14, 136–145.

Ludlow, M.M., 1980b. Adaptive significance of stomatal responses to water stress. In: Turner, N.C., Kramer, P.J. (Eds.), Adaptation of Plants to Water and High Temperature Stress. Wiley, New York, NY, pp. 123–138.

Ludlow, M.M., 1987. Contribution of osmotic adjustment to the maintenance of photosynthesis during water stress. In: Biggins, J. (Ed.), Progress in Photosynthesis Research, vol. 4. Martinus Nijhoff, Dordrecht, The Netherlands, pp. 161–168.

Ludlow, M.M., 1993. Physiological mechanisms of drought resistance. In: Mabry, Nguyen, Dixon (Eds.), Proceedings of Symposium on Application and Prospects of Biotechnology. 5–7 November 1992. Lubbock, TX. IC2 Institute, University of Texas, Austin, pp. 11–34.

Ludlow, M.M., Muchow, R.C., 1990. A critical evaluation of traits for improving crop yields in water-limited environments. Adv. Agron. 43, 107–153.

Ludlow, M.M., Santamareia, F.J., Fakai, S., 1989. Role of osmotic adjustment in reducing the loss of grain yield in sorghum due to drought. Crop Sci. 49, 2386–2392.

Ludlow, M.M., Santamaria, J.M., Fukai, S., 1990. Contribution of osmotic adjustment to grain yield in sorghum under water limited condition II. Water stress after anthesis. Aust. J. Agric. Res. 41, 67–78.

Lux, A., Luxová, M., Hattori, T., Inanaga, S., Sugimoto, Y., 2002. Silicification in sorghum (*Sorghum bicolor*) cultivars with different drought tolerance. Physiol. Plant. 115 (1), 87–92.

Mace, E., Singh, V., Van Oosterom, E., Hammer, G., Hunt, C., Jordan, D., 2012. QTL for nodal root angle in sorghum (*Sorghum bicolor L. Moench*) co-locate with QTL for traits associated with drought adaptation. Theor. Appl. Genet. 124, 97–109.

Madhava Rao, S.S., Sindagi, G., Srinivasalu, 1970. Study of resistance to shoot fly in sorghum. Sorghum Newsl. 13, 37–38.

Madhusudana, R., Umakanth, A.V., Kaul, S., Rana, B.S., 2003. Stability analysis for grain yield in postrainy sorghum (*Sorghum bicolor* (L.) Moench). Indian J. Genet. Plant Breed. 63 (3), 255–256.

Madhusudhan, R., 1993. Combining Ability Studies in Segregating Generations of Sorghum [*Sorghum bicolour* (L.) Moench] (M.Sc. (Agri) thesis). University of Agricultural Sciences, Dharwad.

Madhusudhan, R., 2002a. Heterotic capability of derived inbred lines for grain yield in sorghum [*Sorghum bicolor* (L.) Moench]. Indian J. Genet. Plant Breed. 62, 118–120.

Madhusudhan, R., 2002b. Selection of superior female parents utilizing B x R crosses for A line development in pearl millet. Int. Sorghum Millets Newsl. 43, 81–83.

Madhusudhan, R., Patil, S.S., 1996a. Evaluation of sorghum F3 population for combining ability. New Botanist. 23, 5–11.

Madhusudhan, R., Patil, S.S., 1996b. Combining ability studies in F4 generation of sorghum. New Botanist. 23, 55–61.

Madhusudhan, R., Patil, S.S., 1996c. Relationship between *per se* and test cross performance and combining ability in sorghum. New Botanist. 23, 113–117.

Madhusudhan, R., Patil, S.S., 2000. Use of recurrent recombination to create variability for combining ability. Bioved. 11, 71–74.

Mahajan, R.C., Sudewad, S.M., Jawale, L.N., 1998. Inheritance of seed size and grain yield in sorghum. J. Maharashtra Agric. Univ. 23, 310–311.

Mahalakshmi, V., Bidinger, F.R., 2002. Evaluation of stay-green sorghum germplasm lines at ICRISAT. Crop Sci. 42, 965–974.

Mahdy, E.E., Bakheit, B.R., Ei-Hinnawy, H.H., 1987. Expected and realized gain from S1 and S2 selections in grain sorghum *Sorghum bicolor* (L.) Moench. Assiut J. Agric. Sci. 18, 103–118.

Maiti, R.K., 1980. Role of glossy trichome trait in sorghum crop improvement. Annual Meeting, All India Sorghum Improvement Workshop. 12–14 May 1980, Coimbatore, Tamil Nadu, India, pp. 1–14.

Maiti, R.K., 1996. Root Development and Growth—Sorghum Science. Oxford and IBH Publishing Company Private Limited, New Delhi, pp.183–212.

Maiti, R.K., Bidinger, F.R., 1979. A simple approach to the identification of shoot fly tolerance in sorghum. Indian J. Plant Prot. 7, 135–140.

Major, D.J., 1980. Photoperiod response characteristics controlling flowering of nine crop species. Canadian J. Plant Sci. 60, 777–784.

Malhotra, R.S., Singh, K.B., 1991. Classification of chickpea growing environments to control genotype by environment interaction. Euphytica. 58, 5–12.

Mallick, A.S., Gupta, M.P., 1989. Combining ability for certain physio-morphological traits in grain sorghum. Crop Improv. 16, 57–61.

Mambani, B., Lal, R., 1983. Response of upland rice varieties in drought stress 2: screening rice varieties by means of variable moisture regimes along a topo sequence. Plant Soil. 73, 73–94.

Managoli, S.P., 1973. An attack of shoot bug/Pundalouya-bug (*Peregrinus maidis*) on *rabi* jowar in dry tract of Bijapur district. Farm J. (India).16–17.

Mangush, P.A., Andryushchenko, N.I., 1998. Heterosis for characters in grain sorghum hybrids. Kukuruza-i-sorgo. 3, 10–11.

Mani, N.S., 1981. Heterosis in sorghum [*Sorghum bicolor* (L.) Moench]. Sorghum Newsl. 29, 20.

Manickam, S., Das, L.D.V., 1994. Line x tester analysis in forage sorghum. Int. Sorghum Millets Newsl. 35, 79–80.

Manschadi, A.M., Christopher, J., deVoil, P., Hammer, G.L., 2006. The role of root architectural traits in adaptation of wheat to water-limited environments. Funct. Plant Biol. 33, 823–837.

Markhart, A.H., 1985. Comparative water relations of *Phaseolus vulgaris* L. and *Phaseolus acutifolius* gray. Plant Physiol. 77, 113–117.

Marshall, D.R., 1991. Alternative approaches and perspectives in breeding for higher yields. Field Crops Res. 26, 171–190.

Masferrer, A., Arró, M., Manzano, D., Schaller, H., Fernández-Busquets, X., Moncaleán, P., et al., 2002. Overexpression of Arabidopsis thaliana farnesyl diphosphate synthase (FPS1S) in transgenic Arabidopsis induces a cell death/senescence-like response and reduced cytokinin levels. Plant J. 30, 123–132.

Mastroilli, M., Defilippis, R., Incarnato, D., Katerji, N., Janota Dos Santos, M., Aloni, B., 1992. Consequences of temporary drought at different growth stages on sorghum yield. Proceedings Second Congress of European Society for Agronomy, Warwick University. 23–28 August 1992, pp. 104–105.

Mathews, R.B., Azam-Ali, S.N., Peacock, J.M., 1990a. Response of four sorghum lines to mid-season drought II leaf characteristics. Field Crop Res. 25, 297–308.

Mathews, R.B., Azam-Ali, S.N., Peacock, J.M., 1990b. Response of four sorghum lines to mid-season drought I growth, water use and yield. Field Crop Res. 25 (3-4), 279–296.

Maxson, E.D., Fryar, W.B., Rooney, L.W., Krishnaprasad, M.N., 1971. Milling properties of sorghum grain with different proportions of corneous to floury endosperm. Cereal Chem. 48, 478–490.

Mayee, C.D., Garud, T.B., 1978. An assured method for evaluating sorghum charcoal rot. Indian Phytopathol. 31, 121.

Mc Cree, K.J., Kallisen, C.E., Richardson, S.G., 1984. Carbon balance of sorghum plants during osmotic adjustment water stress. Plant Physiol. 76, 898–902.

McBee, G.G., 1984. Relation of senescence, nonsenescence, and kernel maturity to carbohydrate metabolism in sorghum. In: Mughogho, L.K. (Ed.), Sorghum Root and Stalk Rots: A Critical Review. Proc. Consultative Group Discussion of Research Needs and Strategies for Control of Sorghum Root and Stalk Rot Diseases, Bellagio, Italy. 27 November–2 December 1983. ICRISAT, Patancheru, India, pp. 119–129.

McBee, G.G., Waskom, R.M., Miller, F.R., Creelman, 1983. Effect of senescence and nonsenescence on carbohydrates in sorghum during late kernel maturity states. Crop Sci. 23, 372–376.

McCaig, T.N., Romagosa, I., 1991. Water status measurements of excised wheat leaves: position and age effect. Crop Sci. 31, 1583–1588.

McKersie, B.D., Leshem, Y.Y., 1994. Stress and Stress Coping in Cultivated Plants. Kluwer Academic Publishes, Dordrecht, The Netherlands.

Meckenstock, D.H., 1991 Tropical sorghum conservation and enhancement. In Honduras and Central Amenca In Sorghum Millet Collaborative Research Support Program (CRSP) Annual Report 1991.

Mehetre, S.P., Borikar, S.T., 1992. Combining ability studies involving Maldandi cytoplasm in sorghum. J. Maharashtra Agric. Univ. 17, 247–249.

Miller, F.R., Barnes, D.K., Cruzado, H.J., 1968. Effect of tropical photoperiods on the growth of *Sorghum bicolor* (L.) Moench, when grown in 12 monthly plantings. Crop Sci. 8, 499–502.

Miller, F.R., Kebede, Y., 1981. Genetic contribution to yield gains in sorghum 1950-1980. Genetic Contributions to Yield Gains of Five Major Crop Plants. CSSA, ASA, Madison, WI, CSSA Spec. Publ. 7, pp. 1–14.

Mishra, R.C., Kandalkar, V.S., Chauhan, G.S., 1992. Combining ability analysis of harvest index and its components in sorghum. Indian J. Genet. 52 (2), 178–182.

Mitchell, J.H., Siamhan, D., Wamala, M.H., Risimeri, J.B., Chinyamakobvu, E., Henderson, S.A., et al., 1998. The use of seedling leaf death score for evaluation of drought resistance of rice. Field Crops Res. 55, 129–139. Available from: http://dx.doi.org/10.1016/S0378-4290(97)00074-9.

Mitra, J., 2001. Genetics and genetic improvement of drought resistance in crop plants. Curr. Sci. 80 (6), 758–763.

Mohammed, A.B., 1980. An Evaluation of Eight Female Lines of Grain Sorghum Using Three Groups of Testers (M.Sc. thesis). University of Nebraska, Lincoln.

Mohan, D., 1975. Chemically Induced High Lysine Mutants in *Sorghum bicolor* (L.) Moench (Thesis). Purdue University, West Lafayette, IN, 110 pp.

Moila, S.D., Reddy, C.N.K., 1974. Relationship between biometric characters and grain yield in white seeded sorghum. Sorghum Newsl. 17, 137.

Morgan, J.M., 1984. Osmoregulation and water stress in higher plants. Ann. Rev. Plant Physiol. 35, 299–319.

Morgan, J.M., Condon, A.G., 1986. Water use, grain yield and osmoregulation in wheat. Aust. J. Plant Physiol. 13, 523–532.

Mote, U.N., 1983. Epidemic of delphacids and aphids on winter sorghum. Sorghum Newsl. 26, 76.

Mote, U.N., Kadam, J.R., 1984. Incidence of (*Aphis sacchari* Zehnt) in relation to sorghum plant characters. Sorghum Newsl. 27, 86.

Mote, U.N., Jadhav, S.S., 1993. Seasonal occurrence of flea beetles, delphacids, leaf sugary exudation, and aphids on *rabi* sorghum. J. Maharashtra Univ. Agric. Sci. 18, 133–134.

Mote, U.N., Shahane, A.K., 1993. Studies on varietal reaction of sorghum to delphacid, aphid, and leaf sugary exudation. Indian J. Entomol. 55, 360–367.

Mote, U.N., Shahane, A.K., 1994. Biophysical and biochemical characters of sorghum varieties contributing resistance to delphacid, aphid and leaf sugary exudations. Indian J. Entomol. 56, 113–122.

Mote, U.N., Shinde, M.D., Bapat, D.R., 1985. Screening of sorghum collections for resistance to aphids and oily malady of winter sorghum. Sorghum Newsl. 28, 13.

Muchow, R.C., 1985b. Canopy development in grain legumes grown under different soil water regimes in a semi-arid tropical environment. Field Crops Res. 11, 99–109.

Muchow, R.C., 1989. Comparative productivity of maize, sorghum and pearl millet in a semi-arid tropical environment. I. Yield potential. Field Crop Res. 20 (3), 207–219.

Muchcow, R.C., Coates, D.B., 1986. An analysis of the environmental limitation to yield of irrigated grain sorghum during the dry season in tropical Australia using a radiation interception model. Aust. J. Agric. Res. 37, 135–148.

Muchow, R., Carberry, P.S., 1990. Phenology and leaf area development in tropical grain sorghum. Field Crops Res. 23, 221–237.

Muchow, R., Sinclair, T.R., 1994. Nitrogen response of leaf photosynthesis and canopy radiation use efficiency in field grain maize and sorghum. Crop Sci. 34 (6), 379–389.

Muchow, R.C., Carberry, P.S., 1993. Designing improved plant types for the semiarid tropics: agronomists viewpoints. In: Penning de Vries, F.W.T., Tang, P., Metselaar, K. (Eds.), Systems Approaches for Sustainable Agricultural Development. Kluwer Academic Publishers, Dordrecht, The Netherlands, pp. 37–61.

Mughogho, L.K., Pande, S., 1984. Charcoal rot of sorghum. Proceedings of Consultative Group Discussions on Research Needs and Strategies for Control of Sorghum Root and Stalk Rot Diseases, Italy. 27 November–2 December 1983.

Munamava, M., Riddoch, I., 2001. Response of three sorghum varieties to soil moisture stress at different development stages. S. Afr. J. Plant Soil. 18, 75–79.

Munck, L., Bach Knudsen, K.E., Axtell, J.D., 1982. Milling processes as related to kernel morphology. Proceedings, International Symposium on Sorghum Grain Quality. October 1981, ICRISAT, Patancheru, India, pp. 28–31.

Murphy, L.S., 1975. Fertilizer efficiency for corn and sorghum Pages 49–72. In Proceedings, Thirtieth Annual Corn and Sorghum Research Conference, American Seed trade association, Washington, D.C.

Murthy, U.R., Schertz, K.F., Bashaw, E.C., 1979. Apomictic and sexual reproduction in sorghum. Indian J. Genet. Plant Breed. 39, 271–278.

Murty, U.R., 1991. National programme on sorghum research in India. Paper Presented at the Consultative Meeting to Consider Establishment of Regional Sorghum Research Network for Asia. September, ICRISAT, pp. 16–19.

Murty, U.R., 1992. ICAR_ICRISAT Collaborative Research Projects—Sorghum Progress Report, National Research Center on Sorghum, 27.

Murty, D.S., House, L.R., 1980. Sorghum food quality: its assessment and improvement. Report submitted to the 5th Joint Meeting of the UNDPCIMMYT-ICRISAT Policy Advisory Committee. October, ICRISAT, Patancheru, India, pp. 14–18.

Murty, D.S., Paul, H.D., House, L.R., 1982a. Sorghum *roti* II. Genotypic and environmental variation for *roti* quality parameters. Proceedings, International Symposium on Sorghum Grain Quality. October 1981, ICRISAT, Patancheru, India, pp. 28–31.

Murty, D.S., Patil, H.D., House, L.R., 1982b. Cultivar differences for gel consistency in sorghum. Proceedings, International Symposium on Sorghum Grain Quality. October 1981, ICRISAT, Patancheru, India, pp. 28–31.

Murthy, K.S., Patnaik, R.K., Swan, P., 1986. Net assimilation rate and its related plant characters of high yielding rice varieties. Indian J. Plant Physiol. 29, 5–60.

Murumkar, P.N., 2002. Combining Ability of Newly Established Male Sterile and Restorer Lines in Rabi Sorghum (M.Sc.(Agri) thesis). Punjab Rao Deshmukh Krushi Vishwa Vidyalaya, Akola.

Murumkar, P.N., Atale, S.B., Shivankar, R.S., Meshram, M.P., Bhandarwar, A.D., 2005. Combining ability studies in newly established 'MS' and 'R' lines in *rabi* sorghum. Indian J. Agric. Res. 39 (4), 263–268.

Mutava, R.N., Prasad, P.V.V., Tuinstra, M.R., Kofoid, K.D., Yu, I., 2011. Characterization of sorghum genotypes for traits related to drought tolerance. Fields Crops Res. 123, 10–18.

Mwanamwenge, J., Loss, S.P., Siddique, K.H.M., Cocks, P.S., 1999. Effect of water stress during floral initiation, flowering and podding on the growth and yield of faba bean (*Vicia faba* L.). Eur. J. Agron. 11, 1–11.

Myers, R.J.K., Keefer, G.D., Foale, M.A., 1986. Sorghum growth and development in the tropical and sub-tropical environments growth rate, grain yield and yield components. In: Fale, M.A., Henzel, R.G. (Eds.), Proceedings of the First Australian Sorghum Conference. Gatton, Queensland Australia, pp. 455–463.

Nagabasaiah, K.H.M., 1985. Genetic analysis of ten quantitative characters in F2 generation of a seven-parent diallel set in sorghum [*Sorghum bicolor* (L.) Moench]. Mysore J. Agric. Sci. 19, 135–136.

Nagaraj, N., More, S., Pokharkar, V., Haldar, S., 2011. Impact of potential technologies for post-rainy season sorghum (in Maharashtra) and pearl millet (in Gujarat, Haryana, and Rajasthan) in India. Report on Output 1.2.4 (South Asia) of the HOPE Project, ICRISAT, Patancheru, India.

Nagarajan, K., Saraswati, V., Renfro, B.L., 1970. Incidence of charcoal rot (*Macrophomina phaseolina*) on CSH-1 sorghum. Sorghum Newsl. 13, 25.

Nagur, T., 1971. Studies on Fertility Restoration and Combining Ability in Relation to Genetic Diversity and Cytoplasmic Constitution in Sorghum (Ph.D. thesis). Tamil Nadu Agricultural University, Coimbatore, India.

Naidu, T.C.M., Raju, N., Narayan, A., 2001. Screening of drought tolerance in greengram (*Vigna radiata* (L.) wilczek) genotypes under receding soil moisture. Indian J. Plant Physiol. 6, 197–201.

Naik, V.R., Shivanna, H., Joshi, M.S., Parameshwarappa, K.G., 1994. Heterosis and combining ability analysis in sorghum. J. Maharashtra Agric. Univ. 19, 137–138.

Nandawankar, K.G., 1990. Heterosis studies for grain yield characters in *rabi* sorghum. Indian J. Genet. Plant Breed. 50, 83–85.

Napompeth, B., 1973. Ecology and Population Dynamics of the Corn Plant Hopper, *Peregrinus maidis* (Ashmead) (Homoptera: Delphacidae) in Hawaii (Ph.D. thesis). University of Hawaii, Honolulu, Hawaii, 257pp.

Narayana, D., 1975. Screening for aphids and sooty molds in sorghum. Sorghum Newsl. 18, 21–22.

Narayana, D., Sahib, K.H., Rao, B.S., Rao, M.R., 1982. Studies on the incidence of an aphid (Aphis sacchari) in sorghum. Sorghum Newsl. 25, 72.

Narkhede, B.N., Shinde, M.S., Salunke, C.B., 1999. Phule Yashoda—a new *rabi* sorghum variety of Maharashtra. J. Maharashtra Agric. Univ. 24, 41–45.

Narkhede, B.N., Karad, S.R., Akade, J.H., Kachole, U.G., 2002. Screening of *rabi* sorghum local germplasm of Maharashtra tract against shoot fly reaction. J. Maharashtra Agric. Univ. 27 (1), 60–61.

Navabpour, S., Rezaie, S., 1998. Evaluation of the heterotic patterns and growth indices for grain yield and related traits in sorghum. Seed Plant. 14, 37–47.

Nayakar, N.Y., 1985. Development and Evaluation of New Male Sterile Lines in *Rabi* Sorghum [*Sorghum bicolor* (L.) Moench] (Ph.D. thesis). University of Agricultural Sciences, Dharwad.

Nayakar, N.Y., Gowda, B.T.S., Goud, J.V., 1989. Combining ability analysis for panicle components in *rabi* sorghum. J. Agric. Sci. 23, 306–310.

Nayakar, N.Y., Gowda, B.T.S., Goud, J.V., 1994. Development of new male sterile lines from maintainer x restorer *rabi* sorghum [*Sorghum bicolor* (L.) Moench] crosses. Mysore J. Agric. Sci. 28, 14–18.

Nayeem, K.A., Bapat, D.R., 1984. Heterosis and heterobeltiosis for grain yield and quality in sorghum [*Sorghum bicolor* (L.) Moench]. Indian J. Agric. Sci. 23, 306–310.

Ngugi, K., Kimani, W., Kiambi, D., 2010. Introgression of stay-green trait into a Kenyan farmer preferred sorghum variety. Afr. Crop Sci. J. 18 (3), 141–146.

Nguyen, D.C., Nakamura, S., Yoshida, T., 1997. Combining ability and genotype x environmental interaction in early maturing grain sorghum for summer seeding. Jpn. J. Crop Sci. 66, 698–705.

Nimbalkar, V.S., Bapat, D.R., 1987. Genetic analysis of shoot fly resistance under high level of shoot fly infestation in sorghum. J. Maharashtra Agric. Univ. 12, 331–334.

Nimbalkar, V.S., Bapat, D.R., 1992. Inheritance of shoot fly resistance in sorghum. J. Maharashtra Agric. Univ. 17, 93–96.

Nimbalkar, V.S., Bapat, D.R., Patil, R.C., 1988. Genetic variability, interrelationship and path coefficient of grain yield and its attributes in sorghum. J. Maharashtra Agric. Univ. 13, 207–208.

Nirmal, S.V., Patil, J.V., 2008. A new drought tolerant genotype of rabi sorghum-SPV 1546 (Phule Chitra). Ann. Plant Physiol. 22 (2), 165–168.

Nishiyama, I., 1995. Damage Due to Extreme Temperatures. Food and Agriculture Policy Research Center, Tokyo.

Norem, M.S., Dobrenz, A.K., Voigt, R.L., 1985. Protein and other quality components of drought tolerant sorghum grown with an irrigation gradient system. Sorghum Newsl. 28, 126.

Nouri, M., Stephen, C.M., Lyon, D.J., Prabhakar, D., 2004. Yield components of pearl millet and grain sorghum across environments in the central great plains. Crop Sci. 44, 2138–2145.

Nwanze, K.F., Reddy, Y.V.R., Soman, P., 1990. The role of leaf surface wetness in larval behaviour of the sorghum shoot fly, *Atherigona soccata*. Entomol. Exp. Appl. 56 (2), 187–195.

Obilana, T., 1982. Sorghum traditional foods in Nigeria: their preparation and quality parameters. Proceedings, International Symposium on Sorghum Grain Quality. October 1981, ICRISAT, Patancheru, India, pp. 28–31.

Odvody, G.N., Dunkle, L.D., 1979. Charcoal stalk rot of sorghum: effect of environment on host-parasite relations. Phytopathology. 69, 250–254.

Okiyo, T., Gudu, S., Kiplagat, O., Owuoche, J., 2010. Combining drought and aluminium toxicity tolerance to improve sorghum productivity. Afr. Crop Sci. J. 18 (4), 147–154.

Oliver, S.N., Van Dongen, J.T., Alfred, S.C., Mamun, E.A., Zhao, X.C., Saini, H.S., et al., 2005. Cold-induced repression of the rice anther-specific cell wall invertase gene OSINV4 is correlated with sucrose accumulation and pollen sterility. Plant Cell Environ. 28, 1534–1551.

Oliver, S.N., Dennis, E.S., Dolferus, R., 2007. ABA regulates apoplastic sugar transport and is a potential signal for cold-induced pollen sterility in rice. Plant Cell Physiol. 48, 1319–1330.

Omanya, G.O., Ayieecho, P.I., Nyabundi, J.O., 1997. Variation for adaptability to dryland condition in sorghum. Afr. Crop Sci. J. 5 (2), 127–138.

Ombhako, G.A., Miller, F.R., 1994. Performance, stability and prediction of performance in single, three way and double cross hybrids of sorghum. Int. Sorghum Millets Newsl. 35, 77.

Omer, M.E.H., Frederiksen, R.A., Rosenow, D.T., 1985. Collaborative sorghum disease studies in Sudan. Sorghum Newsl. 28, 93–94.

Omori, T., Agarwal, B.L., House, L.R., 1988. Genetic divergence for resistance for resistance to shoot fly *Atherigorla soccata* Rond. *in* sorghum (*Sorghum bicolor* (L) Moench) and its relationship with heterosis. Insect Sci. Appl. 9, 483–488.

Omra, M.K., Hussein, M.Y., 1987. Genetic variability in root characteristics in relation to yield under drought in early moduring barley. Aust. J. Agric. Sci. 18, 319–332.

O'Neill, M.K., Diaby, M., 1987. Effects of high soil temperature and water stresses on Malian pearl millet and sorghum during seedling stage. J. Agron. Crop. Sci. 159 (3), 192–198.

Oomah, B.D., Reichert, R.D., Youngs, C.G., 1981. A novel, multisample, tangential abrasive dehulling device (TADD). Cereal Chem. 58, 392–395.

Osbourn, A.E., 2001. Plant mechanisms that give defence against soilborne diseases. Aus. Pl. Path. 30, 99–102.

Osterom, E.J.V., Jayachandran, R., Bidinger, F.R., 1996. Diallel analysis of the staygreen trait and its components in sorghum. Crop Sci. 36 (3), 546–555.

Osuna Ortega, J., Endoza, C.M., Mendeza, O.L.E., 2003. Sorghum cold tolerance, pollen production and seed yield in the central high villages of Mexico. Maydica. 48 (2), 125–132.

Paddick, M.E., 1944. Physiological biochemical and genetic basis of heterosis. *Iowa* Agricultural Experiments Station Bulletin, p. 331.

Padhye, A.P., 2006. Inheritance of Grain Yield and Physiological Components in *Rabi* Sorghum (Ph.D. thesis submitted to Mahatma Phule Krishi Vidyapeeth). Rahuri, Maharashtra.

Padmaja, P.G., Woodcock, C.M., Toby, J.A., Bruce, 2010. Electrophysiological and behavioral responses of sorghum shoot fly, *Atherigona soccata*, to sorghum volatiles. J. Chem. Ecol. 36 (12), 1346–1353.

Painter, R.H., 1951. Insect Resistance in Crop Plants. Macmillan & Co., New York, NY.

Palanisamy, S., Prasad, M.N., Rangasamy, S.R., 1983. Heterotic effects of new male sterile lines in sorghum and their combining ability. Proceedings of Scientific Meeting on Genetic and Improvement of Heterotic Systems. School of Genetics, Tamil Nadu Agricultural University, pp. 7–8.

Palanisamy, S., Subramanian, A., 1984. Genetic analysis of harvest index in sorghum. Sorghum Newsl. 28, 39–40.

Pande, S., Karunakar, R.I., 1992. Stalk rots. In: Milliano, W.A.J., Frederiksen, R.A., Bengston, G.D. (Eds.), Sorghum and Millet Diseases: A Second World Review. ICRISAT, Hyderabad, pp. 219–234.

Pandit, V.V., 1989. Genetic Evaluation of Derived B Lines and Induced Mutants of Sorghum (M.Sc. (Agri) thesis). University of Agricultural Sciences, Dharwad.

Pao, C.I., Morgan, P.W., 1986. Genetic regulation of development in *Sorghum bicolor*. I. Role of the maturity genes. Plant Physiol. 82, 575–580.

Pappelis, A.J., Katsanos, R.A., 1966. Effect of plant injury on senescence of sorghum stalk tissue. Phytopathology. 56 (3), 295–297.

Parameshwara, G., Krishnasastry, K.S., 1982. Variability in leaf elongation rate and reduction in green leaf length in sorghum genotypes under moisture stress and on alleviation of stress. Indian J. Agric. Sci. 52 (2), 102–106.

Parameshwarappa, S.G., Karikatti, S.R., 2002. Identification of high yielding *rabi* sorghum varieties for Northern Transition Zone of Karnataka under rainfed conditions. Karnataka J. Agric. Sci. 15 (4), 709–710.

Parameswarappa, R., Kajjari, N.B., Patil, C.S.P., Thimmaiah, H.C., Betsur, S.R., 1976. Charcoal rot incidence recorded at RRS, Dharwad (Karnataka) during *kharif* 1975. Sorghum Newsl. 19, 37.

Pardales Jr., J.R., Kono, Y., 1990. Development of sorghum root system under increasing drought stress. Jpn. J. Crop Sci. 59, 752–761.

Parent, B., Hachez, C., Redondo, E., Simonneau, T., Chaumont, F., Tardieu, F., 2009. Drought and abscicic acid effects on aquaporin content translate into changes in hydraulic conductivity and leaf growth rate: a trans-scale approach. Plant Physiol. 149, 2000–2012.

Parry, M.L., R osenzweig, C., Iglesias, A., Livermore, M., Fischer, G., 2004. Effects of climate change on global food production under SRES emissions and socio-economic scenarios. Glob. Environ. Change. 14, 53–67.

Parsons, L.R., 1982. Plant responses to water stress. In: Christiansen, M.N., Lewis, C.F. (Eds.), Pages 175–192 in Breeding plants for less favourable environments. John Wiley, New York.

Partridge, J.E., Reed, J.E., Jensen, S.G., Sidhu, G.S., 1984. Spatial and temporal succession of fungal species in sorghum stalks as affected by environment. In: Sorghum Shoot and Stalk Rots—A Critical Review: Proceedings of the Consultative Group Discussion on Research Needs and Strategies for Control of Sorghum Root and Stalk Rot Diseases. 27 November–2 December 1983, Beluga, Italy.

Parvatikar, S.R., Hiremath, S.M., 1985. Growth and yield analysis in sorghum—optimizing yield levels in *rabi* sorghum. Sorghum Newsl. 28, 108.

Passioura, J.B., 1977. Grain yield, harvest index and water use of wheat. J. Aust. Inst. Agric. Sci. 43, 117–121.

Passioura, J.B., Angus, J.F., 2010. Improving productivity of crops in water-limited environments. In: Sparks, D.L. (Ed.), Advances in Agronomy, vol. 106. Academic Press, Burlington, pp. 37–75.

Patanothai, A., Atkins, R.E., 1971. Heterosis response for vegetative growth and fruit development in grain sorghum. Crop Sci. 11, 839–843.

Patel, G.M., Sukhani, T.R., 1990. Screening of sorghum genotypes for resistance to shoot fly [*Atherigona soccata* (Rond.)]. Indian J. Entomol. 52, 1–8.

Patel, R.H., Desai, K.S., Kukadia, M.V., Desai, D.T., 1980. Component analysis in sorghum. Sorghum Newsl. 23, 23–24.

Patel, R.H., Kudadia, M.U., Desai, K.B., Patel, S.N., 1985. Inheritance of resistance to sorghum shoot fly in grain sorghum. Sorghum Newsl. 28, 58–59.

Patel, R.H., Desai, K.B., Desai, M.S., 1987. Heterosis in multiple environments for grain sorghum. Indian J. Agric. Sci. 57, 1–4.

Patel, P.I., Patel, R.H., Desai, M.S., 1990. Heterosis and combining ability of grain sorghum [*Sorghum bicolor* (L.) Moench] at different locations in Gujarat. Indian J. Agric. Sci. 60, 382–386.

Patel, D.V., Makne, V.G., Mehta, H.D., Shete, DM., 1993. Genetic architecture of grain yield and related in high energy sorghum. J. Maharashtra Agric. Univ. 18 (2), 261–263.

Patel, D.V., Makne, V.G., Patil, R.A., 1994. Outer—relationship and path coefficient studies in sweet stalk sorghum. J. Maharashtra Agric. Univ. 19, 40–41.

Patel, J.C., Jaimini, S.M., Patel, R.H., Patel, K.G., 1995. Combining ability for early blooming in forage sorghum. Forage Res. 21, 11–14.

Pathak, H.C., Sanghi, A.K., 1992. Combining ability and heterosis studies in forage sorghum [*Sorghum bicolor* (L.) Moench.] across environments. Indian J. Genet. Plant Breed. 52, 75–85.

Patidar, H., Dabholkar, A.R., 1981. 1981, Gene effects for grain size and yield in *rabi* sorghum. Indian J. Genet. 41, 259–263.

Patil, R. P., 1987. Physiological Factors Influencing Growth and Yield of *Rabi* Sorghum Genotypes Under Rainfed Conditions (M.Sc. (Agri.) thesis). University of Agricultural Sciences, Dharwad.

Patil, R.C., 1990. Inheritance of grain yield and its components in grain sorghum. J. Maharashtra Agric. Univ. 15 (1), 54–57.

Patil, S.L., 2002. Sorghum hybrids and varieties suitable for post rainy—season cultivation in Northern Karnataka, India. Int. Sorghum Millets Newsl. 43, 46–48.

Patil, P. R., 2004. Heterosis and Combining Ability Studies for Root, Shoot and Productivity Traits in *Rabi* Sorghum (M.Sc. (Agri) thesis). University of Agricultural Sciences, Dharwad.

Patil, S.L., 2005. Dry matter production, yield, water uses efficiency and economics of winter sorghum varieties under drought conditions in vertisols of South India. Crop Res. 29 (2), 185–191.

Patil, R.C., Thombre, M.V., 1984. Quantitative inheritance of grain yield and its components of F2 diallel cross of sorghum. Indian J. Agric. Sci. 45, 534–537.

Patil, R.C., Thombre, M.V., 1985. Inheritance of shoot fly and earhead midge resistance in sorghum. Curr. Res. Rep. 1, 44–48.

Patil, R.C., Thombre, M.V., 1986. Genetic parameters, correlation coefficient and path analysis in F1 and F2 generations of a 9 × 9 diallel cross of sorghum. J. Maharashtra Agric. Univ. 8 (2), 162–165.

Patil, F.B., Bapat, D.R., 1991. Heterosis in forage sorghum and its relationship with combining ability. J. Maharashtra Agric. Univ. 165, 329–332.

Patil, S.S., Pandit, V.V., 1995. In: Sharma, B. (Ed.), Use of B x R cross for improving combining ability in sorghum. Golden Jubilee Symposium on Genetic Research and Education: Current Trends and the Next Fifty Years. 12–15 February 1991. Indian Society of Genetics and Plant Breeding, New Delhi, pp. 885–888.

Patil, K. D., Prabhakar, M., 2001. Evaluation of diverse source of rabi germplam for physiological traits associated with drought adaptation. National Seminar on Role of Plant Physiology for Sustaining Quality and Quantity of food production in Relation to Environment. UAS, Dharwad, pp. 144.

Patil, B.S., Ravikumar, R.L., 2011. Osmotic adjustment in pollen grains: a measure of drought adaptation in sorghum? Curr. Sci. 10 (3), 377–382.

Patil, R.H., Desai, K.B., Desai, D.T., 1983. Line x Tester analysis for combining ability of new restorers in grain sorghum. Sorghum Newsl. 26, 12–13.

Patil, R.A., Nandanwankar, K.G., Ambekar, S.S., 1985. Combining ability for grain yield in *rabi* sorghum. Sorghum Newsl. 28, 9–10.

Patil, M.S, Mannur, D.M, Patil, J.R., 1998. GRS1 a new *rabi* sorghum variety. Karnataka J. Agric. Sci. 11, 1063–1064.

Patil, S.L., Sheelavantar, M.N., Lamans, V.K., 2003. Correlation analysis among growth and yield components of winter sorghum. Int. Sorghum Millets Newsl. 44, 2003–14–17.

Patil, S.S., Narkhede, B.N., Barhate, K.K., 2005. Genetics of shoot fly resistance in sorghum. Agric. Sci. Dig. 25 (2).

Patil, A., Bashasab, F., Salimath, P.M., Rajkumar, 2012. Genome wide molecular mapping and QTL analysis, validated across locations and years for charcoal rot disease incidence traits in *Sorghum bicolor* (L.) Moench. Indian J. Genet. Plant Breed. 72 (3), 296–302.

Pattanashetti, S. K., 2000. Genetic Studies Involving Milo and Maldandi Sources of Cytoplasmic-Genetic Male Sterility in Sorghum [*Sorghum bicolor* (L.) Moench] (M.Sc. (Agri) thesis). University of Agricultural Sciences, Dharwad.

Pattanayak, C. M., 1977. Annual Report ICRISAT, Kamboinse, Upper Volta.

Pattanayak, C. M., 1978. Annual Report ICRISAT, Kamboinse, Upper Volta.

Patterson, D.P., James, A.B., Radall, S.A., Elizebeth, V.V., 1977. Photosynthesis from controlled and field environment. Plant Physiol. 59, 384–387.

Pawar, S. M., 1996. Physiological Indices for High Yield and Improving Yield Potential in Sorghum (Ph.D. thesis). University of Agricultural Science, Dharwad.

Pawar, S. V., 2000. Heterosis and its Relation to Parental Genetic Divergence in Sorghum [*Sorghum bicolor* (L.) Moench] (M.Sc. (Agri) thesis). University of Agricultural Sciences, Dharwad.

Pawar, K.P., Jadhav, A.S., 1996. Correlation and Path coefficient analysis in *rabi* sorghum. J. Maharashtra Agric. Univ. 32 (3), 344–347.

Pawar, S.M., Chetti, M.B., 1997. Genotypic differences in osmo regulations and their relationship with biomass and grain yield in *rabi* sorghum. Ann. Plant Physiol. 11 (1), 10–14.

Peacock, J.M., Sivakumar, M.V.K., 1987. An environmental physiologist approach to screening for drought resistance in sorghum with particular reference to subsaharan Africa. In: Menyonga, S.M. (Ed.), Proceeding of the International Drought Symposium. May 1986. Kemyatta Conference Centre, Nairobi, Kenya, pp. 19–23.

Peacock, J.M., Soman, P., Howarth, C.J., 1993. High temperature effects on seedling survival in tropical cereals. In: Kuo, C.G. (Ed.), Adaptation of Food Crops to Temperature and Water Stress. Proceedings of an International Symposium. 13–18 August 1992. Asian Vegetable Research and Development Center, Taipei, Taiwan, pp. 106–121.

Pederson, D.G., 1974. Arguments against intermating before selection in a self fertilizing species. Theor. Appl. Genet. 45, 157–162.

Pedgaonkar, S.M., Mayee, C.D., 1990. Stalk water potential in relation to charcoal rot of sorghum. Indian Phytopathol. 43, 2.

Pedrosoperez, R., Rodriguez, F.C., Martin Fagundo, D., Saucedo Castillo, O., 1994. Characterization of the sorghum hybrid 1831 x UDG 110. Centro Agricola. 21, 15–18.

Pereira da Cruz, R., Milach, S.C.K., Federizzi, L.C., 2006. Rice cold tolerance at the reproductive stage in a controlled environment. Sci. Agric. 63, 255–261.

Pereira, M.G., Lee, M., Bramel-Cox, P., Wordman, W., Doebley, J., Whitkus, R., 1994. Construction of an RFLP map in sorghum and comparative mapping in maize. Genome. 37, 236–243.

Perezcabrera, S.I., Miller, R.R., 1985. Combining ability and heterosis studies among newly developed inbreds of sorghum. Sorghum Newsl. 28, 22–23.

Perezcabrero, G. J., 1986. Combining Ability and Heterosis Studies for Adaptation in Newly Developed Male and Female Inbreds of Sorghum [*Sorghum bicolor* (L.) Moench] (Dissertation Abstracts). International Biological Sciences and Engineering, p. 46.

Peterson, G.C., Reddy, B.V.S., Youm, O., Teetes, G.L., Lambright, L., 1997. Breeding for resistance to foliar- and stem-feeding insects of sorghum and pearl millet. In: Proceedings of the International Conference on Genetic Improvement of Sorghum and Pearl Millet. INTSORMIL, Publ. 97-5, pp. 281–302.

Pi, C., Hsieh, J.S., 1982a. Preliminary studies on aphid resistance in sorghum. Natl. Sci. Counc. Month. Repub. China. 10, 153–160.

Pi, C., Hsieh, J.S., 1982b. Studies on grain quality and aphid resistance in sorghum. In: Hsieh, S.C., Liu, D.J. (Eds.), Proceedings of the Symposium on Plant Breeding. Symp. Repub. China Reg. Soc. SABRAO and Agric. Assoc., China, Taiwan, pp. 113–120.

Pierce, M., Raschke, K., 1980. Correlation between loss of turgor and accumulation of abscisic acid in detached leaves. Planta. 148, 174–182.

Pillai, M.A., Rangaswamy, P., Vanirajan, C., Ramalingam, J., 1995. Combining ability analysis for panicle characters in sorghum. Indian J. Agric. Res. 29, 98–102.

Pinjari, M.B., Shindhe, M.S., 1995. Studies on the morpho-physiological traits contributing to the kernel yield in sorghum hybrids. Ann. Plant Physiol. 9, 161–163.

Ponnaiya, B.W.X., 1951. Studies on the genus *Sorghum* I. Field observations on sorghum resistance to the insect pest, *Atherigona indica* (M). Madras Univ. J. 21, 96–117.

Poor, S.N., Rezai, A., 1996. Estimates of genetic parameters for grain yield and related characters in sorghum. Iranian J. Agric. Sci. 27 (2), 77–87.

Prabhakar, M., 1975. Agronomic investigation on rationing of sorghum (Sorghum bicolor (L.) Moench). M.Sc. (Ag) Thesis, University of agricultural Sciences, Dharwad, Karnataka, India.

Prabhakar, 2001a. Heterosis in *rabi* sorghum [*Sorghum bicolor* (L.) Moench]. Indian J. Genet. Plant Breed. 61, 364–365.

Prabhakar, K.M., 2001b. Variability, heritability, genetic advance and character association in *rabi* sorghum. J. Maharashtra Agric. Univ. 26, 188–189.

Prabhakar, Raut, M.S., 2010. Exploitation of heterosis using diverse parental lines in *rabi* sorghum. Electron. J. Plant Breed. 1 (4), 680–684.

Prabhakar, Kannababu, N., Samdur, M.Y., Bahadure, D.M., 2010. Exploitation and assessment of heterosis in *Rabi* sorghum hybrids across environments presented at Global Consultation on Millets Promotion for Health & Nutritional Security, held at Directorate of Sorghum Research. Hyderabad from 18–20, December, 2013, India.

Prasad, P.V.V, Pisipati, S.R, Mutava, R.N, Tuinstra, M.R., 2008. Sensitivity of grain sorghum to high temperature stress during reproductive development. Crop Sci. 48, 1911–1917.

Premachandra, G.S, Saneoka, H., Fujita, K., Ogata, S., 1992. Leaf drought relations, osmotic adjustment, cell membrane stability, epicuticular wax load and growth as affected by increasing drought deficits in sorghum. J. Exp. Bot. 43, 1569–1576.

Quinby, J.R., 1967. The maturity genes of sorghum. In: Norman, A.G (Ed.), Advances in Agronomy, vol. 19. Academic Press, New York, NY, pp. 267–305.

Quinby, J.R., 1973. The genetic control of flowering and growth in sorghum. In: Brady, N. (Ed.), Advances in Agronomy, vol. 25. Academic Press, New York, NY, pp. 125–162.

Quinby, J.R., 1974. Sorghum Improvement and the Genetics of Growth. Texas A&M University Press, College Station, TX.

Quinby, J. R., 1980. Interaction of genes and cytoplasms in male sterility in sorghum. Proceedings of 35th Corn and Sorghum Research Conference. American Seed Trade Association, Chicago, IL.

Quinby, J.R., 1982. Interactions of genes and cytoplasms in sex expression in sorghum. In: House, L.R., Mughogho, L.K., Peacock, J.M. (Eds.), Sorghum in the Eighties. International Crops research Institute for the Semi-Arid Tropics, Patancheru, India, p. 384.

Quinby, J.R., Karper, R.E., 1946. Heterosis in sorghum resulting from the heterozygous condition of a single gene that affects duration of growth. Amer. J. Bot. 33, 716–721.

Quinby, J.R, Hesketh, J.D, Voigt, R.L., 1973. Influence of temperature and photoperiod on floral initiation and leaf number in sorghum. Crop Sci. 13, 243–246.

Quirino, B.F, Noh, Y.S, Himelblau, E, Amasino, R.M., 2000. Molecular aspects of leaf senescence. Trends Plant Sci. 5, 278–282.

Rajasekhar, P., 1989. Studies on the Population Dynamics of Major Pests of Sorghum and Bioecology and Crop Loss Assessment Due to the Shoot Bug, *Peregrinus maidis* (Ashmead) (M.Sc. thesis). Andhra Pradesh Agricultural University, Hyderabad, Andhra Pradesh, India, 230pp.

Rajasekhar, P., 1996. Assessment of crop loss and EIL due to shoot bug, *Peregrinus maidis* Ashmead on sorghum. Ann. Agric. Res. 17, 333–334.

Rajguru, A.B., Kashid, N.V., Kamble, M.S., Rasal, P.N., Gosavi, A.B., 2004. Combining ability analysis for yield and its components in *rabi* sorghum. Crop Improv. 31 (2), 195–200.

Rajguru, A.B., Kashid, N.V., Kamble, M.S., Rasal, P.N., Gosavi, A.B., 2005. Gene action and heritability studies in *rabi* sorghum (*Sorghum bicolor* (L.) Monech. J. Maharashtra Agric. Univ. 30 (3), 367–368.

Rajkumar, 2004. Detection and Mapping of QTLs for Charcoal Rot Resistance and Yield Components in Sorghum (*Sorghum bicolor* (L.) Moench) (M. Sc. (Agri.) thesis). University of Agricultural Sciences, Dharwad.

Rami, J.F., 1999. Etude des facteurs génétiques impliqués dans la qualité technologique du grain chez le maïs et le sorgho (Ph. D. report). Université d'Orsay, France.

Rami, J.F., Dufour, P., Trouche, G., Fliedel, G., Mestres, C., Davrieux, F., et al., 1998. Quantitative trait loci for grain quality, productivity, morphological and agronomical traits in sorghum (*Sorghum bicolor* L. Moench). Theor. Appl. Genet. 97, 605–616.

Rana, B.S., Murty, B.R., 1978. Role of height and panicle type in yield heterosis in some grain sorghums. Indian J. Genet. Plant Breed. 38, 126–134.

Rana, B.S., Rao, V.J.M., Singh, V.U., Indira, S., Rao, N.G.P., 1982a. Breeding for multiple insect and disease resistance in sorghum. Sorghum in the Eighties, Proceedings of the International Symposium on Sorghum. 2–7 November 1981, ICRISAT, Hyderabad, pp. 745.

Rana, B.S., Anahosur, K.H., Rao, V.J.M., Parameshwarappa, R., Rao, N.G.P., 1982. Inheritance of field resistance to sorghum charcoal rot and selection for multiple disease resistance. Indian J. Genetics Plant Breed. 42 (3), 302–310.

Rana, B.S., Anahosur, K.H., Rao, V.J.M., Parameshwarappa, R., Rao, N.G.P., Jayamohan Rao, V., 1982b. Inheritance of field resistance to sorghum charcoal rot and selection for multiple disease resistance. Indian J. Genet. Plant Breed. 42, 302–310.

Rana, B.S., Rao, V.J.M., Reddy, B.B., Rao, N.G.P., Jaya Mohan Rao, V., 1983. Overcoming present hybrid yield in sorghum. In: Proceedings of Precongress Symposium XV International Genetics Congress, 7–9 Dec 1983, Tamil Nadu Agricultural University, Coimbatore, India, pp. 48–59.

Rana, B.S., Singh, B.U., Rao, N.G.P., 1985. Breeding for shoot fly and stem borer resistance in sorghum. Proceedings of the International Sorghum Entomology Workshop. 15–21 July 1984, Texas A&M University, College Station, TX, pp. 347–360.

Rana B.S., Swarnalata, K., Rao, M.H., 1997. Impact of genetic improvement on sorghum productivity in India. Proceedings of an International Conference on the Genetic Improvement of Sorghum and Pearl Millet, held at Lubbock, TX. 22–27 September 1996, International Sorghum and Millet Research (INTSORMIL)—International Crops Research Institute for the Semi-Arid Tropics (ICRISAT), pp. 142–165.

Rana, V.K.S., Ahluwalia, M., 1979. Genetic analysis of characters for dual purpose sorghum. Sorghum Newsl. 22, 17.

Rao, C.L.N., 1997. Effect of water stress during grain filling on growth, yield and yield components of glossy and non-glossy varieties of sorghum (*Sorghum bicolor* L. Moench). Ann. Plant Physiol. 11, 162–179.

Rao, C.L.N., 1999. Effect of water stress on growth, yield and yield components of glossy and non-glossy lines of grain sorghum. J. Res. 27 (4), 38–44, Acharya N.G Ranga Agricultural University.

Rao, C.L.N., Shivaraj, A., 1988. Effect of water stress on grain growth of glossy and non-glossy varieties of grain sorghum. Indian J. Agric. Sci. 58 (10), 770–773.

Rao, D.G., Khanna, S.K.C.S.R., 1999. Comparative performance of sorghum hybrids and their parents under extreme water stress. J. Agric. Sci. 133 (1), 53–59.

Rao, D.N.V., Shinde, V.K., 1985. Inheritance of charcoal rot resistance in sorghum. J. Maharashtra Agric. Univ. 10, 54–56.

Rao, G.K., Balasubramanian, K.A., Sathyanarayan, A., 1989. Contribution of some prime plant characters in relation to sorghum charcoal rot disease. Indian J. Mycol. Plant Pathol. 19, 50–54.

Rao, G.S., Jagadish, C.A., House, L.R., Subba Rao, G., 1978. Combining ability studies in sorghum III. American x African crosses. Indian J. Heredity. 10, 69–75.

Rao, K.N., Reddy, V.S., Williams, R.J., House, L.R., 1980. The ICRISAT charcoal rot resistance program. Sorghum Diseases, a World Review: Proceedings of the International Workshop on Sorghum Diseases. ICRISAT, Hyderabad, India, pp. 315–321.

Rao, M., 1965. Bread or *roti* making quality in sorghum hybrids. Sorghum Newsl. 8, 27–30.

Rao, M.R.G., Patil, S.J., 1996. Variability and correlation studies in F2 population of *kharif* and *rabi* sorghum. Karnataka J. Agric. Sci. 9 (1), 78–84.

Rao, N.G.P., 1970. Genetic analysis of some exotic × Indian crosses in Sorghum I. Heterosis and its interaction with seasons. Indian J. Genet. 30, 347–361.

Rao, N.G.P., 1982. Transforming traditional sorghum in India. Sorghum in the Eighties: Proceedings of the International Symposium on Sorghum. 2–7 November 1981, ICRISAT, vol. 1, Patancheru, India, pp. 39–59.

Rao, N.G.P., Jaya Mohan Rao, V., Reddy, B.B., 1986. Progress in genetic improvement of *rabi* sorghums in India. Indian J. Genet. 46 (2), 348–354.

Rao, N.G.P., Rana, B.S., Balakotaiah, K., Tripathi, D.P., Fayed, M.F.S., 1974. Genetic analysis of some Exotic × Indian crosses in sorghum VIII: F1 analysis of ovipositional nonpreference underlying resistance to shoot fly. Indian J. Genet. Plant Breed. 34, 122–127.

Rao, N.K.S., Singh, S.P., 1998. Contribution of stem sugars and different photosynthetic plant parts into grain development in sorghum. INSA Interdisciplinary Symposium on Photosynthesis and Productivity Abstract Papers, pp. 56–57.

Rao, S.A., Rao, K.E.P., Mengesha, M.H., Reddy, V.G., 1996. Morphological diversity in sorghum germplasm from India. Genet. Resour. Crop Eval. 43, 559–567.

Rao, S.B.P., Rao, D.V.N., 1956. Studies on the sorghum shoot fly, *Atherigona indica* Mallaoch (*Anthomyiidae*: Diptera) at Siruguppa. Mysore Agric. J. 31, 158–174.

Rao, S.B.P., Bhatia, H.P., Dabir, V.N., 1964. Quality tests on grain sorghum hybrids. Sorghum Newsl. 7, 34–35.

Rao, S.K., Gupta, A.K., Baghal, Singh, S.P., 1982. Combining ability analysis of grain quality sorghum. Indian J. Agric. Res. 16 (1), 1–9.

Rao, S.S., Basheeruddin, M., Sahib, K.H., 2000. Genetic variability for shoot fly incidence in sorghum. Crop Res. 19, 485–486.

Rao, S.S., Rana, B.S., Pawar, K.N., Salunke, V.D., Chimmad, V.P., Kusalkar D.V., 2001. Evaluating post rainy-season sorghum genetic resources in multi-environment trials for plant traits conferring adaptation to drought in vertisols of Central India. Proceedings of 22nd Bienn. Grain Sorghum Res. and Util. Conf. 18–20 February 2001, Nashville, TN, 13 p.

Rao, S.S., More P.R., Solunke V.D., Kusalkar D.V., Jirali D.I., Pawar K.N., et al., 2003. Physiological approaches for improving drought tolerance in rabi sorghum. Proceedings of National Seminar on 'Role of Plant Physiology for Sustaining Quality and Quantity of Food Production in Relation to Environment' held at University of Agricultural Sciences, Dharwad, pp. 26–32.

Rao, V.J.M., 1985. Techniques for screening sorghums for resistance to *Striga*. In: Information Bulletin no. 20, International Crops Research Institute for the Semi-Arid Tropics, Patancheru, India, 18.

Rathore, A.S., Singhania, D.L., 1987. Heterosis and inbreeding depression in intervarietal crosses of forage sorghum [*Sorghum bicolor* (L.) Moench]. Indian J. Heredity. 19, 39–49.

Raut, S.K., Patel, P.H., Khorgade, P.W., 1994. Studies on genetic variability in sorghum. Agric. Sci. Dig. 14, 57–59.

Ravikumar, R.L., Patil, B.S., Salimath, P.M., 2003. Drought tolerance in sorghum by pollen selection using osmotic stress. Euphytica. 133, 371–376.

Ravikumar, S., Hammer, G.L., Broad, I., Harland, P., McLean, G., 2009. Modelling environmental effects on phenology and canopy development of diverse sorghum genotypes. Field Crops Res. 111 (1-2), 157–165.

Ravindrababu, Y., Pathak, A.R., 2000a. Gene effects for yield and shoot fly resistance in sorghum. Gujarat Agric. Univ. Res. J. 26, 27–29.

Ravindrababu, Y., Pathak, A.R., 2000b. Gene effects for shoot fly (*Atherigona soccata* Rondani) resistance to sorghum. Indian J. Genet. Plant Breed. 60, 383–385.

Ravindrababu, Y., Pathak, A.R., 2001. Combining ability analysis over environments for yield and shoot fly resistance in sorghum. J. Maharashtra Agric. Univ. 25, 237–239.

Ravindrababu, Y., Patel, N.A., Vekaria, M.V., Pathak, A.R., 1997. Host plant resistance against sorghum shoot fly (*Atherigena soccata* Rondani) in sorghum. Pest Manage. Econ. Zool. 5, 43–46.

Ravindrababu, Y., Pathak, A.R., Tank, C.J., 2001. Studies on combining ability for yield and yield attributes in sorghum [*Sorghum bicolor* (L.) Moench]. Crop Res. 22, 274–277.

Ravindrababu, Y., Pathak, A.R., Tank, C.J., 2002. Studies on heterosis over environment in sorghum. Crop Res. 24, 90–92.

Ravindranath, P., Shivaraj, A.S., 1983. Effect of moisture stress on growth, yield and yield components of field grown sorghum varieties having a glossy and nonglossy leaves. Indian J. Agric. Sci. 53 (6), 428–430.

Rawat, R.R., Saxena, D.K., 1967. Studies on the bionomics of *Peregrinus maidis* (Ashmead) (Homoptera: Araeopidae). JNKVV Res. J. 1, 64–67.

Rawson, H.M., Gardner, P.A., Long, M.J., 1987. Sources of variation in specific leaf area I wheat grown at high temperature. Aust. J. Plant Physiol. 14, 287–298.

Reddy, B.B., Rao, N.G.P., 1978. Time of sowing of rabi jowar. Sorghum News Lett. 21, 17.

Reddy, B.V.S., 1985a. Relatorio final de consultoria. IICA/EMBRAPA/IPA.83.

Reddy, B.V.S., 1985b. Breeding grain sorghums for adaptation to specific drought situations. Relatorio Final De Consultoria. IICA/EMBRAPA/IPA. Jan 84 to Mar 85, pp. 1–83.

Reddy, B.V.S., 1986. Genetic improvement for drought resistance in sorghum: a plant breeder's view point. Genetic Improvement of Drought Resistance, Proceedings of a Discussion Series of the Drought Research Seminar Forums, pp. 28–32.

Reddy, B.V.S., 1997. Development, production, and maintenance of male-sterile lines in sorghum. In: Faujdar Singh, Rai, K.N., Reddy, B.V.S., Diwakar, B. (Eds.), Training manual on development of cultivars and seed production techniques in sorghum and pearl millet. International Crops Research Institute for the Semi-Arid Tropics, Patancheru 502 324, Andhra Pradesh, India, pp. 22–27.

Reddy, B.V.S., Seetharama, N., Maiti, R.K., Bidinger, F.R., Peacock, J.M., House, L.R., 1980. Breeding for drought resistance in sorghum at ICRISAT. Paper presented at All India Coordinated Sorghum Improvement Project Workshop. 12–14 May 1980, Coimbatore, India.

Reddy B.V.S., Rudrappa A.P., Prasada Rao K.E., Seetharama N., House L.R., 1983. Sorghum improvement for *rabi* adaptation: approach and results. Presented at the All India Coordinated Sorghum Improvement Project Workshop. 18–22 April 1983, Haryana Agricultural University, Hisar, India.

Reddy, B.V.S., Sanjana, P., Ramaiah, B., 2003. Strategies for improving post-rainy season sorghum: a case study for landrace hybrid breeding approach. Paper presented in the Workshop on Heterosis in Guinea Sorghum. Sotuba, Mali, pp. 10–14.

Reddy, B.V.S., Ramesh, S., Reddy, P.S., 2006. Sorghum genetic resources, cytogenetics, and improvement. In: Singh, R.J., Jauhar, P.P. (Eds.), Genetic Resources Chromosome Engineering and Crop Improvement, Cereals, vol. 2. CRC Press, Taylor & Francis Group, Boca Raton, FL, pp. 309–363.

Reddy, B.V.S., Ramaiah, B., Ashok Kumar, A., Reddy, P.S., 2007a. Evaluation of sorghum genotypes for staygreen trait and grain yield. An Open Access Journal Published by ICRISAT. 3 (1), 1–4.

Reddy, B.V.S., Ramaiah, B., Ashok Kumar, A., Sanjana Reddy, P., 2007b. Evaluation of sorghum genotypes for stay-green trait and grain yield. E-Journal SAT Agric. Res. 3 (1), 1–4.

Reddy, B.V.S., Ramesh, S., Sanjana Reddy, P., Ashok Kumar, A., 2009. Genetic options for drought management in sorghum. Plant Breed. Rev. 31, 189–222.

Reddy, C., 1980. Studies on Photosynthetic Productivity of Some Sorghum Genotypes Under Rainfed Conditions (Rabi) (M.Sc. (Agri) thesis). University of Agricultural Sciences, Bangalore.

Reddy, J.N., 1993. Combining ability and heterosis of stover yield in sorghum [*Sorghum bicolor* (L.) Moench]. Orissa J. Agric. Res. 6, 164–166.

Reddy, J.N., Joshi, P., 1990. Combining ability analysis for harvest index in sorghum. Crop Improv. 17, 188–190.

Reddy, J.N., Joshi, P., 1993. Heterosis, inbreeding depression and combining ability in sorghum [*Sorghum bicolor* (L.) Moench]. Indian J. Genet. Plant Breed. 53, 138–146.

Reddy, K.A., 2007. Genetic Analysis of Shoot Fly Resistance, Drought Resistance and Grain Quality Component Traits in *Rabi* Sorghum (*Sorghum bicolor* L. Moench) (M Sc (Agri) thesis). University of Agricultural Science, Dharwad, India, 79pp.

Reddy, P.S., Fakrudin, B., Rajkumar, P.S.M., Arun, S.S., Kuruvinashetti, M.S., Das, I.K., et al., 2008. Molecular mapping of genomic regions harbouring QTLs for stalk rot resistance in sorghum. Euphytica. 159, 191–198.

Rego, T.J., Grundon, N.J., Asher, C.J., Edward, D.G., 1988. Comparison of the effects of continuous and a relieved water stress on 'N' nutrition of grain sorghum. Aust. J. Agric. Res. 39 (5), 773–782.

Rekha, B.C., 2006. Genetic Studies on Grain Quality and Productivity Traits in Rabi Sorghum (M Sc (Agri) thesis). University Agricultural Science, Dharwad (India).

Richard, R.A., 2006. Physiological traits used in the breeding of new cultivars for water-scarce environments. Agric. Water Manage. 80 (1-3), 197–211.

Richards, R.A., Passioura, J.B., 1981. Seminal root morphology and water use of wheat. II. Genetic variation. Crop Sci. 21, 249–252.

Robbins, 1941. Genetic basis of heterosis. Am. J. Bot. 28, 216–225.

Robertson, G.W., 1988. Possibilities and limitations of rainfall analysis for predicting crop available water. In: Bidinger, F.R., Johansen, C. (Eds.), Drought Research Priorities for the Dryland Tropics. ICRISAT, Patancheru, India, pp. 3–14.

Rodriguez, F.C., Pedresoperez, R., Martinez, F., Martinfagundo, D., Savcedocastillo, O., 1994. Heterosis for the yield in grains of sorghum hybrids [*Sorghum bicolor* (L.) Moench]. Cent. Agricola. 21, 23–28.

Rooney, L.W., Miller, F.R., 1982. Variation in structure and kernel characteristics of sorghum. Proceedings, International Symposium on Sorghum Grain Quality. 28–31 October 1981, ICRISAT, Patancheru, India.

Rooney, L.W., Murty, D.S., 1982. Color of sorghum food products. In: Proceedings, International Symposium on Sorghum Grain Quality, ICRISAT, 28-31 Oct 1981. Patancheru, A.P., India: ICRISAT.

Rosa, M.D., Maiti, R.K., 1990. Evaluation of glossy sorghum for seedling traits and chlorophyll content under water stress situation. Sorghum Newsl. 31, 36.

Rosenow, D.T., 1983. Breeding for resistance to root and stalk rots in Texas. In: Mughogho, L.K. (Ed.), Sorghum Root and Stalk Rots, a Critical Review. International Crops Research Institute for the Semi-Arid Tropics, Patancheru, India, pp. 209–217.

Rosenow, D.T., 1984. Breeding for resistance to root and stalk rots in Texas. In: Mughogho, L.K. (Ed.), Sorghum Root and Stalk Diseases, a Critical Review. Proc. Consultative Group Discussion of Research Needs and Strategies for Control of Sorghum Root and Stalk Diseases, Bellagio, Italy. 27 November–2 December 1983. ICRISAT, Patancheru, India, pp. 209–217.

Rosenow, D.T., 1987. Breeding sorghum for drought resistance. In: Menyoga, J.M., Bezuneh, T., Youdeowei, A. (Eds.), Food grain Production in Semi Arid Africa. OAU-STRC/SAFGRAD, Ouagadougou, Burkina Faso, pp. 83–89.

Rosenow, D.T., 1994. Evaluation of drought and disease resistance in sorghum for use in molecular marker-assisted selection. In: Witcombe, J.R and Duncan, R.R (Eds.), Use of molecular markers in sorghum and pearl millet breeding for developing countries, Proceedings of an ODA Plant Sciences Research Programs Conference, 29th March- 1st April, 1993, Norwich, U.K. Overseas Development Administration (ODA), UK, pp. 27–31.

Rosenow, D.T., Clark, L.E., 1981. Drought tolerance in sorghum. Proceedings of 36th Annual Corn and Sorghum Industry Research Conference. 9–11 December 1981, Chicago, IL, pp. 18–31.

Rosenow, D.T., Frederiksen, R.A., 1982. Breeding for disease resistance in sorghum. In: Sorghum in the eighties. Proc. Int. Sorghum Workshop, November 1981, Hyderabad, India. Oxford and IBH publication, Co., New Delhi.

Rosenow, D.T., Johnson, J.W., Fredericksen, R.A., Miller, F.R., 1977. Relationship of non-senescence to lodging and charcoal rot in sorghum. Agronomy Abstracts. American Society of Agronomy, Madison, WI, p. 69.

Rosenow, D.T., Schertz, K.F., Solomayor, A., 1980. Germplasm release of three pairs (A and B) of sorghum lines with A_2 cytoplasmic genetic sterility system. Texas Agric. Exp. Station. M.P. 1448, 1–2.

Rosenow, D.T., Quinsenberry, J.A., Wendt, W.C., Clark, L.E., 1983. Drought-tolerant sorghum and cotton germplasm. Agric. Water Manage. 7 (1–3), 207–222.

Rosenow, D.T., Woodfin, C.A., Clarke, L.E., 1988. Breeding for the say-green train in sorghum. Agronomy Abstracts. ASA, Madison, WI, p. 94.

Rosenow, D.T., Ejeta, G., Clark, L.E., Gilbert, M.L., Henzell, R.G., Borell, A.K., et al., 1996. Breeding for pre- and post-flowering drought stress resistance in sorghum. Proceeding of the International Conference on Genetic Improvement of Sorghum and Pearl Millet. 23–27 September 1996, Lubbock, TX, pp. 400–411.

Rosenow, D.T., Ejeta, G., Clark, L.E., Grilbert, M.L., Henzell, R.G., Borrell, A.K., et al., 1997. In: Proceedings of the International Conference on Genetic Improvement of Sorghum and Pearl Millet, September 22–27, 1976, Lubbock, Texas INTSORMIL and ICRISAT, pp. 400–411.

Rosenow, D.T., Dahlberg, J.A., Peterson, G.C., Clark, L.E., Miller, F.R., Sotomayor-Rios, A., et al., 1997a. Registration of fifty converted sorghums from the sorghum conversion program. Crop Sci. 37, 1397–1398.

Rosenow, D.T., Dahlberg, J.A., Stephens, J.C., Miller, F.R., Barnes, D.K., Peterson, G.C., et al., 1997b. Registration of 63 converted sorghums germplasm lines from the sorghum conversion program. Crop Sci. 37, 1399–1400.

Rosenow, D.T., Ejeta, G., Clark, L.E., Grilbert, M.L., Henzell, R.G., Borrell, A.K., et al., 1997c. Breeding for pre- and post-flowering drought stress resistance in sorghum. Proceedings of the International Conference on Genetic Improvement of Sorghum and Pearl Millet. 22–27 September 1976, INTSORMIL and ICRISAT, Lubbock, TX, pp. 400–411.

Rosielle, A.A., Hamblin, J., 1981. Theoretical aspects of selection for yield in stress and non-stress environments. Crop Sci. 21, 943–946.

Rymen, B., Fabio, F., Klaas, V., Dirk, I., Gerritt, Beemster, T.S., 2007. Cold nights impair leaf growth and cell cycle progression in maize through transcriptional changes of cell cycle genes. Plant Physiol. 143, 1429–1439.

Saadan, H.M., Miller, F.R., 1983. Inheritance of endosperm texture in the sorghum kernel. Sorghum News Lett. 26, 45–46.

Sabour, I., Merah, O., El Jaafari, S., Paul, R., Monneveux, P.H., 1997. Leaf osmotic potential, relative water content and leaf excised water loss variations in oasis wheat landraces in response to water deficit. Arch. Int. Physiol. Biochem. Biophys. 105, 14.

Sahib, K.H., Mahaboob Ali, S., Reddy, B.B., 1986. Heterosis and combining ability of yield and yield components in *rabi* sorghum [*Sorghum bicolor* (L.) Moench]. J. Res., Andhra Pradesh Agric. Univ. 14, 134–140.

Sajjanar, G.M., 2002. Genetic Analysis and Molecular Mapping of Components of Resistance to Shoot Fly [*Atherigona soccata* Rond.] in Sorghum [*Sorghum bicolor* (L.) Moench] (Ph.D. thesis). University of Agricultural Sciences, Dharwad, India.

Salem, M.A., Kakani, V.G., Koti, S., Reddy, K.R., 2007. Pollen-based screening of soybean genotype for high temperatures. Crop Sci. 47, 219–231.

Saliah, A.A., Ali, I.A., Lux, A., Luxona, M., Cohen, Y., Sugimoto, Y., et al., 1999. Response of sorghum to water stress under greenhouse conditions. Phenology, water consumption through evapotranspiration and transpiration and yield. Crop Sci. 39, 168–173.

Salunke, C.B., 1995. Heterosis and Combining Ability for Yield and its Components and Study of Morphological Traits in Rabi Sorghum (Ph.D.(Agri.) thesis submitted to M.P.K.V). Rahuri, Maharashtra.

Salunke, C.B., Pawar, B.B., 1996. Heterosis studies for grain yield and physiological traits in *rabi* sorghum. Ann. Plant Physiol. 10, 133–137.

Salunke, C.B., Deore, G.N., 1998. Heterosis and heterobeltiosis studies for grain yield and its component characters in *rabi* sorghum. Ann. Plant Physiol. 12, 6–10.

Salunke, C.B., Pawar, B.B., Deshmukh, R.B., Narkhede, B.N., 1996. Combining ability studies in *rabi* sorghum under irrigated and moisture stress environment. J. Maharashtra Agric. Univ. 21 (3), 426–429.

Salunke, C.B., Deore, G.N., Mate, S.N., 2001. Combining ability studies for physiological traits, harvest index and grain yield in *rabi* sorghum. Ann. Plant Physiol. 14 (2), 190–195.

Salunke, V.D., Deskhmukh, R.V., Aglave, B.N., Borikar, S.T., 2003. Evaluation of sorghum genotypes for drought tolerance. Int. Sorghum Millets Newsl. 44, 88–90.

Sameer, C.V.K., Sreelakshmi, Ch, Shivani, D., 2012. Selection indices for yield in *rabi* sorghum (*Sorghum bicolor* (L.) Moench) genotypes. Electron. J. Plant Breed. 3 (4), 1002–1004.

Sanchez, A.C., Subudhi, P.K., Rosenow, D.T., Nguyen, H.T., 2002. Mapping QTL associated with drought resistance in sorghum (*Sorghum bicolor* L. Moench). Plant Mol. Biol. 48, 713–726.

Sandoval, M.P., Gonzalez-Hernandez, V.A., Mendoza-onofer, L.E., Engleman, E.M., 1989. Drought stress effects on the grain yield and panicle development of sorghum. Can. J. Plant Sci. 69 (3), 631–641.

Sane, A.P., Nath, P., Sane, P.V., 1996. Cytoplasmic male sterility in sorghum: organization and expression of mitochondrial genes in Indian CMS cytoplasms. J. Geneics. 75 (2), 151–159.

Sanghera, G.S., Zarger, M.A., Anwar, A., Singh, S.P., Ahmad, N., Rather, M.A. Studies on spikelet fertility and incidence of leaf blast on certain IRCTN rice genotypes under temperate conditions. Paper presented at National Symposium on plant Protection Strategies for sustainable Agri-Horticulture, SKAUST-Jammu. 2001. Oct 12–13, p. 125.

Sanjay Kumar, Singh, R., Kumar, S., 1996, Combining ability for shoot fly resistance in sorghum. Crop Improvement, 23, 217–220.

Sanjana Reddy, P., Patil, J.V., Nirmal, S.V., Gadakh, S.R., 2012. Improving post-rainy season sorghum productivity in medium soils: does ideotype breeding hold a clue? Curr. Sci. 102, 904–908.

Sankarpandian, R., Bangarusamy, U., 1996. Stability of sorghum genotypes for certain physiological characters and yield under water stress conditions. Crop Improv. 23 (1), 61–65.

Sankarapandian, R., Muppidathi, N., 1995. Evaluation of sorghum genotypes for certain physiological characters with yield and interrelationship analysis under water stress condition. Madras Agric. J. 82 (4), 243–245.

Sankarpandian, R., Krishnadass, D., Muppidathi, N., Chidambaram, S., 1993. Variability studies in grain sorghum for certain physiological characters under water stress conditions. Crop Improv. 20 (1), 45–50.

Sankarapandian, R., Subbaraman, N., Amrithadevarathnam, 1994. Heterosis in grain sorghum. Madras Agric. J. 81, 1–2.

Santamaria, J.M., Fukai, S., 1990. Contribution of osmotic adjustment to kernel yield in *Sorghum bicolor* (L) Moench under water-limited condition. I. Water stress before anthesis. Aust. J. Agric. Res. 41, 51–65.

Santamaria, J.M., Ludlow, M.M., Fukai, S., 1990. Contribution of osmotic adjustment to grain yield in *Sorghum bicolor* (L.) Moench under water-limited conditions. I. Water stress before anthesis. Aust. J. Agric. Res. 41, 51–65.

Santos, F., Dos, B.G., Petrini, J.A., Assis, F.N. De., Morases, D.M.D.E., 1979. Growth analysis of four grain sorghum hybrids in peoltas. In: Sorgo Resultados de Pesquisa Pelotas, Brazil.

Sarkissian, I.V., Srivastawa, H.K., 1967. Heterosis in root growth and nutrient uptake. Genetics. 57, 843–850.

Sarwar, J.A.K., 1983. Assessment of advanced sorghum varieties and hybrids for drought resistance. Sorghum Newsl. 26, 9–10.

Satish, K., Srinivas, G., Madhusudhana, R., Padmaja, P.G., Nagaraja Reddy, R., Murali Mohan, S., et al., 2009. Identification of quantitative trait loci for resistance to shoot fly in sorghum (*Sorghum bicolor* (L.) Moench). Theor. Appl. Genet. 119 (8), 1425–1439.

Saxena, N.P., 1987. Screening for adaptation to drought: case studies with chickpea and pigeonpea. In: Saxena, N. P., Johansen, C. (Eds.), Adaptation of Chickpea and Pigeonpea to Abiotic Stress. ICRISAT, Patancheru, India, pp. 63–76.

Scapim, C.A., Rodrigues, J.A.S., Cruz, C.D., Gomes, J.A., Braccini, L.A., 1998. Gene effects, heterosis and inbreeding depression for grain sorghum characters. Perquisa Agropecuaria Bras. 33, 1847–1857.

Schertz, K.F., 1977. Registration of A2 x 2753 and BT 2753 sorghum germplasm. (Reg. No. GP 30 and 31). Crop Sci. 17, 983.

Schertz, K.F., Ritchey, J.R., 1977. Cytoplasmic genic male sterility systems in sorghum. Crop Sci. 18, 890–893.

Schertz, K.F., Rosenow, D.T., Sotomayor Rios, A., 1981b. Registration of three pairs of (A and B) sorghum germplasm with A2 cytoplasmic genetic male sterility system. Crop Sci. 21, 148.

Scheuring, J.F., Sidibe, S., Kante, A., 1982. Sorghum alkali to: quality considerations. Proceedings, International Symposium on Sorghum Grain Quality. 28–31 October 1981, ICRISAT, Patancheru, India.

Sen, C., Bandopadhyaya, S., 1988. Some aspects of ecological behaviour and disease development and biological inoculums destruction of Macrophomina phaseolina. In: Agnihotri, V.P., Sarbhoy, A.K., Dinesh kumar (Eds.), Perspective in Mycology and Plant pathology. Malhotra Publishing house, New Delhi, pp. 418–443.

Seetharama, N., 1986. Crop physiology and rabi sorghum productivity. Paper presented at 16th Annual Workshop. 14–16 May, AICSIP, Rajendranagar, Hyderabad.

Seetharama, N., 2004. Setting the priorities for postrainy sorghum research. Proceedings of the Brainstorming Session on Managing Biotic Stresses to Increase Postrainy Sorghum Productivity. 11–12 August 2003, Centre on Postrainy Sorghum, National Research Centre for Sorghum, Solapur, Maharashtra, India.

Seetharama, N., Sivakumar, M.V.K., Singh, S., Bidinger, F.R., 1978. Sorghum productivity under receding soil moisture in Deccan plateau. Presented at the International Symposium on Biological Applications of Solar Energy. 1–5 December 1978, Madurai Kamaraj University, Madurai, India.

Seetharama, N., Reddy, B.V.S., Peacock, J.M., Bidinger, F.R., 1982a. Sorghum improvement for drought resistance. pp. 317–338. Drought Resistance in Crops with Emphasis on Rice. International Rice Research Institute, Los Banos, Laguna, Manila, Philippines. pp. 414.

Seetharama, N., Reddy, B.V.S., Peacock, J.K., Bidinger, F.R., 1982b. Sorghum improvement for drought resistance. Drought Resistance in Crop Plants with Emphasis on Rice. IRRI, Philippines, pp. 317–356.

Seetharama, N., Bidinger, F.R., Rao, K.N., Gill, K.S., Mulgund, M., 1987. Effect of pattern and severity of moisture deficit stress on stalk rot incidence in sorghum I. Use of line source irrigation technique, and the effect of time of inoculation. Field Crops Res. 15 (3-4), 289–308.

Seetharama, N., Singh, S., Reddy, B.V.S., 1990. Strategies for improving postrainy sorghum productivity. Proc. Indian Nat. Sci. Acad. 56 (5&6), 455–467.

Seif, E., Evans, J.C., Balaam, L.N., 1979. A multivariate procedure for classifying environments according to their interaction with genotypes. Aust. J. Agric. Res. 30, 1021–1026.

Sekhar, P.R., Kulkarni, N., Reddy, D.M., 1995. Field resistance to shoot bug, *Peregrinus maidis* Ashmead in sorghum. J. Insect Sci. 8 (1), 102–103.

Senthil, N., Palanasamy, S., 1993. Heterosis studies involving diverse cyto-steriles of sorghum. Madras Agric. J. 80, 491–494.

Senthil, N., Palanasamy, S., 1994. Combining ability studies involving diverse cotysteriles of sorghum. Ann. Agric. Res. 15, 339–343.

Setokuchi, O., 1973. Ecology of *Longiunguis sacchari* infesting sorghum. I. Nymphal period and fecundity of apterous viviparous females. Proc. Assoc. Plant Prot., Kyushu. 19, 95–97.

Setokuchi, O., 1976. Ecology of *Longiunguis sacchari* (Zehntner) (Aphididae) infesting sorghums. IV. Varietal difference of sorghums in the aphid occurrence. Proc. Assoc. Plant Prot., Kyushu. 22, 139–142.

Setokuchi, O., 1979. Damage to forage sorghum by *Longiunguis sacchari* (Zehntner) (Aphididae). Proc. Assoc. Plant Prot., Kyushu. 25, 66–70.

Setokuchi, O., 1988. Studies on the ecology of aphids on sugarcane. I. Infestation of *Melanaphis sacchari* (Zehntner) (Homoptera: Aphididae). Jpn. J. Appl. Entomol. Zool. 32, 215–218.

Sezegen, B., Carena, M.J., 2009. Divergent recurrent selection for cold tolerance in two improved maize populations. Euphytica. 167, 237–244.

Shamarao, J., Patil, M.S., Indira, S., 2001. Biological control of charcoal rot of sorghum caused by *M. phaseolina*. Agric. Sci. Dig. 21 (3), 153–156.

Shang, S.P., 1989. Study on the selection index for forage sorghum. J. Taiwan Livestock Res. 22, 59–68.

Sharma, G.C., 1980. Studies on phenotypic stability and genetic architecture in forage sorghum. Thesis Abstr. 6, 109–110.

Sharma, P.S., Kumari, T.S., 1996. Effect of water stress, varieties and potassium on sorghum in vertisols. J. Potassium Res. 12 (1), 96–99.

Sharma, A.D., Singh, P., 2003. Comparative studies on drought-induced changes in peptidyl prolyl *cis-trans* isomerase activity in drought-tolerant and susceptible cultivars of *Sorghum bicolor*. Curr. Sci. 84 (7), 911–918.

Sharma, G.C., Jotwani, M.G., Rana, B.S., Rao, N.G.P., 1977. Resistance to the sorghum shoot fly, *Atherigona soccata* (Rondani) and its genetic analysis. J. Entomol. Res., India. 1, 1–12.

Shawesh, G.A., Voigt, R.L., Dobrenz, A.K., 1985. Stomatal frequency and distribution in drought tolerance and drought susceptible (*Sorghum bicolor* L. Moench) genotypes under moisture stress and non-stress. Sorghum Newsl. 28, 123.

Shekar, B.M.C., Singh, B.U., Reddy, K.D., Reddy, D.D.R., 1993b. Antibiosis component of resistance in sorghum to corn plant hopper, *Peregrinus maidis* (Ashmead) (Homoptera: Delphacidae). Insect Sci. Appl. 14 (5), 559–569.

Shepherd, A.D., 1979. Laboratory abrasive decorticating mill for small grains. Cereal Chem. 56, 517–519.

Sheriff, N.M., Prasad, M.N., 1990. Heterosis in sorghum under different environments. Madras Agric. J. 77, 421–426.

Sheriff, N.M., Prasad, M.N., 1994. Combining ability for yield and yield components in sorghum. Madras Agric. J. 81, 583–585.

Shimono, H., Hasegawa, T., Fujimura, S., Iwama, K., 2004. Responses of leaf photosynthesis and plant water status in rice to low water temperature at different growth stages. Field Crops Res. 89, 71–83. Available from: http://dx.doi.org/10.1016/j.fcr.2004.01.025.

Shimono, H., Okada, K.E., Arakawa, I., 2007. Low temperature induced sterility in rice: evidence for the effects of temperature before panicle initiation. Field Crops Res. 101, 221–231. Available from: http://dx.doi.org/10.1016/j.fcr.2006.11.010.

Shinde, M.S., Narkhede, B.N., 1998. Physiological parameters in relation to grain yield in *rabi* sorghum hybrids. Ann. Plant Physiol. 11 (1), 98–100.

Shinde, M.S., Mutkule, B.R., Gaikwad, A.R., Dalvi, U.S., Gadakh, S.R., 2013. Photoperiod sensitivity studies in sweet sorghum. J. Acad. Indus. Res. 1 (11), 696–699.

Shinde, S.S., Borikar, S.J., 1991. Heterosis studies involving maldandi cytoplasm in sorghum. J. Maharashtra Agric. Univ. 16, 121–122.

Shinde, V.K., 1981. Genetic variability, inter relationship and path analysis of yield and its components in sorghum [*Sorghum bicolor* (L.) Moench]. J. Maharashtra Agric. Univ. 6, 30–32.

Shinde, V.K., Joshi, P., 1985. Genetic analysis of height and maturity in tropical grain sorghum. J. Maharashtra Agric. Univ. 10 (3), 256–259.

Shinde, V.K., Kulkarni, N., 1984. Gene action and combining ability analysis of yield in sorghum. Indian J. Agric. Sci. 54, 955–958.

Shinde, V.K., Sudewad, S.N., 1980. Inheritance of grain yield and seed size in four crosses of sorghum. Sorghum Newsl. 23, 19.

Shinozaki, K., Yamaguchi-Schinozaki, K., 1997. Gene expression and signal transduction in water-stress response. Plant Physiol. 115, 327–334.

Shinde, V.K., Nandwankar, K.C., Ambekar, S.S., 1983. Heterosis and combining ability for grain yield in *rabi* sorghum. Sorghum Newsl. 26, 19.

Shinde, M.S., Narkhede, B.N., Patil, S., 1998. Association of physiological parameters with grain yield of rabi sorghum. Ann. Plant Physiol. 12 (1), 65–66.

Shivalli, S., 2000. Characterization of Morpho-Physiological Traits for Higher Productivity in *Rabi* Sorghum (M.Sc. (Agri.) thesis). University of Agricultural Sciences, Dharwad.

Shivanna, H., 1989. Genetic Studies on Yield Traits and Grain Mold Resistance in Sorghum (Ph.D. thesis). University of Agricultural Sciences, Dharwad.

Shivanna, H., Patil, S.S., 1988. A study of combining combinations for yield and other traits in sorghum. J. Maharashtra Agric. Univ. 13, 70–72.

Shoba, V., Kasturiba, B., Rama, K.N., Nirmala, Y., 2008. Nutritive value and quality characteristics of sorghum genotypes. Karnataka J. Agric. Sci. 20, 586–588.

Shobha Rani, N., Prasad, G.S.V., Singh, S.P., Satyanarayana, K., Kondal Rao, Y., Bhaskar Reddy, P., et al., 2002. Vasumati—a new Basmati variety released in India. Int. Rice Res. Notes. 27 (2), 20.

Shouny, K.A., Yasien, M., Ahmed, I.M., 1990. Combining ability and heterosis in some forage sorghum hybrids. Ann. Agric. Sci., Cairo.(Special Issue), 51–65.

Sinclair, J. (Ed.), 1982. Compendium of Soybean Diseases. The American Phytopathological Society, St. Paul.

Sinclair, T.R., Horie, T., 1989. Leaf nitrogen, photosynthesis and crop radiation use efficiency: a review. Crop Sci. 29, 90–98.

Sinclair, T.R., Ludlow, M.M., 1986. Influence of soil water supply on the plant water balance of four tropical grain legumes. Aust. J. Plant Physiol. 13, 329–341.

Singh, B.L., Chaudhary, 1998. The physiology of drought tolerance in field crops. Field Crops Res. 60, 41–56.

Singh, B.U., 1997. Screening for resistance to sorghum shoot bug and spider mites. In: Sharma, H.C., Singh, F., Nwanze, K.F. (Eds.), Plant Resistance to Insects in Sorghum. International Crops Research Institute for the Semi-Arid Tropics (ICRISAT), Patancheru, India, pp. 52–59.

Singh, B.U., Rana, B.S., 1986. Resistance in sorghum to the shootfly, *Atherigona soccata* Rondani. Insect Sci. Appl. 7, 577–587.

Singh, B.U., Rana, B.S., 1992. Stability of resistance to corn planthopper, *Peregrinus maidis* (Ashmead) in sorghum germplasm. Insect Sci. Appl. 13, 251–263.

Singh, B.U., Padmaja, P.G., Seetharama, N., 2004. Stability of biochemical constituents and their relationships with resistance to shoot fly, *Atherigona soccata* (Rondani) in seedling. Euphytica. 136 (3), 279–289.

Singh, B.V., Rana, B.S., 1994. Influence of varietal resistance on disposition and larval development of stalk borer, *Chilo Partellus* Swinhoe and its relationship to field tolerance in sorghum. Insect Sci. Appl. 5, 287–296.

Singh, R., 1982. Studies on genetic architecture and stability of grain yield and its components in forage sorghum. Thesis Abstr. 8, 251–252.

Singh, R., Axtell, J.D., 1973. High lysine mutant gene (hl) that improves protein quality and biological value of grain sorghum. Crop Sci. 13, 535–539.

Singh, S.K., Nene, Y.L., Reddy, M.V., 1990. Influence of cropping system on *Macrophomina phaseolina* in soil. Plant Dis. 74, 812–814.

Singh, S.K., Kakani, V.G., Brand, D., Baldwin, B., Reddy, K.R., 2008. Assessment of cold and heat tolerance of winter-grown canola (*Brassica napus* L.) cultivars by pollen-based parameters. J. Agron. Crop Sci. 194, 225–236.

Singh, S.P., Jotwani, M.G., 1980a. Mechanisms of resistance in sorghum to shoot fly II. Antibiosis. Indian J. Entomol. 42, 353–360.

Singh, S.P., Jotwani, M.G., 1980b. Mechanisms of resistance in sorghum to shoot fly IV. Role of morphological characters of seedlings. Indian J. Entomol. 42, 806–808.

Singh, S.P., Verma, A.N., 1988. Inheritance of resistance to shoot fly *Atherigona soccata* (Rondani) in forage sorghum. J. Insect Sci. 1, 49–52.

Singh, V., van Oosterom, E.J., Jordan, D.R., Messina, C.D., Cooper, M., Hammer, G.L., 2010. Morphological and architectural development of root systems in sorghum and maize. Plant Soil. 333, 287–299.

Singhania, D.L., 1980. Heterosis and combining ability studies in grain sorghum. Indian J. Genet. Plant Breed. 40, 463–471.

Sinha, S.K., Khanna, R., 1975. Physiological, biochemical and genetic basis of heterosis. Adv. Agron. 27, 123.

Siva Kumar, C., Sharma, H.C., Lakshmi Narasu, M., Pampapathy, G., 2008. Mechanisms and diversity of resistance to shoot fly, *Atherigona soccata* in *Sorghum bicolor*. Indian J. Plant Prot. 36 (2), 249.

Smilovenko, L.A., 2002. Inheritance of quantitative traits in sorghum hybrids. Kukuruza-i-Sorgo. 5, 15–17.

Smith, A., Cullis, B.R., Thompson, R., 2001. Analysing variety by environment data using multiplicative mixed models and adjustments for spatial field trend. Biometrics. 57, 1138–1147.

Smith, C.M., 1989. Plant Resistance to Insects: A Fundamental Approach. Wiley, New York, NY, p. 286.

Smith, C.M., Khan, Z.R., Pathak, M.D., 1994. Techniques for Evaluating Insect Resistance in Crop Plants. CRC Press, Boca Raton, FL, p. 320.

Solomon, S., Qin, D., Manning, M., Alley, R.B., Berntsen, T., Bindoff, N.L., et al., 2007. Technical summary. In: Solomon, S., Qin, D., Manning, M., Chen, Z., Marquis, M., Averyt, K.B., et al., Climate Change: The Physical Science Basis. Contribution of Working Group I to the Fourth Assessment Report of the Intergovernmental Panel on Climate Change. Cambridge University Press, Cambridge, UK/New York, pp. 19–92.

Soman, P., 1990. Development of a technique to study seedling emergence in response to moisture deficit in the field: the seedbed environment. Ann. Appl. Biol. 116, 357–364.

Sprague, G.F., Tatum, L.A., 1942. General and specific combining ability in single crosses in corn. J. Am. Soc. Agron. 34, 923–932.

Sridhar, B., 1991. Studies on combining ability in segregating generations of sorghum (*Sorghum bicolor* (L.) Moench). *M.Sc.(Agri.) Thesis*, University of Agricultural Sciences, Dharwad.

Srihari, A., Nagur, T., 1980. Combining ability studies in sorghum. Sorghum Newsl. 23, 12–13.

Sriram, N., Rao, J.S., 1983. Physiological parameters influencing sorghum yield. Indian J. Agric. Sci. 53, 641–649.

Stack, J., 2000. Sorghum Ergot in the Northern Great Plains. University of Nebraska Cooperative Extension EC00-1879-S, Lincoln, NE.

Stephens, J.C., 1937. Male-sterility in sorghum: its possible utilization in production of hybrid seed. J. Am. Soc. Agron. 29, 690–696.

Stephens, J.C., Holland, R.F., 1954. Cytoplasmic male-sterility for hybrid sorghum seed production. Agron. J. 46, 20–23.

Stephens, J.C., Kuykendall, G.H., George, D.W., 1952. Experimental production of hybrid sorghum seed with a three-way cross. Agron. J. 44, 369–373.

Steponkus, P.L., Shahan, K.W., Cutler, J.M., 1982. Osmotic adjustment in rice. Drought Resistance in Crops with Emphasis on Rice. IRRI, Philippines, pp. 181–194.

Stricevic, R., Caki, E., 1997. Relationships between available soil water and indicators of plant water status of sweet sorghum to be applied in irrigation scheduling. Irrig. Sci. 18 (1), 17–21.

Stickler, F.C., Pauli, A.W., 1969. Influence of date of planting on yield and yield components of grain sorghum. Agron. J. 53, 20–23.

Su, G., Suh, S.O., Schneider, S., Russin, J.S., 2001. Host specialization in the charcoal rot fungus, *Macrophomina phaseolina*. Phytopathology. 91, 120–126.

Subba Rao, G., Jagadish, C.A., House, L.R., 1975. Combining ability studies in sorghum. I. American x African crosses. Indian J. Hered. 7, 5–11.

Subba Rao, G., Jagadish, C.A., House, L.R., 1976a. Combining ability studies in some Indian x exotic crosses of sorghum. Indian J. Hered. 8, 31–38.

Subba Rao, G., Jagadish, C.A., House, L.R., 1976b. Combining ability studies in sorghum. II. American x African crosses. Indian J. Hered. 8, 51–57.

Subramanian, V., Jambunathan, R., 1982. Properties of sorghum grain and their relationship to rati quality. Proceedings, International Symposium on Sorghum Grain Quality. 28–31 October 1981, ICRISAT, Patancheru, India.

Subramanian, V., Murty, D.S., Jambunathan, R., House, L.R., 1982. Boiled sorghum characteristics and their relationship to starch properties. In: Rooney, L.W., Murty, D.S. (Eds.), Proc. International Symposium on Sorghum Grain Quality. International Crops research Institute for the Semi-Arid Tropics, Patancheru, India, pp. 103–109.

Subramanian, et al. 1989.

Subudhi, P.K., Rosenow, D.T., Nguyen, H.T., 2000. Quantitative trait loci for the stay green trait in sorghum (*Sorghum bicolor* (L.) Moench). Consistency across genetic backgrounds and environments. Theor. Appl. Genet. 101 (5-6), 733–741.

Sukhani, T.R., Jotwani, M.G., 1980. Efficacy of mixtures of carbofuran treated and untreated sorghum seed for the control of shoot fly. *Atherigona soccata* (Rondani). J. Entomol. Res. 4, 186–189.

Sukumaran, S., Xiang, W., Bean, S.R., Pedersen, J.F., Kresovich, S., Tuinstra, M.R., et al., 2012. Association mapping for grain quality in a diverse sorghum collection. Plant Genome. 5 (3), 126–135.

Sullivan, C.Y., 1972. Mechanisms of heat and drought resistance in grain sorghum and methods of measurements. In: Rao, N.G.P., House, L.R. (Eds.), Sorghum in Eighties. Oxford and IBH Publishing Co., New Delhi, India, pp. 247–263.

Sullivan, C.Y., Ross, W.M., 1979. Selecting for drought and heat resistance in grain sorghum. In: Mussell, H., Staples, R.C. (Eds.), Stress Physiology in Crop Plants. Wiley–Interscience, New York, NY, pp. 263–281.

Sultan, B., Baron, C., Dingkuhn, M., Sarr, B., Janicot, S., 2005. Agricultural impacts of large-scale variability of the West African monsoon. Agric. For. Meteorol. 128, 93–110.

Surwenshi, A., 1999. Morpho-Physiological Traits Associated with Drought Adaptation and Physiological Basis of Heterosis in Sorghum (M. Sc. (Agri.) thesis). University of Agricultural Sciences, Dharwad.

Swarnalatha, K., Rana, B.S., 1988. Combining ability for biological yield and harvest index in sorghum. Indian J. Genetics. 48 (2), 149–153.

Takele, A., 2000. Seedling emergence and growth of sorghum genotypes under variable soil moisture deficit. Acta Agron. Hung. 48, 95–102.

Takzure, S.C., Phadravis, B.N., Vitkare, D.G., 1998. Effect of winter stress at various stages on yield and its components in grain sorghum. Ann. Plant Physiol. 2 (2), 176–182.

Talwar, H., Vadez, V., 2011. Improving sorghum varieties to meet growing grain and fodder demand in India. South Asia Newsl. 22–25.

Talwar, H.S., Ashok, S., Seetharama, N., 2009a. Use of SPAD chlorophyll meter to screen sorghum (*Sorghum bicolor*) lines for postflowering drought tolerance. Indian J. Agric. Sci. 79 (1), 35–39.

Talwar, H.S., Surwenshi, A., Seetharama, N., 2009b. Use of SPAD chlorophyll meter to screen sorghum (*Sorghum bicolor*) lines for postflowering drought tolerance. Indian J. Agric. Sci. 79 (1), 432–437.

Talwar, H.S., Prabhakar, E.M., Kumari, A., Rao, S.S., Mishra, J.S., Patil, J.V., 2010. Strategies to improve post flowering drought tolerance in rabi sorghum for predicted climate change scenario. Crop Improv. 37 (2), 93–98.

Tan, W.Q., Li, S.M., Guo, H.P., Gao, R.P., 1985. A study of the inheritance of aphid resistance in sorghum. Shanxi Agric. Sci. 8 (12-14).

Taneja, S.L., Leuschner, K., 1985. Resistance screening and mechanisms of resistance in sorghum to shoot fly. In: Kumble, V. (Ed.), Proceedings International Sorghum Entomology Workshop. 15–21 July 1984. Texas A&M University, College Station, TX, International Crops Research Institute for the Semi-Arid Tropics, Patancheru, India, pp. 115–129.

Tangpremsri, T., Fukai, S., Fischer, K.S., Henzell, R.G., 1991a. Genotypic variation in osmotic adjustment in grain sorghum. I. Development of variation in osmotic adjustment under water-limited conditions. Aust. J. Agric. Res. 42, 747–757.

Tangpremsri, T., Fukai, S., Fischer, K.S., Henzell, R.G., 1991b. Genotypic variation in osmotic adjustment in grain sorghum. II. Relation with some growth attributes. Aust. J. Agric. Res. 42, 759–767.

Tangpremsri, T., Fukai, S., Fischer, K.S., Kenell, R.G., 1991c. Genotypic variation in osmotic adjustment in grain sorghum II. Relation with some growth attribution. Aust. J. Agric. Res. 42 (5), 759–767.

Tangpremsri, T., Fukai, S., Fischer, K.S., 1995. Growth and yield of sorghum lines extracted from a population for differences in osmotic adjustment. Aust. J. Agric. Res. 46 (1), 61–74.

Tao, Y.Z., Henzell, R.G., Jordan, D.R., Butler, D.G., Kellu, A.M., McIntyre, C.L., 2000. Identification of genomic regions associated with stay green in sorghum by testing RILs in multiple environments. Theor. Appl. Genet. 100, 1125–1232.

Tarr, S.A.J., 1962. Diseases of Sorghum, Sudangrass and Brown Corn. The Commonwealth Mycological Institute, Kew, Surrey, UK, p. 380.

Tauli, 1964. Growth and Development of Sorghum in Relation to Drought Tolerance. Dissertation Abstraction International (Science and Engineering) 51 (5), 23–28.

Teetes, G.L., 1980. Breeding sorghums resistant to insects. In: Maxwell, F.G., Jennings, P.R. (Eds.), Breeding Plants Resistant to Insects. Wiley, New York, NY, pp. 457–489.

Teetes, G.L., Manthe, C.S., Peterson, G.C., Leuschner, K., Pendleton, B.B., 1995. Sorghum resistant to the sugarcane aphid, *Melanaphis sacchari* (Homoptera: Aphididae), in Botswana and Zimbabwe. Insect Sci. Appl. 16, 63–71.

Tenkouano, A., 1990. Relationships of non-structural carbohydrates to resistance to charcoal rot in sorghum. Summaries of College Station, Texas A&M University, Texas, USA, pp. 129.

Tenkouano, A., Miller, F.R., Frederiksen, R.A., Rosenow, D.T., Cothren, J.T., et al., 1992. Nonstructural carbohydrates and charcoal rot resistance in sorghum. Sorghum Newsl. 33, 33–34.

Tenkouano, A., Miller, F.R., Frederiksen, R.A., Rosenow, D.T., 1993. Genetics of non-senescence and charcoal rot resistance in sorghum. Theor. Appl. Genet. 85, 644–648.

Terry, A.C., 1990. Growth and Development of Sorghum in Relation to Drought Tolerance. Dissertation abstracts International B (Science and Engineering) 51 (5): 213–218.

Thakur, P., Kumar, S., Malik, J.A., Berger, J.D., Nayyar, H., 2010. Cold stress effects on reproductive development in grain crops. Environ. Exp. Bot. 67, 429–443. Available from: http://dx.doi.org/10.1016/j.envexpbot.2009.09.004.

Thawari, S.B., Atale, S.B., Wadhokar, R.S., 2000. Heterosis studies in newly stabilized B lines of sorghum [*Sorghum bicolor* (L.) Moench]. J. Soils Crops. 10, 75–77.

Thomas, H., Howarth, C., 2000. Five ways to stay green. J. Exp. Bot. 51, 329–337.

Thomas, H., Smart, C.M., 1993. Crops that stay-green. Ann. Appl. Biol. 123, 193–219.

Thombre, M.V., Patil, R.C., Hoshi, B.P., 1982. Association of panicle components with grain yield in sorghum. Sorghum Newsl. 25, 17–18.

Thorve, S.B., Upadhye, S.K., Surve, S.P., Kadam, J.R., 2009. Impact of life saving irrigation on yield of rabi sorghum (*Sorghum bicolor* (L.)). Int. J. Agric. Sci. 5 (1), 53–54.

Thul, A.V., 2007. Genetics of Traits Associated with Shootfly and Drought Tolerance in Rabi Sorghum (*Sorghum bicolor* (L.) Moench) (Ph.D.(Agri.) thesis submitted to M.P.K.V). Rahuri, Maharashtra.

Tiryaki, I., Andrews, D.J., 2001. Germination and seedling cold tolerance in sorghum. Agron. J. 93 (6), 1391–1397.

Tiwari, D.K., Gupta, R.S., Mishra, R., 2003. Study of heterotic response for yield and its components in grain sorghum [*Sorghum bicolor* (L.) Moench]. Plant Arch. 3, 255–257.

Tollenaar, M., Daynard, T.B., 1978. Leaf senescence in short-season maize hybrids. Can. J. Plant Sci. 58, 869–874.

Tollenaar, M., Lee, E.A., 2006. Dissection of physiological processes underlying grain yield in maize by examining genetic improvement and heterosis. Maydica. 51, 399–408.

Tourchi, M., Rezai, A.M., 1996. Evaluation of general combining ability of sorghum [*Sorghum bicolor* (L.) Moench] male sterile lines for grain yield andrelated traits. Iran J. Agric. Sci. 27, 37–54.

Traore, M., Sullivan, C.Y., Rosowski, J.R., Lee, K.W., 1989. Comparative leaf surface morphology and the glossy characteristic of sorghum, maize and pearl millet. Ann. Bot. 64 (4), 447–453.

Traore, P.C.S., Kouressy, M., Vaksmann, M., Tabo, R., Maikano, I., Traore, S.B., et al., 2007. Climate prediction and agriculture: what is different about Sudano-Sahelian West Africa? In: Sivakumar, M.V.K., Hansen, J.E. (Eds.), Climate Prediction and Agriculture: Advances and Challenges. Springer-Verlag, Berlin Heidelberg, New York, NY, pp. 188–204. , ISBN-13: 9783540446491.

Trouche, G., Rami, J.E., Chantereau, J., 1998. QTLs for photoperiod response in sorghum. Int. Sorghum Millets Newsl. 39, 94–96.

Tuinstra, M.R., Grote, E.M., Goldsbrough, P.B., Ejeta, G., 1996. Identification of quantitative trait loci associated with pre-flowering drought tolerance in sorghum. Crop Sci. 36, 1337–1344.

Tunistra, M.R., Grote, E.M., Goldsbrough, P.B., Ejeta, G., 1997. Geneticanalysis of post flowering drought tolerance and components of grain development in *Sorghum bicolor* (L.). Moench. Mol. Breed. 3 (6), 439–448.

Tuinsta, M.R., Grote, E.M., Goldsbrough, P.B., Ejeta, G., 1998. Evaluation of near-isogenic sorghum lines contrasting for QTL markers associated with drought tolerance. Crop Sci. 38, 825–842.

Turner, N.C., 1979. Drought resistance and adaptation to water deficits in crop plants. In: Mussell, H., Staples, R. C. (Eds.), Stress Physiology in Crop Plants. Wiley, New York, NY, pp. 343–372.

Turner, N.C., Jones, M.M., 1980. Turgor maintenance by osmotic adjustment: a review and evaluation. In: Turner, N.C., Kramer, P.J. (Eds.), Adaptations of Plants to Water and High Temperature Stress. Wiley, New York, NY, pp. 87–103.

Tysdal, H.M., Kiesselbach, T.A., Westover, H.L., 1942. Alfaalfa breeding. Coll. Agric. Univ. Nebraska Agric. Exp. Stn. Res. Bull. 124, 1–46.

Umakanth, A.V., Madhusudhana, R., Latha, K.V., Kumar, P.H., Kaul, S., 2002. Genetic architecture of yield and its contributing characters in post rainy season sorghum. Int. Sorghum Millets Newsl. 43, 37–40.

Umakanth, A.V., Madhusudhana, R., Madhavi Latha, K., Kaul, S., Rana, B.S., 2003. Heterosis studies for yield and its components in *rabi*sorghum [*Sorghum bicolor* (L.) Moench]. Indian J. Genet. Plant Breed. 63, 159–160.

Umakanth, A.V., Padmaja, P.G., Ashok Kumar, J., Patil, J.V., 2012. Influence of types of sterile cytoplasm on the resistance to sorghum shoot fly (*Atherigona soccata*). Plant Breed. 131 (1), 94–99.

Umakanth, A.V., Rao, S.S., Kuriakose, S.V., 2006. Heterosis in landrace hybrids of post-rainy sorghum {*Sorghum bicolor* (L.) Moench}. Indian J. Agric. Res. 40 (2), 147–150.

Uppal, B.N., 1931. India: *Rhizoctonia bataticola* on sorghum in Bombay Presidency. Int. Bull. Plant Prot. 5, 163.

Usman, S., 1972. Efficacy of granular insecticides for sorghum shoot fly in India. Pesticide. 8, 105–135.

Vadez, V., Deshpande, S.P., Kholova, J., Hammer, G.L., Borrell, A.K., Talwar, H.S., et al., 2011a. Staygreen QTL effects on water extraction and transpiration efficiency in a lysimetric system: Influence of genetic background. Funct. Plant Biol. 38, 553–566.

Vadez, V., Krishnamurthy, L., Hash, C.T., Upadhyaya, H.D., Borrell, A.K., 2011b. Yield, transpiration efficiency and water use variations and their relationships in the sorghum reference collection. Crop Pasture Sci. 62 (8), 1–11.

Vahtin, J.U.V., 1958. The inheritance of combining ability in maize when self pollinated and selected for earliness. Bull. Inat. Biol. Akad. Nauk. Belorusk. SSR.209–216.

Vaksmann, M., Traore, S.B., Niangado, O., 1996. Photoperiodism of sorghum in Africa. Agric. Dev. 9, 13–18.

van den Berg, J., Pretorius, A.J., van Liggerenberg, M., 2003. Effect of leaf feeding by *Melanaphis sacchari* (Zehntner) (Homoptera: Aphididae) on sorghum grain quality. S. Afr. J. Plant Soil. 20, 41–43.

van Oosterrom, E.J., Jayachandran, R., Bidinger, F.R., 1996. Diallel analysis of the stay-green trait and its components in sorghum. Crop Sci. 36, 549–555.

van Oosterom, E., Hammer, G., Kim, H.K., McLean, G., Deifel, K., 2008. Plant design features that improve grain yield of cereals under drought. In: Unkovich, M. (Ed.), Global Issues, Paddock Action. Proceedings of the 14th Australian Society of Agronomy Conference. 21–25 September 2008, Adelaide, South Australia. CD ROM Proceedings (ISBN 1 920842 34 9). The Regional Institute, Gosford, Australia, <www.agronomy.org.au>.

van Oosterom, E., Borrell, A.K., Deifel, K.S., Hammer, G.L., 2011. Does increased leaf appearance rate enhance adaptation to postanthesis drought stress in sorghum. Crop. Sci. 51, 2728–2740.

van Rensburg, N.J., 1973a. Notes on the occurrence and biology of the sorghum aphid in South Africa. J. Entomol. Soc. S. Afr. 36, 293–298.

van Rensburg, N.J., 1973b. Population fluctuations of the sorghum aphid, *Melanaphis* (Longiunguis) *pyrarius* (Passerini) *forma sacchari* (Zehntner). Phytophylactica. 5, 127–134.

Vasal, S.K., Singh, N.N., Dhillon, B.S., Patil, S.J., 2004. Population improvement strategies for crop improvement. In: Jain, H.K., Kharkwal, M.C. (Eds.), Plant Breeding Mendelian to Molecular Approaches. Narosa Publishing House, New Delhi, pp. 391–405.

Vasudev Rao, M.J., 1973. Genetic analysis of eight quantitative characters in a five parent complete diallel of *Sorghum vulgare* pres. Mysore J. Agric. Sci. 7, 657–658.

Vasudev Rao, M.J., Goud, J.V., 1977. Inheritance of plant height and maturity in sorghum and their components. Mysore J. Agric. Sci. 11, 269–275.

Veerabadhiran, P., Palanisamy, S., Palanisamy, G.A., 1994. Heterosis for yield and yield components in sorghum. Madras Agric. J. 81, 643–646.

Veeranna, V.S., 1972. The Effect of Different Levels of Nitrogen on Growth, Nitrogen Uptake and Yield of Sorghum Hybrids and Varieties (M.Sc. (Agri) thesis). University of Agricultural Sciences, Bangalore.

Venkatarao, D.N., Shinde, V.K., Mayee, C.D., 1983. Inheritance of charcoal rot other qualitative characters in sorghum. J. Maharashtra Agric. Univ. 8, 177–178.

Verma, P.K., Eastin, J.D., 1985. Genotypic difference of sorghum (*Sorghum bicolor* L.) Moench in response to environmental stress. Sorghum Newsl. 28, 128.

Verma, P.K., Wade, L.J., Peacock, J.M., Seetarama, N., Prasad, J.S., Huda, K.S., 1983. Leaf area response to water and nitrogen stress in sorghum. Sorghum Newsl. 26, 130–132.

Vietor, D.M., Cralle, H.T., Miller, F.R., 1989. Partitioning of 14C-photosynthate and biomass in relation to senescence characterstics in sorghum. Crop Sci. 29, 1049–1053.

Vijayalakshmi, K., 1993. Study of the inter-relationship of important traits contributing to the resistance of shoot fly in Sorghum bicolor (L.) Moench (En) M.Sc. Thesis, Andhra Pradesh Agricultural University, Hyderabad, Andhra Pradesh. India. 161 pp.

Vinodhana, N.K., Ganesamurthy, K., Punita, D., 2009. Genetic variability and drought tolerant studies in sorghum. Int. J. Plant Sci. 4 (2), 460–463.

Viraktamath, C.S., Raghavendra, G., Desikachar, H.S.R., 1972. Varietal differences in chemical composition, physical properties and culinary qualities of some recently developed sorghums. J. Food Sci. Technol. (India). 9, 73–76.

Virupaksha, P.H., Adiver, S.S., Bhat, R., Narayana, Y.D., Jahagirdar, S., Parameshwarappa, K.G., 2012. Genetic variability in *Macrophomina phaseolina* (tassi.) Goid., causal agent of charcoal rot of sorghum. Karnataka J. Agric. Sci. 25 (1), 72–76.

Voigt, R.L., Borque, P., Dobrenz, A.K., 1983. Photosynthetic stability of sorghum germplasm selected with an irrigation gradient system. Sorghum Newsl. 26, 135–137.

Vong, N.Q., Murata, Y., 1978. Studies on the Physiological Characteristics of C3 and C4 Crop Species II. The Effects of air temperature and solar radiation on the dry matters production of some crops. Jpn. J. Crop Sci. 47 (1), 90–100.

Waghmare, A.G., Varshneya, M.C., Khandge, S.V., Thakur, S.S., Jadhav, A.S., 1995. Effects of meteorological parameters on the incidence of aphids on sorghum. J. Maharashtra Agric. Univ. 20, 307–308.

Wallace, J.R., Ozbun, J.L., Munger, H.M., 1972. Physiological genetics of crop yield. Adv. Agron. 24, 97–146.

Walulu, R.S., Rosenow, D.T., Wester, D.B., Nguyen, H.T., 1994. Inheritance of the stay green trait in sorghum. Crop Sci. 34 (4), 970–972.

Wang, Y.S., 1961. Studies on the sorghum aphid, Aphis sacchari Zehntner. Acta Entomol. Sin. 10, 363–380.

Wang, G.P., Hui, Z., Li, F., Zhao, M.R., Zhang, J., Wang, W., 2010. Improvement of heat and drought photosynthetic tolerance in wheat by over accumulation of glycinebetaine. Plant Biotechnol. Rep. 4, 213–222.

Waniska, R.D., 1976. Methods to Assess Quality of Boiled Sorghum, Gruel and Chapatis from Sorghum with Different Kernel Characters (M.S. thesis). Texas A&M University, College Station, TX.

Watson, D.J., 1947. Comparative physiological studies on growth of field crops I variations in net assimilation rate and leaf area between species and varieties and within and between years. Ann. Bot. 11, 41–76.

Wenzel, W., Ayisi, K., Donaldson, G., 1999. Selection for drought resistance in grain sorghum. Angew. Botanik. 73 (3–4), 118–121.

Wenzel, W., Ayisi, K., Donaldson, G., 2000. Importance of harvest index in drought resistance of sorghum. J. Appl. Bot. 74 (5–6), 203–205.

Wenzel, W.G., 1989. Growth rate and transpirational water loss of cut seedlings as selection criteria for drought resistance in grain sorghum. Angewandtle Botanic. 62 (1-2), 7–15.

Wenzel, W.G., 1990. Inheritance of drought resistance characteristics in grain sorghum seedling. S. Afr. J. Plant Soil. 8 (4), 169–171.

Whaley, W.G., 1952. In: Gowen, J.W. (Ed.), Heterosis. Iowa State University Press America, pp. 98–113.

White, J.W., 1988. Preliminary results of the bean international drought yield trial (BIDYT). In: White, J.W., Hoogenboom, G., Ibarra, F., Singh, S.P. (Eds.), Research on Drought Tolerance in Common Bean. CIAT, Cali, Colombia.

Willey, R.W., Basiime, D.R., 1973. Studies on the physiological determinants of grain yield in five varieties of sorghum. J. Agric. Cambridge. 81, 537–548.

Wilson, D.R., Van Bavel, C.H.M., McCree, K.J., 1980. Carbon balance of water deficient grain sorghum plains. Crop Sci. 20, 145–153.

Witcombe, J.R., 2002. A Mother and Baby trial system. In: Breeding Rainfed Rice for Drought-Prone Environments: Integrating Conventional and Participatory Plant Breeding in South and Southeast Asia. Proceedings of a DFID Plant Sciences Research Programme/IRRI Conference. 12–15 March 2002, IRRI, Los Baños, Laguna, Philippines. Department for International Development (DFID) Plant Sciences Research Programme, Centre for Arid Zone Studies (CAZS) and International Rice Research Institute (IRRI), Bangor and Manila. Appendix, pp79.

Witcombe, J.R., Packwood, A.J., Raj, A.G.B., Virk, D.S., 1998. The extent and rate of adoption of modern cultivars in India. In: Witcombe, J.R., Virk, D.S., Farrington, J. (Eds.), Seeds of Choice. Making the Most of New Varieties for Small Farmers. Oxford IBH: Intermediate Technology Publications, New Delhi and London, pp. 53–68. Published by.

Wong, R.R., Munoz Orozco, A., Mondozaonfre, L.E., 1983. Effect of drought on vegetative, reproductive and efficiency characteristics of sorghum varieties. Agociencia. 51, 101–114.

Woo, H.R., Chung, K.M., Park, J.-H., Oh, S.A., Ahn, T., Hong, S.H., et al., 2001. ORE9, an F-box protein that regulates leaf senescence in Arabidopsis. Plant Cell. 13, 1779–1790.

Wood, A.J., Saneoka, H., Rhodes, D., Joly, R.J., Goldsbrough, P.B., 1996. Betaine aldehyde dehydrogenase in sorghum. Molecular cloning and expression of two related genes. Plant Physiol. 110 (4), 1301–1308.

Worstell, J.V., Kidd, H.J., Schertz, K.C., 1984. Relationships among male sterility inducing cytoplasms of sorghum. Crop Sci. 24, 186–189.

Wright, G.C., Smith, R.C.G., 1983. Differences between two sorghum genotypes in adaptation to drought stress. II. Root water uptake and water use. Aust. J. Agric. Res. 34, 627–636. Available from: http://dx.doi.org/10.1071/AR9830627.

Wright, G.C., Smith, R.C.G., McWilliam, J.R., 1983. Differences between two grain sorghum genotypes in adaptation to drought stress I Crop growth and yield responses. Aust. J. Agric. Res. 34 (6), 615–626.

Xin Z., Wang M., Barkley Jr., N., et al., 2007. Development of a tilling population for sorghum functional genomics. Proceedings of the 15th International Plant & Animal Genome Conference. San Diego, CA.

Xu, G.W., Magill, C.W., Schertz, K.F., Hart, G.E., 1994. A RFLP linkage map of *Sorghum bicolor* (L.) Moench. Theor. Appl. Genet. 89 (2-3), 139.

Xu, W.W., Subudhi, P.K., Crasta, O.R., Rosenow, D.T., Mullet, J.E., Nguyen, H.T., 2000. Molecular mapping of QTLs conferring stay-green in grain sorghum (*Sorghum bicolor* L. Moench). Genome. 43, 461–469.

Yadav, S., Jyothi lakshmi, N., Maheshwari, M., Venkateswarlu, B., 2003. Influence of water deficit at vegetative, anthesis and grain fillings stages on water relation and grain yield in sorghum. Indian J. Plant Physiol. 10 (1), 20–24.

Yang, J.C., Zhang, J.H., Wang, Z.Q., Zhu, Q.S., Wang, W., 2001. Remobilization of carbon reserves in response to water deficit during grain filling of rice. Field Crops Res. 71, 47–55. Available from: http://dx.doi.org/10.1016/S0378-4290(01)00147-2.

Young, N.D., 1996. QTL mapping and quantitative disease resistance in plants. Annu. Rev. Phytopathol. 34, 479–501.

Yark, J.O., 1977. Proposed reclassification of certain genes that influence pericarp and testa colour in sorghum. Agron. Abstr. 78, 13–18.

Yin, Y., Wang, Z.Y., Mora-Garcia, S., Li, J., Yoshida, S., et al., 2002. BES1 accumulates in the nucleus in response to brassinosteriods to regulate genes expression and promote stem elongation. Cell. 109, 181–191.

Yoshida, S., 1972. Physiological aspects of grain yield. Ann. Rev. Plant Physiol. 23, 437–464.

Yoshida, S., 1976. Physiological consequences of altering plant type and maturity. Proceedings of International Rice Research Conference. IRRI, Las Banos, Philippines.

Yoshida, S., 2003. Molecular regulation of leaf senescence. Curr. Opin. Plant Biol. 6, 79–84.

Young, W.R., 1970. Sorghum insects. In: Wall, J.S., Ross, W.M. (Eds.), Sorghum Production and Utilization. AVI Publishing Co., Westport, CT, pp. 235–287.

Youngquist, J.B., Calter, D.C., Youngquist, W.C., Clegg, M.D., 1993. Phenotypic and agronomic characteristics associated with yield and yield stability of grain sorghum in low rainfall environments. Agron. Trends Agric. Sci. 1, 25–32.

Youngquist, J.R., Calter, D.C., Youngquist, W.C., Clegg, M.O., 1990. Phenotypic and agronomic characteristics associated with yield and yield stability of grain sorghum in low rainfall environments. Bulletin of Agricultural Research, Boswana, pp. 21–23.

Younis, M.E., Shahaby, E.I., Hamed, Ibrahim, A.H., 2000. Effect of water stress on growth, pigments and CO_2 assimilation in three sorghum cultivars. J. Agron. Crop. Sci. 1875, 73–82.

Yu, J., Tuinstra, M.R., 2001. Genetic analysis of seedling growth under cold temperature stress in grain sorghum. Crop Sci. 41, 1438–1443.

Zaman-Allah, M., Jenkinson, D.M., Vadez, V., 2011a. Chickpea genotypes contrasting for seed yield under terminal drought stress in the field differ for traits related to the control of water use. Funct. Plant Biol. 38, 270–281.

Zaman-Allah, M., Jenkinson, D.M., Vadez, V., 2011b. A conservative pattern of water use, rather than deep or profuse rooting, is critical for the terminal drought tolerance of chickpea. J. Exp. Bot. 62 (12), 4239–4252.

Zhang, J., Kirkham, M.B., 1994. Drought-stress induced changes in activities of superoxide dismutase, catalase and peroxidases in wheat leaves. Plant Cell Physiol. 35, 785–791.

Zhang, J., Kirkham, M.B., 1996. Enzymatic responses of the ascorbate-glutathione cycle to drought in sorghum and sunflower plants. Plant Sci. 113, 139–147.

Zhao, T.Y., Shi, P., Guo, J.A., 1983. Relationship between heterosis of photosynthetic characters and yield components in sorghum. Shanxi Agric. Sci. 9, 18–21.

Zimmerman, E.C., 1948. Insects of Hawaii, Homoptera: Auchenorhyncha, vol. 4. University of Hawaii Press, Honolulu, Hawaii, p. 268.

Zinn, K.E., Ozdemir, M.T., Harper, J.F., 2010. Temperature stress and plant sexual reproduction: uncovering the weakest links. J. Exp. Bot. 61, 1959–1968.

CHAPTER

8

Industrial or Alternate Uses

OUTLINE

Sorghum (*Sorghum bicolor*), is classified under coarse grain millet because of its hard grain texture (Murty and Kumar, 1995). It complements well with lysine-rich vegetables (leguminous) and animal proteins and forms nutritionally balanced composite foods of high biological value.

Most of *rabi* sorghum grain is utilized for food since the grain is of superior quality and hence preferred for consumption. *Rabi* sorghum prices are higher by 20–40% compared to *kharif* sorghum grain, thus making their use uneconomical in alternative uses like poultry feed and alcohol manufacture in relation to other substitute cereals. Besides its use as a staple at the household level, small quantities of *rabi* sorghum are used in the processed food industry (personal communication with experts and field surveys). On an average about 3–5% of *rabi* sorghum is used in the processed food industry, while another 5% is used to prepare *rotis* sold at restaurants (through field surveys and consultation with processors). For *rabi* sorghum, it is assumed that in the 1970s, about 5% of *rabi* production was being used by the processed food industry, and more recently, in 2005, about 10% of the *rabi* sorghum production is used for alternative uses like processed food, seed, and *rotis* in restaurants (Parthasarathy Rao et al., 2010).

Alternative uses of sorghum encompass the utilization of grain and sweet stalk in food and nonfood sectors for the production of commercial products, such as alcohol (potable and industrial grade), syrups (natural and high-fructose), glucose (liquid and powder), modified starches, maltodextrins, jaggery, sorbitol, and citric acid (downstream products from starch). The food products include baked goods and industrial products. Various food products are traditionally prepared from winter sorghum harvested over centuries in the Indian subcontinent. These include unleavened pancakes/flatbread from fermented or unfermented dough, mainly in the states of Maharashtra, Karnataka, and parts of Andhra Pradesh in India. It can be stored for several days in the crisp dehydrated

Genetic Enhancement of rabi sorghum – Adapting the Indian Durras.
DOI: http://dx.doi.org/10.1016/B978-0-12-801926-9.00008-X

form without losing quality (Subramanian and Jambunathan, 1982) Other forms of food includes stiff or thin porridges, snack foods, deep-fried products, sweet or sour opaque beer, nonalcoholic beverages, and boiled decorticated grains. The traditional preparations like *annam* (similar to cooked rice), *sankati* (toh) and *ganji* (gruel) are popular with farmers (Murty and Subramanian, 1981). New value added/processed food products for human consumption are emerging such as popped sorghum, *papad*, porridge, *rava*, and as an ingredient for Indian dishes like *dosa* and *khichdi*, which, though in the nascent stage, are likely to be significant avenues for diversifying utilization trends of sorghum (Parthasarathy Rao et al., 2010).

Bakery products: Common bakery products include bread, cakes, and biscuits. Sorghum flour does not contain proteins similar to wheat gluten; therefore, yeast-leavened products from 100% sorghum flour are difficult to use for baking. However, sorghum and wheat flour blends have been used to produce many baked products, such as yeast-leavened pan bread, hearth bread, and flatbreads, cakes, muffins, cookies, biscuits, and flour tortillas (Rooney et al., 1980; Morad et al., 1984a,b; Badi et al., 1990). Quintero-Fuentes et al. (1999) found that 15–20% waxy sorghum improved the flexibility of tortillas significantly during storage.

Tortillas: Serna-Saldivar et al. (1988) produced acceptable tortilla chips from blends containing 50% sorghum/50% maize. Flour from white sorghum can replace up to 20% of wheat flour for the production of flour tortillas (Torres et al., 1993).

Extrusion, micronization, and related processes: Sorghum grain, grits, and meal can be easily extruded (MacLean et al., 1983; Gomez et al., 1988; Almeida-Dominguez et al., 1996; Floyd, 1996), flaked (McDonough et al., 1998a), puffed (Suhendro et al., 1998), micronized (Cruzy Celis et al., 1996), and shredded to produce a wide array of ready-to-eat breakfast foods, snacks, and other products. The grain of waxy sorghum has excellent properties for some food systems, including steam flaking (McDonough et al., 1998a), micronizing for granolas (Cruzy Celis et al., 1996), inhibition of tortilla staling, and improvement in the texture of baked tortilla chips (Quintero-Fuentes et al., 1999). Waxy sorghum has a lower yield and would be a specialty crop.

Noodles and pasta: Sorghums with soft texture, yellow endosperm, white pericarp, and no pigmented testa produced the best pasta products. Hard endosperm sorghum produced the best noodles in terms of texture after cooking and reduced dry matter losses.

8.1 INDUSTRIAL PRODUCTS

Different industrial products that can be prepared from sorghum grain include potable alcohol, glucose, high-fructose syrup, modified starches, maltodextrins, and starch downstream products like sorbitol and citric acid (Ratnavathi et al., 2004).

Sorghum varieties with high starch and moderate protein content are preferred by industries for alcohol production. Sorghum grain contains 63.4–72.5% starch, 17.8–21.9% amylase, 7.9–11.5% protein, 1.86–3.08% fat, and 1.57–2.41% fiber (Ratnavathi and Bala Ravi, 2000). Starch from sorghum grain is comparable to maize in quality, though it has 5–8% less quantity. Starch can be used in the preparation of liquid glucose, high-fructose syrup, and maltodextrins. Maltodextrins are used in the preparation of low-calorie and low-fat cookies.

Sorbitol is used as a syrup base in the pharmaceutical industry. The conversion percentage of glucose to sorbitol is 90. The malt prepared from sorghum grain is used in the preparation of baby food and beverages. Milo, a sorghum drink, is prepared from sorghum malt. The diastatic activity of sorghum is found to be 80% of that of barley (*Hordeum vulgare*). Grain sorghum flakes are used as an adjunct in the brewing industry.

References

Almeida-Dominguez, H.D., Ku-Kumul, M.M., McDonough, C.M., Rooney, L.W., 1996. Effect of extrusion parameters on physic-chemical and pasting characteristics of a food type white sorghum. In: Walker, C.E., Hazelton, J.L., (Eds.), Proceedings 1994 AACC RVA Symposium: Applications of the Rapid Visco Analyser. Neuport Science Pty. Ltd., Warriewood, New South Wales, Australia, pp. 31*–35.

Badi, S., Pedersen, B., Manowar, L., Eggum, B.O., 1990. The nutritive value of new and traditional sorghum and millet foods from Sudan. Plant Foods Hum. Nutr. 40, 5–19.

CruzyCelis, L.P., Rooney, L.W., McDonough, C.M., 1996. A ready-to-eat breakfast cereal from food-grade sorghum. Cereal Chem. 73 (1), 108–114.

Floyd, C.B., 1996. Affect of aging of endosperm characteristics and processing functionality of sorghum and corn. Ph.D. Dissertation. Texas A&M University, College Station, Tx, 88p.

Gomez, M.H., Waniska, R.D., Rooney, L.W., 1988. Extrusion-cooking of sorghum. J. Food Sci. 53, 1818–1822.

MacLean, W.C., Lopez de Romana, G., Gastanaday, A., Graham, G.G., 1983. The effect of decortications and extrusion on the digestibility of sorghum by pre-school children. J. Nutr. 113, 2071–2077.

McDonough, C.M., Anderson, B.J., Acosta-Zuleta, H., Rooney, L.W., 1998a. Steam flaking characteristics of sorghum hybrids and lines with differing endosperm characteristics. Cereal Chem. 75 (5), 634–638.

Morad, M.M., Doherty, C.A., Rooney, L.W., 1984a. Effect of sorghum variety on baking properties of U.S. conventional bread, Egyptian pita "balady" bread and cookies. J. Food. Sci. 49, 1070–1074.

Morad, M.M., Doherty, C.A., Rooney, L.W., 1984b. Utilization of dried distiller grain from sorghum in baked food systems. Cereal Chem. 61 (5), 409–414.

Murty, D.S., Kumar, K.A., 1995. Traditional uses of sorghum and millets. In: Dendy, D.A.V. (Ed.), Sorghum and Millets: Chemistry and Technology. American Association of Cereal Chemists, St. Paul, MN, pp. 185–221.

Murty, D.S., Subramanian, V., 1981. Sorghum *roti*: I. Traditional methods of consumption and standard procedures for evaluation. In: Proceedings of the International Symposium on Sorghum Grain Quality, 28-31 Oct 1981, ICRISAT Center, Patancheru, India. International Crops Research Institute for the Semi-Arid Tropics, Patancheru, Andhra Pradesh, India, pp. 73–78.

Parthasarathy Rao, P., Basavaraj, G., Wasim Ahmad, Bhagavatula, S., 2010. An analysis of availability and utilization of sorghum grain in India. E-journal SAT Agric. Res. 8, 1–8.

Quintero-Fuentes, X., McDonough, C.M., Rooney, L.W., Almeida-Dominguez, H., 1999. Functionality of rice and sorghum flours in baked tortilla and corn chips. Cereal Chem. 76 (5), 705–710.

Ratnavathi, C.V., Bala Ravi, S., 2000. A study on the suitability of unmalted sorghum as a brewing adjunct. J. Inst. Brewing. 106 (6), 383–387.

Ratnavathi, C.V., Biswas, P.K., Pallavi, M., Maheswari, M., Vijay Kumar, B.S., Seetharama, N., 2004. Alternative uses of sorghum—methods and feasibility. Indian Perspective in Alternative Uses of Sorghum and Pearl Millet in Asia. Proceedings of the Expert Meeting, ICRISAT. 1–4 July 2003, International Crops Research Institute for the Semi-Arid Tropics, Patancheru, India. CFC Technical Paper No. 34, Common Fund for Commodities, Amsterdam, The Netherlands, pp. 188–200.

Rooney, L.W., Khan, M.N., Earp, C.F., 1980. The technology of sorghum products. In: Inglett, G.E., Munck, L. (Eds.), Cereals for Food and Beverages: Recent Progress in Cereal Chemistry and Technology. Academic Press, New York, NY.

Serna-Saldivar, S.O., Cannett, R., Vargas, J., Gonzalez, M., Bedolla, S., Medina, C., 1988. Effect of soyabean and sesame addition on nutritional value of maize and decorticated sorghum tortillas produced by extrusion cooking. Cereal Chem. 65, 44.

Suhendro, E.S., McDonough, C.M., Rooney, L.W., Waniska, R.D., Yetneberk, S., 1998. Effects of processing conditions and sorghum cultivar on alkaline-processed snacks. Cereal Chem. 75 (2), 187–193.

Subramanian, V., Jambunathan, R. 1982. Properties of sorghum grain and their relationship to rati quality. In: Proceedings, International Symposium on Sorghum Grain Quality, ICRISAT. 28–31 October 1981, Patancheru, India.

Torres, P.I., Ramirez-Wong, B., Serna-Saldivar, S.O., Rooney, L.W., 1993. Effect of sorghum flour addition on the characteristics of wheat flour tortillas. Cereal Chem. 70 (1), 8–13.

Index

Note: Page numbers followed by "*f*" and "*t*" refer to figures and tables, respectively.

D

E

F

T

Printed in the United States
By Bookmasters